SAMPLING AND ANALYSIS OF INDOOR MICROORGANISMS

The Wiley Bicentennial–Knowledge for Generations

Each generation has its unique needs and aspirations. When Charles Wiley first opened his small printing shop in lower Manhattan in 1807, it was a generation of boundless potential searching for an identity. And we were there, helping to define a new American literary tradition. Over half a century later, in the midst of the Second Industrial Revolution, it was a generation focused on building the future. Once again, we were there, supplying the critical scientific, technical, and engineering knowledge that helped frame the world. Throughout the 20th Century, and into the new millennium, nations began to reach out beyond their own borders and a new international community was born. Wiley was there, expanding its operations around the world to enable a global exchange of ideas, opinions, and know-how.

For 200 years, Wiley has been an integral part of each generation's journey, enabling the flow of information and understanding necessary to meet their needs and fulfill their aspirations. Today, bold new technologies are changing the way we live and learn. Wiley will be there, providing you the must-have knowledge you need to imagine new worlds, new possibilities, and new opportunities.

Generations come and go, but you can always count on Wiley to provide you the knowledge you need, when and where you need it!

William J. Pesce
President and Chief Executive Officer

Peter Booth Wiley
Chairman of the Board

SAMPLING AND ANALYSIS OF INDOOR MICROORGANISMS

CHIN S. YANG
P&K Microbiology Services, Inc.
Cherry Hill, New Jersey

PATRICIA A. HEINSOHN
Micro Bios Pacifica, California

WILEY-INTERSCIENCE
A JOHN WILEY & SONS, INC., PUBLICATION

Published by John Wiley & Sons, Inc., Hoboken, New Jersey.
Published simultaneously in Canada.

For general information on our other products and services or for technical support, please contact our Customer Care Department within the United States at (800) 762-2974, outside the United States at (317) 572-3993 or fax (317) 572-4002.

Wiley publishes in a variety of print and electronic formats and by print-on-demand. Some material included with standard print versions of this book may not be included in e-books or in print-on-demand. If this book refers to media such as a CD or DVD that is not included in the version you purchased, you may download this material at http://booksupport.wiley.com. For more information about Wiley products, visit www.wiley.com.

Wiley Bicentennial Logo: Richard J. Pacifico

Library of Congress Cataloging-in-Publication Data:

Sampling and analysis of indoor microorganisms/[edited by] Chin S. Yang, Patricia A. Heinsohn.
p. cm.
ISBN-13: 978-0-471-73093-4
ISBN-10: 0-471-73093-9
1. Buildings. 2. Microbial ecology. 3. Molds (Fungi). 4. Environmental sampling. I. Yang, Chin S. II. Heinsohn, Patricia A.

QR100.S26 2007
579′.17--dc22 2006024643

10 9 8 7 6 5 4 3 2 1

CONTENTS

PREFACE

Over the last two decades (since the mid-1980s), concerns of human exposures to microbiologicals originating in the indoor environment due to water intrusion or moisture control problems have been centered on fungi, which include molds. It is, however, well understood that fungal growth is not the only group of microorganisms thriving in such conditions. Bacteria are particularly common in the indoor environment, with growth possibly resulting from water damage. Because of their ubiquity and small size, bacteria react to water intrusion or excessive moisture much faster than do fungi. Bacterial growth can produce endotoxins and other biogenic toxins. Some environmental bacteria are also infectious. In fact, two legionelloses are associated with waterborne bacteria. Legionnaire's disease was first reported in the outbreak in the 1976 American Legion convention in Philadelphia. Pontiac fever is a nonpneumonic legionellosis, believed to represent reaction to inhaled *Legionella* antigens rather than bacterial invasion. *Legionella* bacteria are now known to thrive in building water systems such as cooling towers or hot-water systems.

Although fungal spores are ubiquitous, fungal growth is considered unusual in a properly maintained building. Because of the ubiquitousness, fungal spores can germinate and grow into colonies when there is access to moisture. During their lifecycle, fungi produce a wide variety of chemicals, including secondary metabolites, some of which are toxic to animals and humans; hence the term *mycotoxins*. Fungal spores are known to contain allergens and fungal glucans. Exposure to fungal allergens and mycotoxins in a water-damaged mold-infested indoor environment is generally believed to pose an increased risk over that posed by outdoor exposures.

Because of the concerns and interest in indoor fungal contamination issues, several guidelines and reference books have been published. Conventional and newly invented sampling and analytical methodologies have routinely been employed in the assessments and investigations of microbiologics in the indoor environment. Some of the methodologies have never been properly evaluated and validated by scientific methods. Furthermore, results and data derived from the methodologies are often misused or misinterpreted. In 2004, the Institute of Medicine (IOM) of the National Academies reported that currently used sampling and analytical methodologies were inadequate for microbial exposure assessment in the indoor environment because of widespread uncertainty from potential errors and bias in the methodologies as well as complexities from large temporal and spatial variability and diverse microbial species and metabolites that are produced in such environments. The IOM further recommended that "existing exposure assessment methods for fungal and other microbial agents need rigorous validation and further refinement to make them more suitable for large-scale epidemiologic studies." Before such recommendations are fulfilled, it is our desire and opinion that investigators and laboratory managers be thoroughly familiar with the details of investigation, including sampling, and analytical methodologies so that they understand the advantages and limitations of sampling and analytical methods, utilize methods properly and appropriately, and interpret the results accordingly. The approach to assessment and investigation is toward evaluating the conditions of growth and contamination in the indoor environment rather than assessment of human exposure.

In 2004, a group of scientists at the National Institute for Occupational Safety and Health (NIOSH) studied employees in 13 college buildings. They reported that building-related respiratory symptoms can be predicted with semiquantitative indices of exposure to dampness and mold by classifying water stains, visible mold, mold odor, and moisture using semiquantitative scales. They then estimated individual exposure indices weighted by the time spent in specific rooms. The use of surface coverage area of visible fungal growth to guide and design mold remediation requirements for human and environmental protections was detailed in guidelines and documents published by the New York City Department of Health, the U.S. Environmental Protection Agency (USEPA), and Health Canada. The New York City Guidelines available in 1993 and 2000 and the USEPA's *Mold Remediation in Schools and Commercial Buildings* published in 2001 used square footages to define requirements for personal protective equipment and isolation containment. This implies the relationship between visible mold area and exposure. In 2004, Health Canada published a document titled *Fungal Contamination in Public Buildings: Health Effects and Investigation Methods*, in which it suggested that there is dose–effect relationship in that more visible mold yielded more symptoms. Although the relationship requires further refinement, investigators should always keep this in mind when conducting an assessment, an investigation, or an inspection of fungal growth.

Although we understand the complexity of microbiological problems in a water-damaged environment and the knowledge gap in practices of sampling and analysis, we think there is sufficient information and knowledge in microbiological sampling and testing available to properly assess and inspect an environment for microbiological contamination. Clearly, any new advance will greatly add to the tools and knowledge available for better methodology to sampling and testing for microbes indoors. In light of the IOM publication, *Damp Indoor Spaces and Health* and the University of Connecticut's *Guidance for Clinicians on the Recognition and Management of Health Effects Related to Mold Exposure and Moisture Indoors* (http://www.oehc.uchc.edu/clinser/MOLD%20GUIDE.pdf), the impact of microbial growth and contamination in a water-damaged indoor environment on human health is minimally discussed in this book. Readers are referred to the documents cited above as well as those published by governmental agencies, such as New York City Department of Health, USEPA, NIOSH, or Health Canada. Readers are also encouraged to stay abreast of the scientific literature in this area as it is published.

Finally, we would like to thank chapter authors for their efforts and contributions to this book. We are also very grateful for Drs. Francis Harrington, De-Wei Li, Michael Berg, and Philip R. Morey for reviewing manuscripts. Dr. Keith Wheatstone of Severn Trent Laboratories is thanked for his support to Chin S. Yang.

CHIN S. YANG

Cherry Hill, New Jersey

PATRICIA A. HEINSOHN

Pacifica, California

CONTRIBUTORS

SUSAN E. ANAGNOST, Ph.D., Faculty of Construction Management and, Wood Products Engineering, State University of New York College of Environmental Science and Forestry, Syracuse, NY 13210 (email: seanagno@esf.edu)

FRANCIS HARRINGTON, Ph.D., 15 Hinman Street, Meriden, CT 06450 (email: odyharrington@aol.com)

RICHARD A. HAUGLAND, Ph.D., National Exposure Research Laboratory, U.S. Environmental Protection Agency, Cincinnati, OH 45268 (email: haugland.rich@epamail.epa.gov)

Patricia Heinsohn, Ph.D., CIH, Micro Bios, 2309 Palmetto Avenue, C-1, Pacifica, CA 94044 (email: phmicrobios@aol.com)

DE-WEI LI, Ph.D., The Connecticut Agricultural Experiment Station, Valley Laboratory, 153 Cook Hill Road, P.O. Box 248, Windsor, CT 06095 (email: dewei.li@po.state.ct.us)

PHILIP R. MOREY, Ph.D., CIH, Director of Microbiology, Boelter & Yates, Inc., 2235 Baltimore Pike, Gettysburg, PA 17325-7015 (email: pmorey@boelter-yates.com)

LINDA D. STETZENBACH, Ph.D., Professor, School of Public Health, University of Nevada, Las Vegas, NV 89154-4009 (email: stetzenl@unlv.nevada.edu)

STELLA M. TSAI, Ph.D., CIH, Research Scientist, New Jersey Department of Health and Senior Services, P.O. Box 369, Trenton, NJ 08625-0369 (email: stella.tsai@doh.state.nj.us)

STEPHEN J. VESPER, Ph.D., National Exposure Research Laboratory, U.S. Environmental Protection Agency, Cincinnati, OH 45268 (email: vesper.stephen@epamail.epa.gov)

FLORENCE Q. WU, Ph.D., Aemtek Inc., 46309 Warm Springs Blvd., Fremont, CA 94539 (email: florencewu@aemtek.com)

CHIN S. YANG, Ph.D., P&K Microbiology Services, Inc., 1936 Olney Ave, Cherry Hill, NJ 08003 (email: cyang1000@comcast.net)

CHAPTER 1

INTRODUCTION TO MICROBIOLOGICAL GROWTH AND CONTAMINATION INDOORS

CHIN S. YANG and PATRICIA A. HEINSOHN

1.1. INTRODUCTION

Humans in modern society have been reported to spend 80% or more of their time indoors, whether at work or at home.[1] Indoor air quality has become an important public health concern since the mid-1970s, at least. Among various indoor pollutants, microbiologicals are one of the most important.[2] It has been estimated that one-third of indoor air quality (IAQ) complaints may be due to microbial contamination.[3]

Among the many microbial contaminants, fungal growth and contamination in the indoor environment has been the focus of news media reports since the mid-1990s as cases of infant death from acute pulmonary hemorrage/hemosiderosis in Cleveland, Ohio were associated with mold growth, specifically *Stachybotrys chartarum*.[4–7] Later a review by a panel of outside experts convened by the Centers for Disease Control and Prevention (CDC) found shortcomings in the implementation and reporting of the investigation. CDC concluded, on the basis of the review, that a possible association between acute pulmonary hemorrage/hemosiderosis in infants and exposure to molds, specifically *S. chartarum*, was not proved.[8] Unfortunately, there was no explanation for the cause(s) of the infant's death in the CDC review, and there has been no cause offered so far.

Indoor microbiological contamination is not new. Leviticus 14:33–45 of the *Old Testament* is often cited as the earliest known reference to a "mold problem" in the indoor environment. Although it is reasonable to believe that the reference of "plague" means mold, it can also be fungi, mildew, bacteria, algae, or any combination of these microorganisms due to moisture problems. Straus described it as putting forth a detailed protocol for the remediation of contaminated structures.[9]

Sampling and Analysis of Indoor Microorganisms, Edited by Chin S. Yang and Patricia A. Heinsohn

It did imply a wet or damp condition for microbial growth. In the East, the Chinese have traditionally associated "mushroom growing and becoming moldy" (*sen gu fa mei*; 生菇發黴 or 生菇发霉) on materials or in the living environment as an unhealthy sign. These references are significant because they indicate microbial growth due to wet and damp environments as unhealthy.

The interest in and concern for indoor microbiological growth and contamination has led to many books and publications, including reference books from the American Industrial Hygiene Association and American Conference of Governmental Industrial Hygienists.[10–13] There have been other reference books and conference proceedings related to the topic published since the mid-1980s. A list of selected books includes Refs. 2 and 14–24. The scope of these books varies widely. Some are texts with certain scientific depth and were written by research or practicing scientists. Some are how-to books for practicing professionals. In addition, most or all of the books offer basic, standardized approaches to assessing mold or microbial growth and contamination indoors. Furthermore, some books focus exclusively on mold even though bacterial growth and other biological contaminants do occur in water-damaged conditions. Chapters of this book are written by several scientists with advanced academic degrees and extensive practical experience in areas of indoor microbial contamination. The book intends to bridge the gap between the sciences and reality.

In additional to reference books, substantial numbers of publications (e.g., reviews, position papers, peer-reviewed original articles) have accumulated in the literature. This book is not designed to extensively review the entire literature. Instead, useful and relevant references with thoughts and depth that are authored by reputable and credible scientists or professionals are selected. Governmental and professional guidance documents are referenced where they are appropriate. These documents are often guidelines but not standards, even though some are titled standards. Such guidance documents may be designed to serve certain constituents and can become out-of-date rather quickly because it often takes a long time to prepare and produce such documents.

In 2004, the Institute of Medicine (IOM) published a report titled *Damp Indoor Spaces and Health* based on "a comprehensive review of the scientific literature regarding the relationship between damp or moldy indoor environments and the manifestation of adverse health effects, particularly respiratory and allergic symptoms."[25] The primary backdrop for the review was public health concerns due to fungal growth from damp environments as well as visible mold in homes with recent water damage. The report, however, emphasized the importance of a comprehensive approach toward microbial assessment in the indoor environment. This approach would include bacteria, fungi, and components of microbial agents, such as allergens of microbial origin, structural components of fungi and bacteria, microbial volatile organic compounds (MVOCs), and potentially toxic products of microbial secondary metabolism. Allergens of dust mites and cockroaches can also be important factors in a damp environment. Two previously published books by IOM[2,26] also emphasized the importance of various microbiologicals as allergens and asthma triggers in the indoor environment.

It is extremely important to emphasize that fungi are never the only group of organisms to proliferate in an indoor environment in which moisture is not controlled. This agrees with the three IOM reports as well as basic principles of biology. Various microbes, from bacteria, fungi, and slime molds, to algae and protozoa, have been associated with such an environment.[2,25,26] Therefore, fungi and other microbes, particularly bacteria, and their byproducts should be taken into consideration when conducting an assessment of a water-damaged environment. Furthermore, some environmental bacteria, including *Legionella* species and *Pseudomonas aeruginosa*, are common in water associated with building systems, such as hot water, sewage, or cooling water of the HVAC system, and are opportunistic pathogens that can cause death on infection without proper treatment. Reports of outbreaks of sporadic and nosocomial infections of Legionnaire's disease are not uncommon. Nosocomial infection by *P. aeruginosa* is one of the major infection control issues in healthcare facilities.[27]

In a water-damaged environment, bacterial, fungal, and other biological (e.g., insects and mites) growth is likely to occur depending on the duration of water damage. Expertise and experience from a variety of scientists and professionals are necessary to address the complex problem. This book is intended to be multi- and interdisciplinary, including authors who have the best and most thorough knowledge as well as extensive experience in diagnosing complex indoor microbial growth and contamination issues. The authors include microbiologists, mycologists, public health scientists, environmental professionals, and certified industrial hygienists with advanced academic degrees.

In a modern building, whether residential, commercial, or industrial, construction is complex and contains many components and different systems for various purposes, from functionality and comfort to necessity. These components and systems may be microbiological reservoirs or may become the cause of or a contributor to microbial exposure. For example, heating–ventilating–air-conditioning (HVAC) systems have been known to contain mold growth, requiring remediation.[28–30] Materials used in a building can also be part of the problem because of their nutritional value and hygroscopicity, or ability to absorb moisture. Wood and paper products are notoriously susceptible to moisture, leading to problems of fungal infestation and decay (see also Chapters 8 and 10).

1.2. HEALTH EFFECTS OF INDOOR FUNGAL AND BACTERIAL GROWTH

It is well established that fungi and bacteria have known health effects in humans.[3,12,25] Many species of fungi and bacteria can cause infections. They also produce a wide range of chemical byproducts, from microbial volatile organic compounds (MVOCs), endotoxins, and fungal glucans, to mycotoxins. Fungi include allergens and triggers of asthma. Fungal glucans are inflammatory in the lung and have been reported to be associated with headaches. Mycotoxins have a wide range of health effects, primarily in ingestion exposures. Although airborne mycotoxin exposures and health effects have been widely reported,[31] their impact is controversial within the medical community.[25] Gram-negative bacteria are producers of endotoxins, which have a wide range of health effects, from mild fever and flulike symptoms to death in extreme exposure conditions. Some medical clinicians question health effects other than allergy-related diseases from such exposures. Nonetheless, the American Industrial Hygiene Association took the position that the significant presence of fungi in indoor air that are not present or are a minor component of the outdoor air is unacceptable from a health (and building performance) perspective.[10,11]

In addition to the health impact on humans, fungal and bacterial growth can cause unsightly stains on building materials, degradation of building materials, and wood decay.[31–34] In fact, fungi are well documented biodeteriorating agents of many foodstuffs and paper and wood products.[34–36]

1.3. TEAM AND INDIVIDUAL EXPERTISE

A modern indoor environment is designed to accommodate many human needs and comfort, and has many built-in components and systems to fulfill such needs. Because of the complexity of such an environment, a team of professionals with various areas of expertise is often necessary in a comprehensive investigation and assessment, including sampling and analysis for microbiological organisms. Individual professionals and their expertise and experience in such activities are discussed below.

Many assessments and investigations are often initiated by complaints or reports of illness related to the indoor environment. More recently, surveys for fungal and microbial contamination have become routine as a part of due diligence in real estate transactions, whether residential, commercial, or industrial. If health complaints, personal injury, or health-related issues are alleged, medical and public health professionals, including physicians, must be included in the team. Specialty board-certified physicians, such as occupational health physicians, allergists, immunologists, pulmonolgists, infectious disease specialists, and epidemiologists, are some of the medical experts considered. Most physicians have limited experience in diagnosing patients with mold or microbial exposure. In fact, the University of Connecticut Health Science Center published a document titled *Guidance for Clinicians on the Recognition and Management of Health Effects Related to Mold Exposure and Moisture Indoors* in 2004, with a grant from the U.S. Environmental Protection Agency.[37] The title of the document unambiguously states that it is designed to assist clinicians to recognize and manage health effects related to mold exposure. A more recent position paper published in 2006 by the American Academy of Allergy, Asthma and Immunology (AAAAI) reaffirms that the allergy community requires better professional education and training on this topic.[38]

The 2006 AAAAI position paper on the medical effects of mold exposure intended to provide a state-of-the-art review of the role that molds are known to play in human disease.[38] The paper includes a section on measurement of molds and mold product exposure in the patient's environment and makes the following three conclusions: (1) sampling of both indoor and outdoor air for mold spores provides a measure of potential exposures and can be useful in certain conditions but has many shortcomings; (2) bulk, surface, and within-wall cavity measurement of mold or mycotoxins, although having potential relevance for other purposes, cannot be used to assess exposure; and (3) testing for airborne mycotoxins in nonagricultural environments cannot be used to diagnose mold exposure. Because physicians lack the education, understanding, training, and experience necessary to assess the extent of microbial growth within buildings, the conclusions beg further discussion and clarity than this book intends to provide. The results of properly conducted sampling and testing can be used to identify and determine the extent of fungal growth and contamination indoors. Such information can be used to evaluate and index occupants' exposures qualitatively as well as to determine the environmental control treatment of patients' living and work environments. The results typically consist of qualitative and quantitative data. Qualitative data include the identification of fungal taxa. By expert and logical analysis of the data, one can determine whether fungi are actively growing indoors. Clearly, the allergy community has very little training and background in mycology, sampling, and testing for fungi, or in result interpretation of data derived from such testing. Allergists could benefit from learning how to work with environmental professionals and use such information and evaluations in their practice.

Spores of *Alternaria alternata* are often cited as evidence of sensitization to the fungus[39,40] and links to the presence, persistence, and severity of asthma.[38,41] It was the only fungal allergen included in the National Health and Nutrition Examination Surveys (NHANES) II and III studies and reported by Arbes et al.[40] The report indicated that 54.3% of the U.S. population is sensitive to one or more allergens by skin testing 10 common allergens.[40] Prevalences to indoor allergens are as follows: dust mite 27.5%, German cockroach 26.1%, cat 17.0%, and at least one indoor allergen 43%.[40] *Alternaria alternata*, which was considered to be one of the outdoor allergens in the study, was the only fungal allergen tested. It is known to grow on water-damaged indoor environments. Among the population, 12.9% are allergic to *A. alternata*.

Spores of *A. alternata* are considered ubiquitous but as a minor component in abundance in both outdoor and indoor air. Because it is possible to measure its allergens in dust, it has been widely measured indoors[42] and because it is always present, although typically at low numbers compared with other funal spores, questionable epidemiological conclusions have been drawn and permeate the literature. Furthermore, the spore size of *A. alternata* varies from 7–18 × 16–63 μm[43], 7–18 × 18–83 μm,[15] 8–12 × 20–40 μm,[36] 8.5–14 × 18–66 μm,[44] to 9–18 × 20–63 μm.[45] The average size is 13 × 37 μm according to Ellis.[45] These spore sizes and ranges are inhalable but not respirable and, therefore, unlikely to reach deep into the respiratory system and the lungs. Andersen et al. studied and concluded that *A. alternata, A. longipes*, and *A. gaisen* are different species but often mis-identified as *A. alternata*.[46] In addition, Nielsen believed that growth and spores of *A. tenuissima* are far more common indoors than *A. alternata*.[47] *Alternaria alternata* is considered a rare species and is often misidentified simply because it manifests black *Alternaria*-like spores.[48] This raises the issue regarding the significance and importance of the spores as a human allergen. This further suggests that this type of study is multidisciplinary and calls for the input and involvement of mycologists in such important studies.

Fungi grow and reproduce by spore production. In their lifecycle, fungi also produce chemical byproducts, many of which are associated with fungal spores. Fungal spores are biologically designed for easy dispersal by various means, such as minute air movements and wind, water, insects, small rodents, and tiny creatures. Airborne dispersal is the primary dissemination route for most fungal spores. The presence of active fungal growth indoors is correlated with human exposures,[49] although the exposure dosages or quantities are difficult to measure because of sampling and testing difficulties, the biological complexity of fungi, and spatial and temporal variations.

The selection of physicians who are open-minded and diligent in conducting a review of accurate and up-to-date medical literature and research is important. If a physician, in examining a patient, determines that a specific group of infectious or allergic agents, their byproducts, or both is responsible for the patient's conditions or symptoms, this information will enable the investigator to design a sampling–testing strategy to maximize the probability of finding those etiologic agents in the patient's environment. A physician may also review medical opinions offered by opposing medical professionals.

Architects and various engineering professionals may play an important role in a team that is investigating a building for moisture leading to microbial problems. These professionals may include architects, civil engineers, mechanical engineers, structural engineers, roofing specialists, and geotechnical engineers. A team of architects and engineers are often involved in building design and construction. It is, therefore, important to have architects and engineers with background and training in design, construction, and maintenance of buildings on the team participating in the assessment and investigation of water-damaged buildings. Sometimes, architects and engineers with experience and acumen in forensic investigations play pivotal roles. Although an architect or an engineer with a professional license (such as AIA, PE, or SE) is important, it does not necessarily guarantee that this individual is the most competent professional available. Experience and knowledge are critically important. The same architects and engineers or other professionals may also be asked to provide solutions to the structural problems and assist in designing a microbial remediation project.

Industrial hygienists and environmental professionals usually play an important role in microbial assessments indoors. They may include certified industrial hygienists (CIHs), other environmental consultants, and possibly home and building inspectors. They are often called on to investigate, sample, and test for microbial contaminants indoors. They should review all the relevant facts pertinent to the building, perform an independent assessment of the subject building, and render an opinion on the likelihood of the cause of the mould and bacteria as it overlaps with engineers also involved. They should include an opinion on the implications of their findings on occupant health, and offer specifications to remediate the property and restore it to its previous condition to the extent possible. Although many environmental consultants are very competent in providing these services, a CIH with experience in microbial assessment is often preferred. Certified industrial hygienists are certified by a process that requires educational prerequisites, professional experience, and an examination administered by the American Board of Industrial Hygiene (ABIH). It is a professional certification that has gained wide recognition by many, including local and federal governmental agencies. On the other hand, a CIH may be educated and trained in chemistry, engineering, public health, or areas of environmental science but not in biology, mycology, or microbiology. They may practice broad general industrial hygiene, which may include ergonomics, noise, asbestos, lead, hazardous-site assessment, indoor air quality assessment and mould assessment. They may acquire skill and knowledge in environmental microbial sampling and analysis from literature, references, or by attending seminars and professional development courses. Other consultants may include home and building inspectors, registered sanitarians, or other individuals, all of whom provide such services for a fee. There are few regulations, professional requirements, certifications, or licensing procedures for these consultants. Their competency is usually highly variable and depends heavily on the individual's educational background, training, and experience. Several trade and professional organizations have set up "certifications" for residential mold inspectors or indoor air quality specialists. The credibility and usefulness of such "certifications" are questionable and highly uncertain at this time. It is strongly recommended that potential candidates, whether they are CIHs or other environmental consultants, should be interviewed and evaluated for their competence and experience in microbial assessment and sampling as well as their ethics.

"Microbiologist" is a collective term used to include scientists who are specialists in various disciplines of virology, bacteriology, mycology or parasitology. Some microbiologists may specialize in the biology of a single microbial species, such as *Escherichia coli*. On the other hand, there are microbiologists who have a broad interest in all aspects of microbiology. It is important to understand and evaluate the background and research interest of individual microbiologists.

Bacteriologists are the scientists who study bacteria. Bacteria are *prokaryotic*, which means that their cellular organization is structurally simpler than eukaryotic organisms, such as fungi, plants, animals, and humans, and their cellular functions are simpler also. Bacteria do not have a nucleus, mitochondria, or other organelles. Their DNA, RNA, and enzymes are dispersed in the cell. In evolution, eukaryotic organisms are considered much more advanced than prokaryotes. The differences between prokaryotes and eukaryotes are many and profound. This underscores the importance of using mycologists in analyzing fungi and bacteriologists for analyzing bacteria.

Mycologists are the scientists who study fungi. Mycology is a unique branch of microbiology and biology, and only mycologists have a true understanding of the science. There are very few colleges and universities offering degreed study in mycology. Therefore, there are not many trained, degreed mycologists.

Furthermore, very few mycologists have expertise in the group of molds and fungi that are found in water-damaged moldy environments. For example, there are mycologists who specialize in wild mushrooms, which are seldom found growing indoors. The mycologist who specializes in basidiomycetes, including mushrooms and wood-decaying bracket fungi and polypores, may not be familiar with ascomycetes and deuteromycetes, which include most microfungi growing in water-damaged environments. It is important to find the right expertise. The expertise of mycologists is different from that of virologists, bacteriologists, parasitologists, or microbiologists. Mycologists may play a role in assisting with planning a field investigation, sampling, laboratory analysis of samples, assisting with interpretation of the data, remediation, or any combination of these actitivies because they know the biology of fungi. On the other hand, mycology is a very broad field. It covers five major groups of fungi: zygomycetes, ascomycetes, basidiomycetes, deuteromycetes, and myxomycetes. Mycologists may offer information and guidance about the identification, detection, health effects, eradication, and control of fungi. On a few occasions, biologists with various areas of expertise may be called on to address issues caused by other biological agents, such as dust mites, cockroaches, pollens, and insects.

Microbiologists, bacteriologists, and mycologists are scientists who are seldom subject to certification or licensing requirements. However, there are certifications required for practicing public health and medical microbiologists, bacteriologists, and mycologists. They are usually affiliated with public health and medical laboratories. Their expertise is in medically important microbial species. Their practices are often of limited use in the study of microorganisms from the indoor environment.

Other professionals who may be helpful in sampling and testing for microbes and their byproducts can include biochemists, some of whom specialize in mycotoxins or other biochemical toxicants, and toxicologists, some of whom specialize in the toxic effects of mycotoxins and other biochemical toxicants. However, the biochemistry and toxicology of microbial byproducts are highly specialized. There are very few scientists and professionals who have expertise in these areas. A careful selection of experts is critical.

1.4. APPROACH OF THIS BOOK

In an assessment for fungal growth and contamination indoors, it is important to start with the building structure and the water history of the building. An engineer or a professional with such training and experience should conduct a detailed inspection and survey of the building and its systems for design, construction, installation, and maintenance issues that can lead to leaks, floods, and condensation. A building inspector should understand the control of moisture in three different forms: liquid water, water vapor, and condensation. Liquid water includes water from pipe bursts, overflow, flood, or leaks. Water vapor in a building may travel with the airflow and by diffusion. In a hot, humid environment, outdoor airflow into a building can carry large amounts of water vapor and increase both the moisture content or a_w of building materials and the likelihood of condensation. A high dew-point temperature is an indication of high moisture load in the air, which can increase the possibility of condensation on cold surfaces. In cold-climate buildings, high humidity load, particularly in late fall and early winter, can also result in condensation on and around cold surfaces, such as windows. If an elevated humidity load maintains steady through the entire winter as a result of human activities, such as cooking and washing, condensation can continue. The importance of moisture to indoor microbial growth is such that readers will find discussions of moisture in several chapters in the book. It is recommended that readers of this book further consult the references listed at the end of this chapter, including those by L'stiburek and Carmody[50] and by Harriman et al.[51] for more information on understanding and dealing with moisture indoors.

In a building with moisture issues, fungal growth is often the observed result. However, whenever fungal growth is observed and identified, bacterial growth is likely to occur also. In general, bacterial cells are much smaller than fungi and react much more quickly in response to water. In fact, the 2004 IOM report points out the likelihood of bacterial growth in a damp space and the importance of a comprehensive evaluation of bacteria, fungi, and components of microbial agents in such an environment.[25] A chapter on airborne bacteria in indoor environments is included in this book (Chapter 6).

The focus on fungal growth and contamination in water-damaged indoor environments is justifiable because the likelihood of fungal growth is minimal if there are no water damage or humidity control problems. Spores of common types of fungi of outdoor origin are ubiquitous indoors.[11,52] These spores will not germinate and grow without the required a_w or moisture content in the substrates. Fungal growth indoors is a very strong indication of water damage or long-term excessive humidity. On the other hand, bacterial growth can be from humans, or on continually rewetted surfaces such as in the kitchen, bathroom, laundryroom, and toilets. Assessing fungal growth and contamination is the better approach than determining bacterial growth indoors. The primary nutrients available indoors for fungal growth are paper and wood products. In most inspections, fungal growth on lumber and wood products is often overlooked. A chapter on fungal growth on wood and wood decay (Chapter 8) provides important information on fungal growth and wood decay. Several chapters in the book look into sampling and laboratory analysis of fungi. Common and appropriate sampling techniques and strategies are discussed. Emphasis is also placed on conventional and modern laboratory analytical methodologies, statistical analysis of data, and understanding of the ecology of fungi and bacteria found indoors. It is difficult to evaluate the quality of laboratory data without an understanding of how samples are handled, processed and analyzed in the laboratories. The importance of result interpretation is emphasized here. Several chapters include result interpretation as well as how to use laboratory test results. Recipients of the test results should take full advantages of the information contained in the laboratory reports. A chapter on the retrospective and forensic approach to the assessment of fungal growth in the indoor environment

(Chapter 11) is also included. The ultimate goal and purpose of assessment, sampling, and analysis is to detect, identify, and estimate the extent of fungal growth indoors for remediation and cleanup. Therefore, a chapter on proper professional remediation and cleanup (Chapter 12) concludes the main text of this book. Each topic covered is important in the diagnosis and mitigation of moisture and fungal problems in buildings.

1.5. CONCLUSIONS

There has been tremendous interest in the effects of microbial contamination in the indoor environment. It is well understood that bacteria, fungi, and possibly other biologics, such as mites and insects, grow in water-damaged indoor environments. An understanding of the moisture dynamics is the crucial step in dealing with microbial growth and contamination indoors. An investigation without an understanding of moisture issues in the building and their impact on microbial growth and contamination cannot address and solve the problems. Sampling and analysis for fungi, bacteria, and their byproducts are often requested, required and performed during an investigation. Sampling and analysis without proper planning, sampling, laboratory analytical methodology, and statistical analysis can, in fact, lead to erroneus conclusions. The importance of using the information and data in a specific environment with focused objectives in formulating opinions is also emphasized. In most building construction, paper and wood products are widely used. They are nutrients for bacterial and fungal growth. The use of such products is selective for cellulolytic fungi and possibly wood decay fungi when they become wet and remain wet long enough. To properly address fungal growth problems, an understanding of indoor fungal ecology is very important. In situations where responsibilities are shared, an understanding of the chronology of water damage and fungal growth is also important.

REFERENCES

1. Spengler, J. D. and K. Sexton, Indoor air pollution: A public health perspective, *Science* **221**:9–17 (1983).
2. Pope, A. M., R. Patterson, and H. Burge, *Indoor Allergens*, National Academy Press, Washington, DC, 1993.
3. Lewis, F. A., Regulating indoor microbes, the OSHA proposed rule on IAQ a focus on microbial contamination, in *Fungi and Bacteria in Indoor Air Environments, Health Effects, Detection and Remediation*, E. Johanning and C. Yang, eds., Eastern New York Occupational Health Program, Albany, NY, 1995, pp. 5–9.
4. Centers for Disease Control (CDC), Acute pulmonary hemorrhage/hemosiderosis among infants—Cleveland, January 1993–November 1994, *Morb. Mort. Wkly. Rep.* **43**: 881–883 (1994).
5. CDC, Update: Pulmonary hemorrhage/hemosiderosis among infants—Cleveland, Ohio, 1993–1996, *Morb. Mort. Wkly. Rep.* **46**:33–35 (1997).

6. Montana, E., R. Etzel, T. Allan, T. Horgan, and D. Dearborn, Environmental risk factors associated with pediatric idiopathic pulmonary hemorrhage and hemosiderosis in a Cleveland community, *Pediatrics* **99**:117–124 (1997).
7. Etzel, R., E. Montana, W. Sorenson, G. Kullman, T. Allan, and D. Dearborn, Acute pulmonary hemorrhage in infants associated with exposure to Stachybotrys atra and other fungi, *Arch. Pediatr. Adolesc. Med.* **152**:757–762 (1998).
8. CDC, Update: Pulmonary hemorrhage/hemosiderosis among infants—Cleveland, Ohio, 1993–1996. *Morb. Mort. Wkly. Rep.* **49**:180–184 (2000).
9. Straus, D. C., ed., *Sick Building Syndrome*, Vol. 55, *Advances in Applied Microbiology*, Elsevier Academic Press, San Diego, 2004.
10. Dillon, H. K., P. A. Heinsohn, and J. D. Miller, *Field Guide for the Determination of Biological Contaminants in Environmental Samples*, American Industrial Hygiene Assoc., Fairfax, VA, 1996.
11. Hung, L.-L., J. D. Miller, and H. K. Dillon, *Field Guide for the Determination of Biological Contaminants in Environmental Samples*, 2nd ed., American Industrial Hygiene Assoc., Fairfax, VA, 2005.
12. ACGIH, *Guidelines for the Assessment of Bioaerosols in the Indoor Environment*, American Conf. Governmental Industrial Hygienists, Cincinnati, OH, 1989.
13. ACGIH, *Bioaerosols, Assessment and Control*, American Conf. Governmental Industrial Hygienists, Cincinnati, OH, 1999.
14. Morey, P., J. Feeley, and J. Otten, *Biological Contaminants in Indoor Environments*, ASTM, Philadelphia, 1990.
15. Gravesen, S., J. C. Frisvad, and R. A. Samson, *Microfungi*, Munksgaard, Copenhagen, Denmark, 1994.
16. Rylander, R. and R. R. Jacobs, *Organic Dusts: Exposure, Effects, and Prevention*, Lewis Publishers, Boca Raton, FL, 1994.
17. Samson, R. A., B. Flannigan, M. E. Flannigan, A. P. Verhoeff, O. C. G. Adan, and E. S. Hoekstra, eds., *Health Implications of Fungi in Indoor Environments*, Elsevier, Amsterdam, 1994.
18. Burge, H. A., ed., *Bioaerosols*, Lewis Publishers, Boca Raton, FL, 1995.
19. Johanning E. and C. S. Yang, *Fungi and Bacteria in Indoor Air Environments, Health Effects, Detection and Remediation*, Eastern New York Occupational Health Program, Albany, NY, 1995.
20. Hurst, C. J., G. R. Knudsen, M. J. McInerney, L. D. Stetzenbach, and M. V. Walter, *Manual of Environmental Microbiology*, American Society for Microbiology, Washington, DC, 1996.
21. Hurst, C. J., G. R. Knudsen, M. J. McInerney, L. D. Stetzenbach, and M. V. Walter, *Manual of Environmental Microbiology*, 2nd ed., American Society for Microbiology, Washington, DC, 2002.
22. Johanning, E., *Bioaerosols, Fungi and Mycotoxins: Health Effects, Assessment, Prevention and Control.* Eastern New York Occupational and Environmental Health Center, Albany, NY, 1999.
23. Samson, R., E. S. Hoekstra, J. C. Firsvad, and O. Filtenborg, *Introduction to Food- and Airborne Fungi*, 6th ed., CBS, Utrecht, Netherlands, 2000.
24. Flannigan, B., R. A. Samson, and J. D. Miller, *Microorganisms in Home and Indoor Work Environments*, Taylor & Francis, London, 2001.

25. Institute of Medicine (IOM) of the National Academies, Committee on Damp Indoor Spaces and Health, Board on Health Promotion and Disease Prevention, *Damp Indoor Spaces and Health*, The National Academies Press, Washington, DC, 2004.
26. Institute of Medicine (IOM) of the National Academies, Committee on the Assessment of Asthma and Indoor Air, *Clearing the Air: Asthma and Indoor Air Exposures*, National Academies Press, Washington, DC, 2000.
27. Botzenhart, K. and G. Doring, Ecology and epidemiology of *Pseudomonas aeruginosa*, in *Pseudomonas aeruginosa as an Opportunistic Pathogen*, M. Campa, M. Bendinelli, and H. Friedman, eds., Plenum Press, New York, 1993, pp. 1–18.
28. Yang, C. S., Fungal colonization of HVAC fiber-glass air-duct liner in the USA, *Proc. Indoor Air Conf. 1996*, Vol. 3, pp. 173–177.
29. Yang, C. S. and P. J. Ellringer, Evaluation of treating and coating HVAC fibrous glass liners for controlling fungal colonization and amplification. *Proc. Indoor Air Conf. 1996*, 1996, Vol. 3, pp. 167–172.
30. Yang, C. S. and P. J. Ellringer, Antifungal treatments and their effects on fibrous glass liner, *ASHRAE (Am. Soc. Heat. Refrig. Air-Condit. Eng. Assoc.) J.* **46**(4): 35–40, (2004).
31. Yang, C. S. and E. Johanning, Airborne fungi and mycotoxins, in *Manual of Environmental Microbiology*, 2nd ed., C. J. Hurst, G. R. Knudsen, M. J. McInerney, L. D. Stetzenbach, and M. V. Walter, eds., American Society for Microbiology, Washington, DC, 2002, pp. 839–852.
32. Flannigan, B. and J. D. Miller, Microbial growth in indoor environments, in *Microorganisms in Home and Indoor Work Environments*, B. Flannigan, R. A. Samson, and J. D. Miller, eds., Taylor & Francis, London, 2001.
33. Li, D.-W. and C. S. Yang, Fungal contamination as a major contributor of sick building syndrome, in *Sick Building Syndrome, Vol. 55, Advances in Applied Microbiology*, David Straus, ed., Elsevier Academic Press, San Diego, 2004, pp. 31–112.
34. Zabel, R. A. and J. J. Morrell, *Wood Microbiology*, Academic Press, San Diego, 1992.
35. Morey, P. R., Remediation and control of microbial growth in problem buildings, in *Microorganisms in Home and Indoor Work Environments*, B. Flannigan, R. A. Samson, and J. D. Miller, eds., Taylor & Francis, London, 2001, pp. 83–99.
36. Pitt, J. I. and A. D. Hocking, *Fungi and Food Spoilage*, Blackie Academic & Professional, London, 1997.
37. Storey, E., K. H. Dangman, P. Schenck, R. L. DeBernardo, C. S. Yang, A. Bracker, and M. J. Hodgson, *Guidance for Clinicians on the Recognition and Management of Health Effects Related to Mold Exposure and Moisture Indoors*, Univ. Connecticut Health Center, Division of Occupational and Environmental Medicine, Center for Indoor Environments and Health, Farmington, CT (http://www.oehc.uchc.edu/clinser/MOLD%20GUIDE.pdf), 2004.
38. Bush, R. K., J. M. Portnoy, A. Saxton, A. I. Terr, and R. A. Wood, The medical effects of mold exposure. *J. Allergy Clin. Immunol.* **117**:326–333 (2006).
39. Anderrson, M., S. Downs, T. Mitakakis, J. Leuppi, and G. Marks, Natural exposure to *Alternaria* spores induces allergic rhinitis symptoms in sensitized children. *Pediatr. Allergy Immunol.* **14**:100–105 (2003).
40. Arbes, S. J., P. J. Gergen, L. Elliott, and D. C. Zeldin, Prevalences of positive skin test responses to 10 common allergens in the US population: Results from the Third National

Health and Nutrition Examination Survey, *J. Allergy Clin. Immunol.* **116**:377–383 (2005).

41. Bush, R. K. and J. J. Prochnau, *Alternaria*-induced asthma, *J. Allergy Clin. Immunol.* **113**:227–234 (2004).
42. Gergen, P. J. and P. C. Turkeltau, The association of individual allergen reactivity with respiratory disease in a national sample: Data from the second National Health and Nutrition Examination Survey, 1976–1980 (NHANES II), *J. Allergy Clin. Immunol.* **90**:579–588 (1992).
43. Domsch, K. H., W. Gams, and T.-H. Anderson, *Compendium of Soil Fungi*, Vol. 1, reprinted by IHW-Verlag with supplement by W. Gams, 1993.
44. Wang, C. J. K. and R. A. Zabel, *Identification Manual for Fungi from Utility Poles in the Eastern United States*, American Type Culture Collection, Rockville, MD, 1990.
45. Ellis, M. B., *Dematiaceous Hyphomycetes*, CAB International, Commonwealth Mycological Institute, Kew, Surrey, UK, 1971, pp. 608.
46. Andersen, B., E. Kroger, and R. G. Roberts, Chemical and morphological segregation of Alternaria alternata, A. gaisen and A. longipes, *Mycol. Res.* **105**:291–299 (2001).
47. Nielsen, K. F., *Mould growth on building materials: Secondary Metabolites, Mycotoxins and Biomarkers*, Ph.D. thesis, BioCentrum-DTU, Technical Univ. Denmark, Lyngby, Denmark, 2002.
48. Nielsen, K. F., Mycotoxin production by indoor molds, *Fung. Genet. Biol.* **39**:103–117, (2003).
49. Park, J.-H., P. L. Schleiff, M. D. Attfield, J. M. Cox-Ganser, and K. Kreiss, Building-related respiratory symptoms can be predicted with semi-quantitative indices of exposure to dampness and mold, *Indoor Air* **14**:425–433 (2004).
50. L'stiburek, J. and J. Carmody, *Moisture Control Handbook: Principles and Practices for Residential and Small Commercial Buildings*, Wiley, New York, 1994.
51. Harriman, L., G. Brundrett, and R. Kittler, *Humidity Control Design Guide for Commercial and Institutional Buildings*, ASHRAE, Atlanta, GA, 2001.
52. Horner, W. Elliott, A. G. Worthan, and P. R. Morey, Air- and dustborne mycoflora in houses free of water damage and fungal growth. *Appl. Environ. Microbiol.* **70**:6394–6400, (2004).

CHAPTER 2

CONDUCTING BUILDING MOLD INVESTIGATIONS

PATRICIA A. HEINSOHN

2.1. INTRODUCTION

A building investigation for mold is warranted if a building becomes water-damaged by water intrusion through the building envelope, leaks occur within the building envelope, or the building suffers from chronic excessive humidity. Water can penetrate the building envelope by diverse means, such as natural disasters, construction defects, or lack of building maintenance. Multiple sources may contribute to the total moisture burden. Leaks within the building envelope such as plumbing leakage can result in fungal growth that may ultimately affect interior gypsum wallboard or plaster and interior finishes. Fungal growth can also result from excessive and chronic humidity. This is commonly seen on the window caulking and window sills at metal single-pane windows and on gypsum wallboard ceilings in insufficiently ventilated bathrooms. Excessive humidity may also result in large areas of visible growth on interior wall surfaces.

In addition to investigation for mold due to water intrusion and humidity control problems, a building inspection or a baseline investigation for mold and microbiological contamination is recommended as a part of the due-diligence process in a real estate transaction, in the approval of a mortgage application, or in the issuance of an insurance policy. An inspection as a part of due diligence is usually much more limited in scope than a building investigation for mold due to moisture problems. In that destructive sampling and testing are usually not conducted.

The importance of conducting informed building investigations of buildings subject to the above-mentioned sources of moisture cannot be overstated. Effective mold remediation cannot be done if biased, flawed, or inadequate building investigations precede remediation. Building investigations involve many different kinds of investigations. The type of investigation conducted depends upon the goals of the investigator. Table 2.1 describes commonly performed investigations conducted

Sampling and Analysis of Indoor Microorganisms, Edited by Chin S. Yang and Patricia A. Heinsohn

TABLE 2.1. Types of Building Investigations

Type of Investigation	Description	Goal
Baseline	Consists of an informed visual inspection, baseline air sampling for culturable and total fungal spores, surface/source sampling for fungi using a variety of sampling methods Consists of moisture mapping	Determine nature and extent of visible fungal growth Determine whether indoor air quality is affected by visible and/or hidden fungal growth
Destructive testing	Consists of surface sampling on surfaces hidden from view in the baseline investigation May consist of air sampling if documentation is required to verify that there was no escape of fungal spores disturbed during the destructive testing process	Determine whether hidden fungal growth exists Determine whether hidden fungal growth is degrading the indoor air quality; if so, determine extent of growth, i.e., amount and location of materials to be remediated
Contents	Consists of sampling of contents, and possible finishes (such as carpeting)	Determine whether contents are contaminated by species of fungal growth documented in baseline and destructive investigations, respectively
Mold remediation oversight	Consists of sampling of documented moldy surfaces Consists of sampling of fungal growth newly revealed	Determine efficacy of remediation, i.e., whether surface treatment can cease or must be repeated Determine species of fungal growth on surfaces newly revealed through remediation
Mold postremediation sampling	Consists of surface and air sampling	Determine that air inside containment is not degraded with species of fungi being remediated from surfaces Determine whether newly revealed surfaces are effectively remediated

to investigate buildings. Perhaps the most critical investigation is the baseline investigation. Others will be discussed in less detail.

2.2. BASELINE INVESTIGATION

Building investigations do not start with sampling for mold. They start with getting as much relevant background information as possible. Possible sources of relevant information could include the following:

- Building owner
- Building manager
- Building maintenance personnel
- Building architect or architectural plans
- Building occupants
- Other consultants, including various engineers or environmental consultants

The kind of information needed may vary somewhat depending on the nature of water intrusion but commonly includes the following:

- *The Leak History (i.e., a Chronology of Water Intrusion Events) Including Known or Suspected Sources, Location in the Building, and Duration of the Leak and Wetness.* It is rare that an entire structure is water-damaged, although there have been such instances (e.g., complete submersion in a flood, during construction before the roof assembly is complete). The length of time that a material remains wet can predict the kinds of fungi that colonize it.
- *Building Plans and Drawings Such as Floorplans (with Scales), Mechanical Plans, and Plumbing Plans.* Floorplans are needed to estimate the number of air sampling locations needed to appropriately characterize the indoor air quality.
- *Reports from Consultants (e.g., Architects, Engineers) Who Have Conducted Forensic Water Intrusion Investigations.* These reports can be very helpful because fungal growth can occur in the presence of sufficient moisture on a substrate capable of supporting its growth. The opposite is also true; fungal growth does not occur in the absence of sufficient moisture. Documented fungal growth in locations not explained by water intrusion experts begs explanation. The absence of fungal growth in locations known to have been wetted is most likely due to a lack of adequate nutrients and/or lack of the continued moisture due to drying. However, the mold investigator must verify the absence of growth in known or suspected wetted areas.
- *Reports from Other Environmental Consultants Who Have Conducted Independent Investigations in the Past.* Laboratory data from samples collected by others can serve as the reference to expand an independent baseline survey. For example, given the following (1) *Penicillium/Aspergillus* spores excessively

dominate indoor spore trap samples, (2) airborne culture samples are not dominated by undifferentiated *Penicillium* species on malt extract agar (MEA) media, and (3) tape lift samples indicate growth of undifferentiated *Penicillium* species, it is prudent to conduct a second survey including culturable air samples using various fungal isolation media, including a selective medium for xerophilic fungi, and surface samples submitted for culture with species identification on MEA and DG-18 media. This sampling design will likely be sufficient to support or refute the hypothesis that the likely source of growth was due to the flooding incident. If the species of surface *Penicillium* growth differs from the *Penicillium* degrading the indoor air quality, then additional investigation that may involve destructive testing is required to find the source of that *Penicillium* species. (Caution is urged if the original spore trap data were collected with an Air-O-Cell cassette; see Section 2.2.2.2). That growth site may still be related to crawlspace flooding or not, depending on location of the growth site and remainder of the leak history. But the obvious fact remains, without identifying growth sites, mold remediation will fail.

- *Age and Use of the Building.* The older a building is, the more likely that it has a water damage history. An investigator needs to know whether the building is in use and whether occupancy might restrict the investigation. Is the building furnished, or have contents been removed? If yes, why? Does the building have electrical power, and is adequate lighting available? What information has been shared with occupants? Are there any access issues with respect to occupants? The use of many commercial and institutional buildings can predict the amount of building occupant generated sources of moisture, such as, laundry facilities. It is also very important that qualified consultants conduct asbestos and lead surveys if the age of the building indicates that these contaminants are likely in various building construction or finish materials. This information must be collected prior to destructive testing and preparing written technical specifications for mold remediation to ensure that the appropriate steps are taken to minimize remediation worker exposure and prevent atmospheric contamination of all contaminants that will become airborne, not just fungal. Investigators should be thoroughly familiar with all applicable federal and state regulations involving asbestos and lead or must depend on investigations of those who are qualified. An example of an appropriate step would be the inclusion of using wet methods (required in OSHA asbestos regulations[1]) to remove asbestos containing-materials that are rarely included in mold remediation protocols.
- *Occupant Health Status.* In general, the mold investigator needs to know whether building occupants include infants, persons recovering from recent surgery, people with immune suppression, hypersensitivity pneumonitis, severe allergies, sinusitis, or other chronic inflammatory lung diseases.[2,3] Health status should be taken into account when interpreting laboratory data relative to occupancy. A knowledgeable physician may be consulted for making a final decision on occupancy.

- *Building Design and Construction Including Accurate Information on the Building Envelope and Foundation.* It is impossible to specify materials to be remediated unless those materials are known. It is also impossible to specify the method of remediation unless construction and finish materials are known.
- *A Chronology of Water Loss Restoration (if Any) or Prior Mold Remediation (if Any).* Some important questions to be answered include the following:

  Where was the initial water intrusion observed?

  What steps were taken and when to repair the leak and dry out the building?

  What mold remediation steps were taken at what locations, and when?

  An example follows that illustrates why this information is critical:

  Baseline air sampling was conducted in a building that was extensively water-damaged by a high-pressure pipe leak onto crawlspace framing for as much as 2 weeks. The crawlspace framing was never dried out or completely remediated of fungal growth. However, water-damaged and moldy gypsum wallboard and hardwood flooring were removed during mold remediation without containment of affected areas followed by 30 h of detailed cleaning of interior surfaces from which contents had been removed. Baseline air sampling after these activities indicated the presence of low concentrations of *Penicillium* species in the indoor air not found in the outdoor air that was documented growing on crawlspace framing. More likely than not, detailed cleaning reduced levels of airborne *Penicillium* based on these and additional data and information. Additional investigation revealed extensive growth not remediated in the crawlspace and elsewhere. In other circumstances, low concentrations of airborne *Penicillium*, particularly in the absence of documented growth sites of *Penicillium*, might not have been significant.

This example demonstrates the importance of detailed, thorough informed investigations and the importance of good air sample data.

2.2.1. Physical Inspection

Physical inspection requires access to all areas of the subject building unless there is a clearly defined, justifiable reason not to inspect one or more areas. The underlying reason for this is that a moisture problem in one area of a building can impact another. This includes occupied areas and unoccupied areas such as attics, hot-water-heater closets, storage rooms, garages, and crawlspaces. In part, all of the relevant areas investigated depend on the type and use of the building, such as, single-family residence, hospital, hotel, school, office building, and retail commercial space. Gloves should be used when actually handling suspect moldy materials.

The investigator should have all of the appropriate equipment for personal protection and investigation immediately available as needed. Construction areas may require a hardhat, reflective vest, and steel toe footwear. If significant visible mold growth is evident or independently collected air sample data indicate that the indoor air quality is degraded, then respiratory protective equipment is needed. Boots or walking shoes should be worn to climb ladders and access rooftop areas.

It is important to take the right tools, equipment, and supplies to the site for inspection. Please refer to Table 2.2 for a list of these and comments on their use.

TABLE 2.2. Inspection Tools, Equipment, and Supplies

Item	Comment
Generator	Needed if electrical power is not provided
Ladder	Needed to access some areas, including ceilings
Extension cords	—
Plug conversion adapter	Older buildings do not accept two-prong male ends
Flashlight or other auxiliary lighting	Often necessary for attics and crawlspaces
Pocket knife, utility knife, or other sharp edge for cutting	Often needed for collecting bulk samples
Screwdriver	Needed to access components of a heating, ventilating, or air conditioning system
Extra batteries	—
Tape measure	Needed to measure distances to relevant architectural details Needed to measure the square footage of suspect fungal growth
Reflecting mirror	Often needed to view inaccessible areas
Clip chart	Needed to hold inspection forms in an orderly fashion
Forms and plans	Floorplans are very useful for orientation and making notes relevant to architectural details
Graph paper	—
Smoke tube or tissue	Qualitative assessment of working order of exhaust fans
Camera and film	—
Moisture content meter	—
Compass	—
Air sampling equipments with media, supplies, and tripod	Supplies include alcohol wipes, parafilm, labeling pens
Coolers with blue ice	—
Surface sampling equipment	Including, but not limited to, sterile swabs, transparent tape, Rodac surface contact plates

2.2.1.1. Visual Inspection. The primary objectives in a visual inspection portion of a building mold investigation are to note the presence or absence of suspect water staining, suspect water damage, and suspect fungal growth relative to architectural details first, and finishes and contents, second. Therefore, these will be discussed in depth. However, it is impossible to mention all conditions here that are critical to note because they are as diverse as are the buildings investigated. For example, the conditions inside single-family homes are unique to each home. A few of these will be discussed. In addition, the reader is provided with forms appended to this chapter (see Figs. 2.1, 2.2, and 2.6) that may be useful and can be adapted or revised as needed. They illustrate the level of detail that is necessary to conduct professional, competent investigations.

One of the most important steps to take initially is to stand outside the building and determine the orientation of the building on the lot, its relationship to surrounding buildings, and the direction it faces. Building use and design of adjacent buildings may be important, such as the location of a dumpster beside a lower-story window of the subject building that contributed to atypical concentrations of *Aspergillus fumigatus.* Activities on or adjacent to the property may also affect sample data and therefore are important to note during the visual inspection portion of a baseline investigation. Examples are grass-cutting and weed-eating or earthmoving equipment preparing ground for a new housing development.

An investigator should inspect all walls, the ceiling, and the floor for suspect water stains, water damage, and fungal growth in every room inspected. Either the presence or absence should be recorded. One such description could read "Suspect water stain noted on the ceiling of the dining area, round in shape, located approximately one foot from the exterior west wall. No suspect fungal growth seen." This location should also be photographed and the photograph number clearly recorded with the description.

Sometimes what appears to be a water stain, water damage, or fungal growth is not; thus the description as "suspect." Suspect water stains should be documented by forensic water intrusion experts. Suspect water damage may or need to be documented by others, especially if damage is severe and moisture content readings are elevated. Suspect fungal growth should be described by color, coverage area, location relative to architectural details, and general appearance. One such description might be "16 ft^2 of suspect visible growth on painted gypsum wallboard at a height of approximately two feet above the finished floor running the entire length of the bed placed approximately 3 in. from the east exterior wall. Visible after bed was pulled out from the wall. Growth is dark and spotty. There is no suspect growth above the level of the bed. Moisture content readings are all normal, $n = 32$ . . . ." All of these observations are critical in terms of causation, remediation, and implications of the growth.

It is important to note that *visible* fungal growth sometimes is hidden and not apparent in the inspection. In the preceding example, although the top edge of suspect growth was visible without moving the bed out from the wall, it would not have been possible to describe the amount and appearance of growth or take moisture content readings without moving the bed. This suspect growth can correctly be called suspect *visible* growth even though occupants placed the bed too close to the wall to allow for ventilation and rendering the area somewhat inaccessible. Suspect growth behind a refrigerator due to a water supply line leak is normally hidden from view and therefore should not be described as visible. Suspect fungal growth within a wall cavity on any surface that can be seen only after destructive testing or demolition is seldom visible or accessible. Thus, it is correctly described as "hidden" growth. Said another way, just because you can see growth, this does not mean that it is visible growth. *Visibility* to relates with location and accessibility. As long as investigators *define* their terms and are *consistent*, their documentation is clear. This also serves to emphasize the importance of detailed descriptions; there cannot be too much detail. When it comes time to interpret all the findings, taking into account all the background information gathered, a complete physical inspection, and air and surface sampling, you will find this level of detail and documentation valuable.

Suspect fungal growth is verified by surface sampling using an appropriate method after air sampling is completed to avoid potential bias of the air sample data. It is strongly recommended that surface sampling be done for two major reasons: (1) to confirm and determine the species of fungal growth and (2) to determine whether vegetative growth only or if vegetative and reproductive growth exists on the surface sampled (please refer to Table 2.3).

Investigators should inspect interiors to the level of detail described in Table 2.4. Observations should then be recorded on field notes sheets or photograph logs (see Figs. 2.1 and 2.2).

Everything that is noteworthy, literally, should be photographed with the photograph number and a description of why the photo was taken. Photographic documentation will be discussed in greater detail in Section 2.2.1.2, below.

Investigators should inspect exteriors and nonoccupied areas to the level of detail described in Section 2.2.1.2 and record on a combination of field notes sheets, photograph logs, and photograph depictions (see Figs. 2.1–2.3).

2.2.1.2 Documentation. All suspect water stains, suspect water damage, and suspect visible fungal growth must be documented in writing, drawings, or photographs, and via surface sample data. The previous section discussed written descriptions and subsequent sections, and Chapters 4 and 5 discuss documentation verification of the presence or absence of fungal growth through surface sampling and laboratory testing. This section presents some principles of photographic documentation and recommends good practices.

Examples of written descriptions in Section 2.2.1.1 use compass directions: north, south, east, and west. Another acceptable wall designation utilizes numbering walls 1, 2, 3, and 4 coinciding with the building elevation entered, the left side, the opposite side, and the right side of entry, respectively. Figure 2.4 exemplifies the use of this designation scheme. As long as the method used is consistent, either can be used.

All photographs must be physically labeled with the photograph number used on all field documentation. Professional photographers label the reverse side of photographs.

Basic photo documentation involves, to the extent possible, taking three perspectives: overview (the entire view), midrange (the specific object or condition), and

TABLE 2.3. Importance and Use of Species Data

Use	Comments	Other Data Needed
Determine whether indoor air quality is degraded in a baseline survey	Species identification is required to determine the biodiversity between indoor and outdoor air	Both indoor and representative outdoor air sample species data are necessary for comparison
Determine likely cause of degraded indoor air quality in a baseline survey	Not all fungal growth results in degrading the indoor air quality Very small areas of growth can have little or no impact on indoor air quality Hidden fungal growth may not degrade indoor air quality	Species identification in both air and surface samples is required to "connect the dots"
Verify that surface remediation was effective	Goal of mold remediation is to remove fungal growth; visual inspection is the first step but must be followed by surface sampling to verify the absence of growth—this is particularly important for wood and wood products remediated in place because they are porous Surface sampling methodology must be used and same area should be resampled	Initial baseline or destructive test survey data from the same surface are necessary for comparison
Verify that air within containment is not degraded by fungal growth discovered during remediation	Containment barriers should not be removed until air quality is known	Surface sample species and clearance air sample species data are required for comparison

closeup (the specific information about the object or condition). One such example relative to a building mold investigation follows:

- *Overview Photo*—the entire exterior wall cavity revealed through destructive testing
- *Midrange Photo*—portion of wall cavity indicating the area of wet exterior sheathing and suspect fungal growth
- *Closeup*—the location in the cavity where a fastener penetrates the exterior cladding and sheathing

TABLE 2.4. Visual Inspection Checklist

Interior Inspection
Crevices between refrigerators and counters or walls
Headers of windows, window frames, window sills, window caulking
Condensation on inside surface of windows
Condensation (known as "fogging") between window dual panes
Sink cabinet floors, sides, and back and at sink plumbing lines
Underside of toilet tanks and at plumbing lines
Whether windows are fixed, i.e., nonopening and their location
Whether doors or windows were open or closed on arrival at the site
Housekeeping practices, e.g., visibly dusty surfaces, food left on plates in sinks, mildew in bathroom tile grout and caulking, dirty and stained carpet, dust bunnies
In general, the amount of contents
Placement of furniture adjacent to walls
Presence of an exhaust fan in bathroom and whether it actually exhausts
Presence and location of houseplants
Condition of the floor surface underneath plant containers
Presence of location of silk and dried plants
Presence and location of aquariums
Presence and location of wood stored inside
Presence of cats, dogs, birds, or other pets, or evidence of pets (food bowls, cages, etc.)
Presence of smokers or evidence of smoking
Whether clothes driers are exhausted outdoors
Clotheslines indoors for hanging wet clothes to dry
Presence, location, and type of insects
Whether the range is gas
Presence, use, and location of dehumidifiers and humidifiers; record the manufacturer and model
Presence, use, and location of air purifiers; record the manufacturer and model
Presence, location, type, and condition of wallpaper
Presence and location of warping of hardwood flooring
Condition of "drip edge" of shower curtain liners
Exterior Inspection
Amount of vegetation and proximity to building
Presence and location of citrus trees with fruit
Efflorescence on concrete masonry
Cracks in concrete slabs
Noticeable negative soil grade toward building and location
Presence and location of standing water
Presence and location of algal growth on concrete surfaces
Presence or absence of gutters
Termination characteristics of downspouts
Presence or absence of scuppers
Condition of exterior cladding, e.g., peeling, cracking paint on wood siding, termination of stucco (presence/absence of weep screed)
Presence and location of suspect dry rot
Presence and location of suspect wood stain fungi

Page_of_

MICRO BIOS FIELD NOTES

Client/Project:	Project No:	Date:

Signature: ______________________ Date: ______________

Forms/Field Notes Sheet

Fig. 2.1. An example of field notes.

Photograph Log

Acme Warehouse				
123 Elm Street, Utopia, CA		Micro Bios Project No.: 06-004		
Baseline Investigation				April 13, 2006
Photo No.	Room/Area	Sample No.	Sample Type	Description
1.24	Shipping	1.24-S	Swipe	SVMG on GWB of exterior west wall to left of receiving door #6 from warehouse floor to height of 2' and 1' width.
1.23	Employee's bathroom			Unaffected carpet tack strip under bathroom window on east wall
1.22	Employee's lunchroom			Overview of window on east wall
1.21	Employee's lunchroom			Water damaged carpet tack strip under east-facing window, MC = 28.7%. GWB MC = 6.9%, off scale, 3.1%, and 3.9%. See moisture map.

Descriptions:
SVMG: Suspect visible mold growth
MC: Moisture content
GWB: Gypsum wallboard

Fig. 2.2. An example of a photograph log.

Sometimes it is useful to take multiple photographs constituting a 360° circumnavigation of the building. An "establishing shot" is a photograph that shows the front of the building, which documents the size, shape, location, or address. This photograph establishes the north zero degrees (0°). Sometimes an overhead view is appropriate. An overhead photograph is a view from a safe position on a ladder or other relevant elevation. The height of the photographer should be documented in writing. One such photograph description might read "overhead view of suspect preexisting blue stain fungi on roof joists taken standing on second from top rung of 10-ft ladder inserted into above ceiling access hatch in hallway."

If possible, a familiar object, or ruler should be placed in closeup shots to gage the scale involved. Consider using a color chart in a photo other than reference photos. If the color chart is not true to colors in the developed photos, then ask the developer to adjust the color.

All sketches, such as a moisture map described in Section 2.2.1.3, should have a legend that contains the following information:

Date

Author

Photos #–#

[ ] = 1 foot (unit of measurement, e.g., one square = 1 ft)

N (direction north)

Photograph Depiction

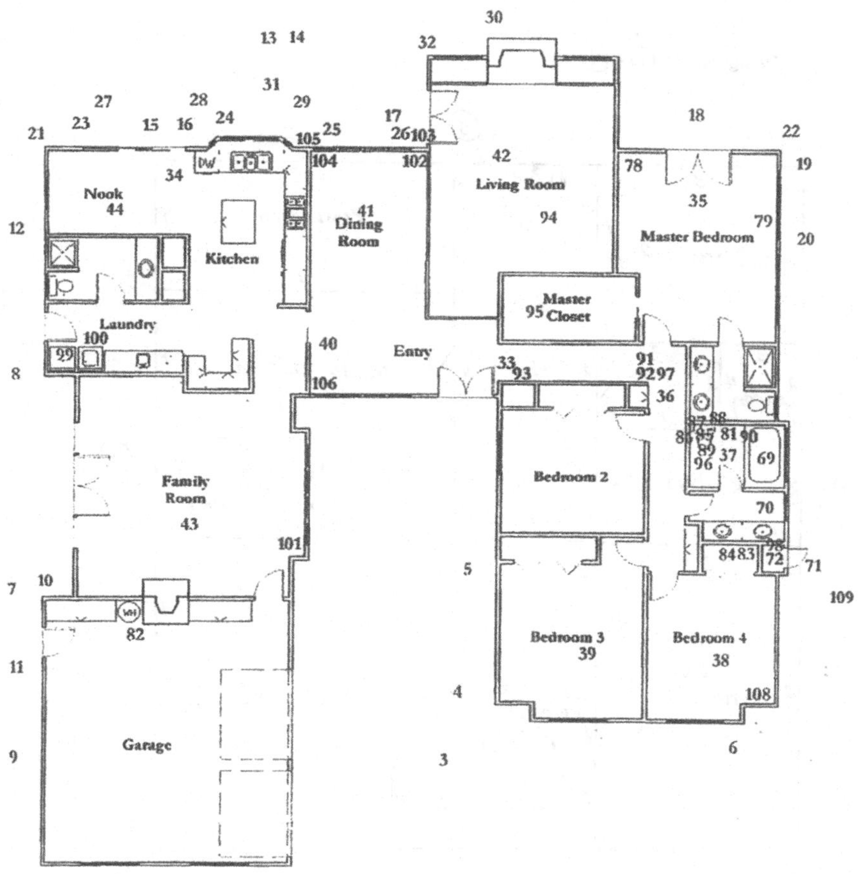

1 2

Case Name: Harold	Client: Daniel King
Affiliated Professional Services, Inc.	Photograph Location Survey

Fig. 2.3. Photograph depiction.

Wall Designations Using Point of Entry as Reference 1

Wall Designations ①–④

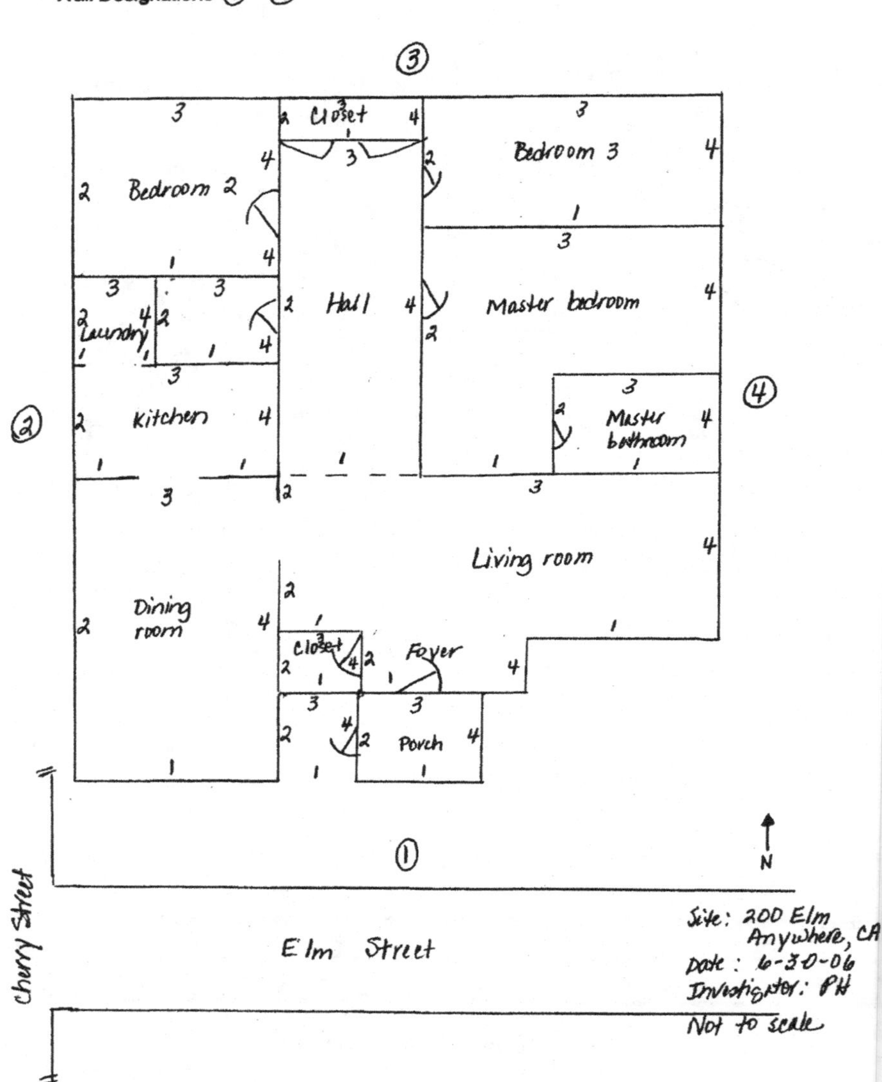

Fig. 2.4. Wall designation using point of entry as reference.

Photo logs should contain at least the following information:

- Photograph number (this is the permanently assigned number that is used in a unique label placed on the back of the photograph)
- Camera shot (if different from the photograph number)
- Photographer
- Date
- Photo description
- Samples collected from a surface in the photograph (if any)

Some of this information can be combined in a header if applicable to all photographs (see Fig. 2.2 for an example of a photo log). Typical photograph descriptions include why the photograph was taken, the subject or object of the photograph, and its significance.

A common and useful practice is to depict photograph numbers on a floorplan. The photograph number and direction is given on the floorplan at the location where the camera was held with an arrow pointing in the direction of the photograph (see Fig. 2.3 for an example). Of course, a written photograph log must accompany it.

Photographs and all documents generated as a result of building investigations are subject to chain-of-custody procedures and should be treated as evidence. This is no less important than following chain-of-custody procedures for samples.

2.2.1.3. Moisture and Moisture Mapping. Sufficient moisture in and on building materials, finishes, and other contents is the most important factor that leads to growth of fungi within buildings. Therefore, surveying for moisture is a very important part of conducting building investigations. Moisture surveys must be done by those conducting mold investigations independent of investigations of a forensic water intrusion expert, such as a qualified and experienced civil or structural engineer, whose job is to identify all sources of water intrusion through or within the building envelope and make recommendations for mitigating it. The mold investigator's first job is to determine the nature and extent of fungal growth (if any) and contamination (if any) resulting from all sources of moisture in the building. Remediation of fungal growth without mitigating the underlying moisture source is doomed to failure. As long as sufficient moisture is present for a long enough period of time, growth will reestablish in the building even if the original water-damaged building materials, finishes, or contents are remediated by treatment in place or removal and replacement.

The source of moisture and type of building material in part predict the location, amount, and species of fungi (if possible) growing. In other words, the mold investigator can assist in determining causal sources of moisture depending on the location, amount, and species of fungi (if possible) growing on or in a building material. Furthermore, if the leak history and the findings of a forensic water intrusion expert are not consistent with the mold investigator's finding of the location, amount, and species of fungi (if possible) growing on building materials, then additional investigation should be done to find the causal source of moisture. Thus, the mold investigator can assist other experts investigating the building. However, if claims are made of chronic water intrusion through the envelope onto building materials that would likely support fungal growth, and the mold investigator finds that the materials are dry and not supporting fungal growth, then these data will serve to refute these claims.

Conducting moisture surveys requires a basic understanding of moisture in buildings and how to use moisture measuring instruments as well as their limitations. The following are basic concepts regarding moisture in buildings:

- All buildings "breathe" moisture. The predominant flow of moisture is determined by geographic location (climate) and a building designer/architect's choice and placement of building envelope materials. The reader should consult texts (e.g., Ref. 4) in building science for in-depth discussion of building science as it relates to these topics.
- Building materials can get wet and dry without supporting fungal growth.
- Building materials are not sterile at time of construction from the manufacturer.
- Occupants themselves, appliances and equipment used inside buildings, and occupant activities generate moisture in all buildings. Some of these and often overlooked non-occupant-related sources of indoor moisture are listed in Table 2.5. Quantitative measurement of the amount of moisture generated has been determined by others.[4]

(*Note*: All sources of moisture should be noted and documented in a mold investigation to the extent possible. For example, nine people living in a two-bedroom one-bath apartment is significant in terms of the amount of moisture generated through respiration, bathing, and gas cooking.)

Air contains varying amounts of moisture in the gas or vapor form depending on the temperature. Cooler air holds less moisture, and warmer air can hold more moisture. The absolute humidity is the actual amount of moisture contained in air. The absolute humidity is the ratio of the mass of water vapor to the mass of dry air. The relative humidity is the ratio of the actual amount of moisture in the air to the maximum amount of moisture the air can hold at a given temperature. Taking air temperature and relative humidity readings in multiple locations in buildings is an important part of a building investigation and is seldom done by mold investigators. There have been general recommendations that indoor humidity should be maintained at 30–50%, is considered excessive at 60–70%, and is likely to promote microbial growth above 70%. However, the dewpoint is the temperature at which humidity in the air reaches saturation (100%) and will condense from that air to form condensation or "dew" on surfaces. Thus, reaching dewpoint resulting in condensation on building materials and finishes is the determining factor in establishing fungal growth.

Many temperature and humidity instruments are commercially available. Some instruments can easily be attached to a tripod on which fungal air samplers are mounted. The dewpoint is easily determined by reading it off a psychometric chart

TABLE 2.5. Occupant-Related and Often Overlooked Non-Occupant-Related Indoor Moisture Sources

Occupant-Related Sources
Aquariums by evaporative loss
Bathing by tub and shower
Clothes washing by hand
Clothes drying: indoor line-dried clothes, nonvented dryers
Cooking by gas, uncovered simmering, uncovered boiling
Hand dishwashing
Firewood storage
Gas range pilot
Floor mopping
House plants
Humidifier use
Refrigerator defrost
Saunas, steam baths, whirlpools
Vegetable storage
Non-Occupant-Related Sources
Desorption of materials, both seasonal and new construction
Ground moisture migration
Plumbing leaks
Rain or snowmelt penetration
Seasonal high outdoor absolute humidity

using the measured temperature and relative humidity. Some temperature/humidity instruments also display the calculated dewpoint. It is important to remember that these instruments need to be calibrated according to the manufacturer's specifications. Materials for conducting in-house calibration can be purchased from the manufacturer.

Although severe, chronic indoor humidity resulting in condensation can actually make a building material wet, the primary use of moisture content readings is to assess wetness due to liquid water from water intrusion and leakage. Moisture content is the percentage or weight of moisture in a material, as compared to the weight of that material when completely dry. It can be viewed as that amount of water available for fungal growth. Moisture content readings are used in moisture mapping. They can be very useful in locating sources of moisture relative to the presence or absence of fungal growth.

To do moisture content mapping of a target area, multiple measurements are taken at incremental spacing to cover the suspect area and beyond. Measurements must be recorded on a form standardized by the investigator. Ideally, measurements are written on or accompany a sketch of the subject area. For example, readings are taken at 6-in. increments to form a square grid. The final shape of the grid may be dictated by objects such as fixtures. Consider the following example. Water staining is noted along the edge of a tub and the edge of resilient flooring has delaminated from the sub floor curling slightly upward in a bathroom currently in use. Mildew discoloration is observed in the caulking along the tub for about a distance of one foot from the plumbing wall (at head of the tub) and in the caulking at the leg of the tub up about 3 in. above the finished floor. Mildew is noted in caulking at the back of the toilet at the edge of the flooring.

A claim was made that all the water staining and mildew observed is due to shower use; that is, water escapes the shower/tub to wet the flooring despite the presence of a shower curtain. The claimant denies a plumbing leak anywhere along the plumbing wall (at the sink, toilet, or shower/tub).

Moisture content readings were taken of gypsum wallboard using a Delmhorst Contractor's Moisture Meter BD-2100 on scale 3 with 2941/A-100 contact pins with $\frac{5}{16}$-in. penetration according to the manufacturer's recommendations. (Delmhorst products are mentioned here by way of example and because they are in widespread use by building investigators.) Moisture content readings were taken at 6-in. increments of the entire accessible plumbing wall. The first set of measurements was taken around the sink plumbing, including above, below, and both sides of the cold- and hot-water supply lines and drainline. No signs of water staining directly below the sink drainline were observed. The second set of measurements was of the entire area between the edge of the sink cabinet and tub/shower fixture, including above, below and to the sides of the water supply line to the toilet tank. It is important to survey this entire area because it can eliminate one or more leaking areas. The third set of measurements was taken at the side of the tub/shower enclosure at the plumbing wall from the floor level to the height of the shower head. Please refer to Fig. 2.5 for a sketch of the moisture mapping area and the recorded readings. Note that all the relevant information for the survey is provided, including the name of the investigator, date, manufacturer and model of meter, electrode attached, scale used, room, orientation, and exact location for each reading.

As can be seen from Fig. 2.5, the visual observations and moisture content readings clearly eliminate leaks at the sink plumbing lines. The majority of moisture content readings between the sink cabinet and shower/tub clearly indicate the absence of any sink plumbing leak that migrated to this area or plumbing leaks associated with the toilet. The only elevated moisture content readings appeared to begin about 10 in. above the level of the tub (which is about 2 in. above the tub plumbing lines) and extending downward in a conical fashion with the widest portion at floor level. Importantly, none of the readings 2 in. above the tub's faucet were elevated, thus discrediting the claim that damage was due to occupant shower use. In this example, moisture mapping combined with visual observations were powerful tools in determining whether the claim had merit. Of course, a plumber or a water intrusion expert with expertise in plumbing is needed to confirm that the shower/tub plumbing was the source of the leak.

There are essentially three methods for determining the moisture content of a material. These are based on oven-dry mass measurement,[5] measurement based on electrical resistance using a pin meter, and measurement based on capacitance using a pinless meter. While an exact method, the oven-dry mass method is impractical and rarely used for building investigations. For pin meters, the electrical resistance is measured between two physical points in the subject material along the

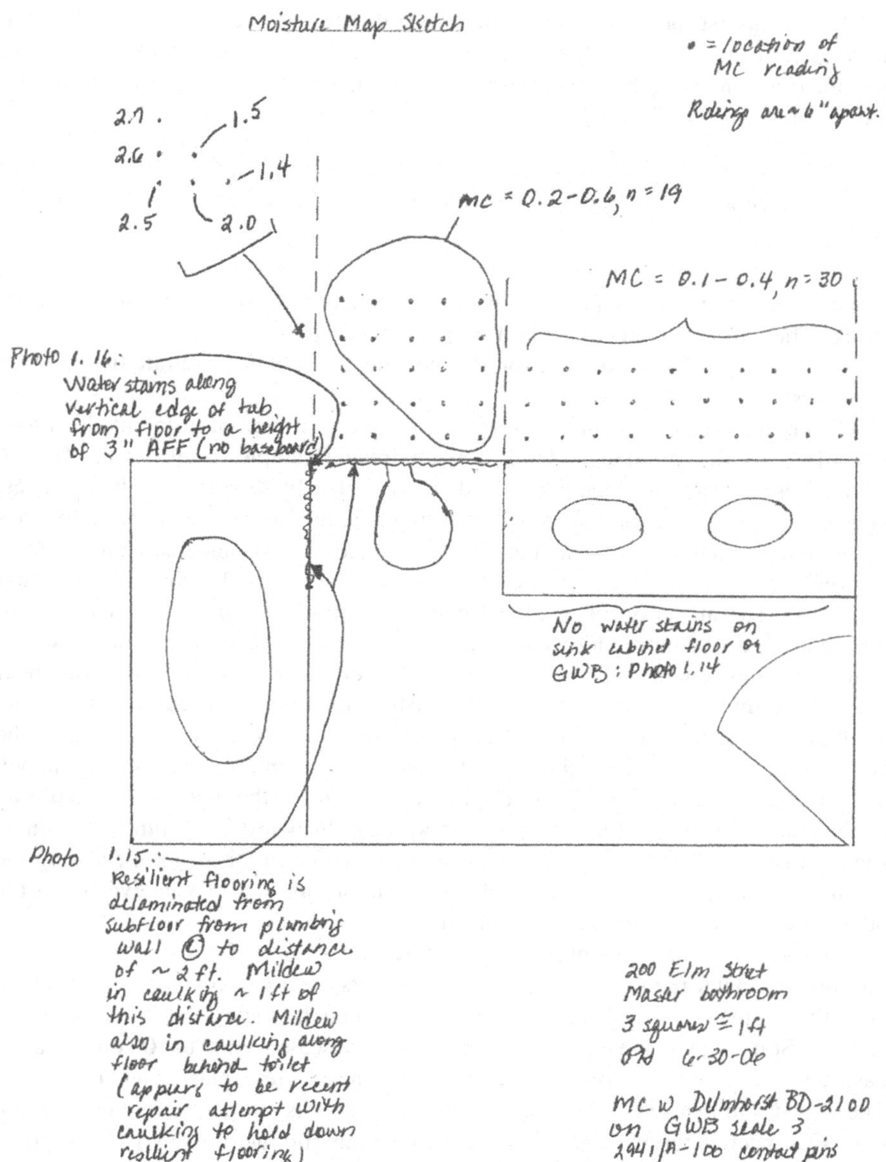

Fig. 2.5. A moisture map sketch.

length of the pin at the point of highest moisture content (for noninsulated pins). For pinless meters, moisture content is measured at a fixed depth below the sensor specified by the model's manufacturer as a function of electromagnetic waves that penetrate the subject material. These differences must be understood in order to select and use a meter correctly.

The Delmhorst meter described above is a direct-read meter with pins that are inserted into the material of interest. The BD-2100 model is capable of reading the moisture content of three categories of material, namely, wood, plaster/concrete, and gypsum (sheetrock). Their scales vary as follows:

- Wood scale—15%
- Plaster/concrete reference scale—85%
- Gypsum—1%

Moisture meters usually require factory calibration. Some have a built-in calibration check. The calibration check should be performed before, during, and after every moisture survey. The contact pins or attached electrode must penetrate the material of interest.

Most moisture meter manufacturers provide an assortment of different electrodes that allow for measurement of hard-to-reach areas. Examples of electrodes for use with different Delmhorst models include pins, rolled blades, and flat bent blades. Pins can be insulated or not and of varying length for varying penetration. As stated above, noninsulated pins measure the electrical resistance at a pin depth at the highest content. This content may not be at the tip. In contrast, moisture content is read at the tip of insulated pins. This allows for insertion of the tip to different depths to more closely ascertain the origin of the water. In other words, is the moisture on the surface, in the middle, or the far side of plywood shear wall? Determining the location of the moisture requires knowledge of the exact thickness of all materials that will be penetrated and a means of measuring the depth penetrated. The investigator must select the manufacturer, meter, model, and electrodes according to the user's needs and then use them properly in building investigations. The sensor of pinless meters usually penetrates a depth of $\frac{3}{4}$–1-in. If such a meter is placed on $\frac{3}{8}$-in. plywood in a wall assembly, it is not measuring the moisture content of the plywood at all. It is measuring that of whatever is on the other side of it unless the other side is air.

Although moisture content readings can be extremely useful, caution is urged in interpreting meter readings. Scales between meters are not necessarily interchangeable; that is, the wood scale on one meter may not be the same as the wood scale for another. Some meters are equipped with a reference scale from 0 to 100 on a relative basis and can be used on nonwood materials such as concrete, plaster, and insulation. The user has to measure reference, nonsuspect materials and compare them to readings of suspect materials in order to interpret the measurements. Clearly, one cannot make a direct comparison between readings taken of the same subject material with different meters without knowing exactly how the measurements were taken and how each meter operates.

The Delmhorst BD-2100 model wood scale is factory-calibrated against Douglas fir, the USDA standard. However, the electrical characteristics of different species vary, and some species may read differently at the same moisture content. For most situations, the Douglas fir reading can be used for construction lumber (e.g., joists, studs). If the user is measuring a different solid wood, the Douglas fir reading can be converted by using a species conversion chart. If the user needs an exact reading for the following wood product, the following specified conversion factor is recommended by Delmhorst:

- *Plywood*—use the SPF (spruce pine fir) correction factor.
- *Oriented Strand Board* (OSB)—use the basswood correction factor.
- *Medium-Density Fiberboard* (MDF)—use the Philippine mahogany correction factor.

If relative readings of these wood products are sufficient, then conversions are not necessary. A reading can be more precise by making temperature conversions as well. However, this conversion is usually considered insignificant if the temperature ranges within 50–90°F.

Moisture meters are used by contractors for their particular trade. Moisture content readings of wood are useful to a painter because wood must begin dry and remain dry after the application of paint. Outdoor wood can be safely painted without danger of peeling if the moisture content is 15% or less (BD-2100). In drier climates, the maximum reading should be 10–11%. In fact, there are recommended average moisture content averages for interior use of wood products in various areas of the United States.[7] Indoor wood should be 7–8% prior to painting. A similar scenario could be described for testing concrete slabs before finished floors are applied. For gypsum wallboard, the readings are useful to know if gypsum wallboard is too wet for painting or hanging wallpaper.

Contractors' uses contrast starkly with those of building mold and water intrusion investigators. While tracing leaks is a valid application of these meters, care must be taken not to overinterpret moisture content readings. For example, gypsum wallboard readings of 0.5–1% do not mean that the gypsum wallboard is necessarily "wet" or that the reading is necessarily associated with a leak. The same can be said of readings at 1%, 1.2%, and 1.4%. The latter means that the gypsum wallboard contains more moisture than it should to paint or hang wallpaper. Like all other parts of building mold investigations, moisture content readings are viewed in light of all other findings, information, and data. If, on the other hand, gypsum wallboard moisture content readings were much higher (e.g., 5.7%) in the absence or presence of suspect visible fungal growth, the readings can be immediately interpreted as indication of an active leak. Sources of active leaks must be terminated and repaired as soon as possible. Air and surface sampling should be performed.

Other survey instruments can also be useful in locating moisture in buildings to greater depths, such as those based on infrared thermography. Generally, these are used by the water loss restoration industry and therefore are mentioned only briefly here. One application might be infrared scanning an entire kitchen cabinet wall to determine whether drying needs to be done behind the cabinetry.

So-called normal moisture content does not mean the absence of unseen fungal growth. The subject material could have become wet and dried, even more than once. If it was sufficiently wet for some time, nonvisible fungal growth could have resulted. However, normal readings certainly argue against chronic, continual water intrusion.

2.2.2. Sampling Design

After obtaining all relevant information, the mold investigator develops a sampling design that will fulfill the goal to determine the nature and extent of fungal growth and contamination. Execution of the sampling design follows the physical inspection.

This chapter includes recommendations for various investigation sampling designs. Consult other chapters in this book that specifically address statistics, sample analysis, and fungal ecology to complement discussion here. A recommended sampling design follows that includes selection of air sampling methods, selection of air sampling instruments, air sampling flowrate and air sampling duration, the number of indoor and outdoor air sampling locations, sampling techniques, surface sampling and analysis methods, laboratory selection, and recommended prior notification to occupants. All of these parameters directly affect the resulting data on which interpretation relative to the building and subsequent recommendations are based.

These decisions are not trivial and predict the quality of data collected and all subsequent decisions. At the time of writing, it has been a decade since the American Industrial Hygiene Association (AIHA) published the first edition of the *Field Guide for the Determination of Biological Contaminants in Environmental Samples*[6] and 7 years since the American Conference of Governmental Industrial Hygienists (ACGIH) published *Bioaerosols: Assessment and Control.*[2] Yet even the issue of whether to conduct air sampling is still debated. Results of air sampling provides empirical evidence on which to make informed decisions about indoor air quality that necessarily impacts decisions regarding occupancy and relating that to growth sites. If a decision is made not to conduct air sampling, then vacating the building is not justifiable because fungal growth can potentially exist without significantly degrading the indoor air quality. Likewise, leaving immune-suppressed individuals in the presence of air degraded, for example, by pathogenic fungi, would likely be considered negligent. Also, without air sample data, it is also not possible to determine the effect, if any, of hidden fungal growth. Baseline data also serve as the basis of comparison for postremediation sample data. However, caution is urged. Poorly conceived sampling designs, poorly executed sampling designs, poor-quality or inaccurate laboratory analysis, and/or over- or underinterpreted sample data also may not fulfill the investigator's goals and have unsupportable ramifications with respect to disruption of life and/or business, and the liability of many more parties than is first apparent.

2.2.2.1. Air Sampling Methods. Air sampling for fungal spores should include sampling both total fungal spores, also called "spore trap sampling," and sampling for culturable fungi (sometimes erroneously referred to as "viable sampling"). Both the AIHA[6,8] and ACGIH pubications[2] cited above encourage species identification.

The latter unambiguously states that indoor and outdoor air comparisons cannot be made "unless the genera and species found indoors and outdoors have been identified."[2]

Many more air sampling and analytical methods are available than those for total fungal spores and culturable fungi, including but not limited to, microbial volatile organic compounds (MVOCs), polymerase chain reaction (PCR), $(1 \rightarrow 3)$-β-D-glucan (polyglucose compound in the cell wall of fungi), and ergosterol (the most common sterol of fungi). Currently, none of these surpasses the quality of information provided by culture with proper and accurate species identification accompanied by spore trap.

2.2.2.2. Selection of Air Sampling Instruments. For a thorough understanding of the advantages and disadvantages of various bioaerosol samplers or the mechanisms of aerosol collection, the reader is referred to a comprehensive discussion of inertial sampling instruments.[9] Some reference sources do not clarify whether sampling bioaerosols is species-dependent;[10] however, it has been known for decades that sampling efficiency with culturable analysis is species-dependent.[11,12] More recently, sampling efficiency with non-culture-based analysis has also been shown to be species-dependent.[12]

New sampling instruments are continuously introduced on the market with and without published validation data pertaining to sampling efficiency under controlled test atmospheric conditions. For example, the Zefon Air-O-Cell spore trap air sampling cassette is widely used. No studies with regard to sampling efficiency were available when the cassette was first marketed. On review of the cassette design, it seemed that its collection characteristics should compare well with the Allergenco or Burkard model because the slit dimensions were almost identical. Side-by-side comparisons of the Allergenco and Air-O-Cell in field investigations revealed that the data were not entirely comparable; it appeared that higher percentages of *Penicillium/Aspergillus* were recovered with the Air-O-Cell, which led to reaching different conclusions as to the quality of indoor air (P. Heinsohn, unpublished data). Although it yielded useful data, this study design could not assign the amount of variability due to the instrument itself without knowing how much analytical variability there is in the data, or how much of the difference was attributable to environmental variability. Subsequently, the Air-O-Cell cassette was shown under controlled laboratory conditions to result in an underestimation of spores of *Cladosporium cladosporioides*,[13] which is one of the most common fungi in indoor and outdoor air. In otherwise nondegraded indoor air, its underestimation could provide an incorrect rank order for fungi that it more efficiently recovers from the air, including *Penicillium/Aspergillus*.[13,14] This does not mean that the Air-O-Cell cannot or should not be used. Its usefulness outweighs this pitfall. However, because *C. cladosporioides* spores are undersampled, the investigator should use results of Air-O-Cell if this differential sampling efficiency is factored in when interpreting the data. The side-by-side culturable air sample data with species identification are usually more important, especially at low spore concentrations.

Proper selection of air sampling instrumentations requires that investigators have a rigorous foundation with regard to the underlying principles of aerosol technology and mycology and that they continually read the peer-reviewed scientific literature. One should be wary of any air sampling instrument for which no peer-reviewed published data independently collected exist. Investigators should avoid using samplers that lack data from laboratory-controlled test bioaerosol challenges demonstrating sampling efficiency.

The use of the Andersen N-6 for culturable air samples side-by-side with the Air-O-Cell spore trap samples has become a common practice in mold investigations.[14] The Andersen N-6 is a single stage sieve impactor that is the sixth stage from the Andersen six-stage sampler. The sampling efficiency for this sampler is perhaps the best known of all of the bioaerosol samplers[15] and has long been a "standard" bioaerosol sampler.[8,16] Collection of side-by-side Andersen culturable samples with species identification and spore trap samples is recommended.

2.2.2.3. Air Sampling Flowrate, Pump Calibration, and Sampling Duration. The Andersen N-6 sampler is designed to be operated at its optimal flowrate to ensure collecting a particle size distribution with the known cutpoint. The Allergenco and Burkard samplers for total fungal spores should also be operated at their design flow rates, and the Air-O-Cell at the manufacturer's current recommended flowrate of 15 L/min. Thus, it is important to calibrate pump flowrate to ensure particle collection according to design specifications. However, the Air-O-Cell manufacturer's recommended flowrate is somewhat artificial and arbitrary. The cassette was found to have better collection efficiencies for two fungal spore types at higher flowrates in research laboratory testing.[17] A disadvantage of the Allergenco and Burkard samplers is that investigators can only *measure* the flowrate. These samplers have to be returned to the factory to be calibrated.

Ideally, pump flowrate should be calibrated using a primary flow standard such as the Bios DryCal, which is traceable to the standards established at the National Institute of Standards and Technology (NIST). Pump calibration should be done prior to use in the field and the flowrate measured at the conclusion of sampling to obtain the average flowrate over the entire sampling period. Periodic field checks can also be performed. Flowrate checks can be performed with a rotameter if the rotameter has been calibrated against a primary standard, such as a bubble flowmeter or DryCal. Temperature and pressure corrections should be done if necessary. Figure 2.6 is an example of the kind of documentation necessary for pump flowrate calibration. Note the level of information that should be included with all pump calibration. The average flowrate over the entire sampling period is used to calculate the air volume sampled. A difference of more than 5% of the calibration flowrate is considered unacceptable.

The sampling period or duration needs to be sufficiently long enough to meet the limit of quantification for the sampler chosen, such as the Andersen N-6 sampler.[8] It is clear that 1- and -2 min samples are often too short to meet the limit of quantification for most indoor environments.[12] A 3-min sampling period for culturable

Pump Flow Calibration

Micro Bios Pump flowrate calibration

Micro Bios Project No.

Sampler	Pump	Calibrator
Andersen N6, SN L337	Gast Model 10-709 SN 0500-2867	DryCal DC-Lite Model DCL-H SN 3131

Date: Initials: Date: Initials:

Calibration flowrate, n =	Post-calibration flowrate, n =	Average flowrate

Sampler	Pump	Calibrator
Zefon Air-O-Cell	Gast Model 10-709 SN 2866	DryCal DC-Lite DCL-H SN 3131

Date: Initials: Date: Initials:

Calibration flowrate, n =	Post-calibration flowrate, n =	Average flowrate

Fig. 2.6. An example of an air pump flowrate calibration record.

air sampling using the Andersen N-6 is usually sufficient. Culturable air sampling for longer periods of time always has to be balanced with the knowledge that desiccation can occur as more air passes over propagules already impacted on the media resulting in a loss of culturability. In other words, lengthier sampling periods may fail to meet the statistical limit of quantification not because the airborne concentration is too low but because fewer propagules remain viable and culturable. In addition, sieve impactors such as the Andersen N-6 has an upper limit of collection controlled by the number of holes.

Some manufacturers recommend different sampling periods for different built environments. While this makes sense for industrial work environments such as cotton mills or grain elevators where concentrations of fungi as high as 1×10^4 CFU/m^3 (colony-forming units per cubic meter of air) have been documented,[18] it does not make sense for nonindustrial indoor environments when sampling as part of a baseline survey. In the unlikely event that baseline survey samples are overloaded, a shorter sampling time can be used in a second survey. Samples collected indoors and outdoors for the same sampling duration can be directly compared, whereas those with disparate sampling times cannot.

2.2.2.4. Number of Indoor and Outdoor Air Samples, Air Sampling Locations, and Order of Sampling. Although there are no generally recommended guidelines, at least two spore trap samples and two culturable air samples should be collected at each location. The underlying reason for collecting two of each kind of sample is to have more data from the same location to assess the true mean of the airborne concentration at that location at that point in time. The recommendation is to collect a pair of samples, namely, one spore trap and one culturable air sample simultaneously, followed immediately by a second such pair. This approach, rather than collecting two samples for total fungi collected simultaneously followed by two samples for culturable fungi collected simultaneously, allows for direct comparison between the two methods collected at the same time for that air sampling location that is needed for a direct comparison between the two kinds of samples for the same time period. Trends in the data will be obvious when samples are collected this way.

Some investigators collect air samples in a "lap" fashion, where a spore trap–culturable air sample pair is collected in one location and the second sample of each method at that location is not collected until after the first pair is collected at all other air sampling locations. Other investigators may separate pairs of samples with an arbitrary time period, such as 2 h. The inherent temporal environmental variability may or may not be detectable in the "lap" sampling.

The decision of how many air samples of the same type at the same location can be addressed from either a statistical or practical standpoint. Other sources should be consulted regarding statistical calculations of how many samples are needed on the basis of the goal of sampling.[2] In general, collecting the number of samples required to meet statistical significance is simply not logistically or economically feasible. Consideration should also be given as to the possibility of actually resampling room air using multiple high-volume air sampling pumps in small rooms, such as typical bathrooms and laundryrooms. It would be difficult if not impossible to accommodate multiple sampling trains with tripods in most typically furnished rooms. More than one operator would be required to operate and move multiple sampling trains efficiently from one location to another. This increases labor costs to the high capital expense of multiple sampling trains and many times the analytical cost.

One sampling location per 500 ft^2 is usually adequate unless there is an area for which no information is desired or unless the floorplan dictates otherwise. Fungal bioaerosols and the concentration of these particles does not reach an equilibrium concentration as would a gas or vapor in a defined space given sufficient time. In fact, as do all particles, they settle out of the air as a function of their particle aerodynamic diameter,[19] unless the air is purposely moved from one location to another and sufficient transport velocity is maintained to keep them from settling out.[19,20] Also, the aerosol concentrations are highest at the point of generation prior to settling out of the air. The goal of most air sampling is to determine whether fungi that are growing on building materials are degrading the air quality.

Reference outdoor air samples should be collected at the outdoor air intake of the air handling unit(s) serving the building, whenever possible. This is not always possible since in many high-rise buildings the outdoor air intakes can be several stories high with no access. The mechanical plans will reveal how the building is serviced, the locations of the air-handling unit serving each section of the building, the location of the outdoor air intake, mixing plenums, and building exhaust locations. Based on review of the mechanical and architectural plans, the investigator must choose appropriate outdoor air sampling locations that would be indicative of the outdoor air infiltrating into occupied areas. For example, consider the appropriate outdoor air locations for a 15-story apartment building that is designed to mechanically ventilate common-use areas such as hallways leading to apartments that have balconies. Appropriate locations include the air intake of the air handler supplying the hallway outside the apartment entry door and on the balcony of that apartment. In other words, all outdoor air sources for the apartment should be characterized. Natural ventilation occurs through open doors to the balcony and/or open windows and through window weep holes to a lesser degree. The fact that the higher above the ground a story is, the lower the concentration of phylloplane and soil fungi is likely to be should be considered when planning the number and location of outdoor air samples in high-rise buildings. It is recommended that the investigator assess the pressure relationship of the hallway and apartment and what the conditions are at the time. In other words, that pressure relationship depends on other factors such as whether bathroom or kitchen exhaust fans are operating and elevator use.

2.2.2.5. Sampling Techniques. An investigator has the option of using techniques to influence the air sample data by creating air currents to resuspend settled fungal spores, so-called aggressive air sampling. Some investigators attempt to maximize airborne concentrations of fungi by purposefully stirring up dusts through activities such as stomping on carpet, banging on furniture, or waving a clipboard in the air. Aggressive air sampling cannot be reliably reproduced. It may produce positively biased data. The more appropriate sampling method for determining whether surface concentrations of fungi constitute background levels or contamination above background include microvacuum dust samples or surface contact samples, rather than air samples.

Please refer to Table 2.6 for a summary of previous sections.

2.2.2.6. Surface Sampling Techniques and Analysis. There are five basic surface sampling methods: wipe (swipe, swab), tape lift, bulk, surface contact plate, and settled dust cassette. There are advantages and disadvantages of each of these sampling methods. Additional information on surface sampling may be found in Chapter 5.

In simplest terms, all sampling methods for which the analysis is culture allow for identification of fungi to the species level. The importance of species information

TABLE 2.6. Recommended Air Sampling Design

	General
Number of indoor air sampling locations	One location per 500 ft^2 of floor space unless room sizes are smaller
Placement of air sampling equipment within room	Central to room to the extent possible; ≥ 1 m away from an exterior wall; away from plants; on a tripod approximately 4 ft above finished floor
Number of outdoor air sample locations for single-story, nonmechanically ventilated buildings	At least two, front and back of the building; tripod should sit on concrete or asphalt, if possible, and away from plants
Number of outdoor air sample locations for multiple-story, nonmechanically ventilated buildings	Two-grade level at front and back of building and on a balcony or deck for at least the first five stories; tripod should sit on concrete or asphalt, if possible, and away from plants
Number of outdoor air sample locations for multiple-story, mechanically ventilated buildings	At air intake of the air handling unit supplying the area(s) of interest; tripod should sit on concrete or asphalt, if possible, and away from plants
Air sampling methods	Total fungal spores and airborne culturable fungi
Laboratory analysis of air samples	—
Media for buildings without immune-suppressed occupants and incubation temperature	MEA (see text discussion) and 25°C
Media for buildings with immune-suppressed occupants and incubation temperature	MEA at 25°C and 37°C
Laboratory analysis of culturable air samples	Isolation, enumeration, and identification to the species level
Laboratory analysis of total fungi	See Ref. 13 for discussion
Air sampling instruments, flow rate, and time	Andersen N-6 at 28.3 L/min for culturable air samples for 3 min Allergenco, Burkard, Air-O-Cell
Sample identification	Blinded

cannot be overstated for air or surface samples and has been already been discussed in this chapter and is also partially summarized in Table 2.3. Wipe, surface contact plate, and settled dust cassette surface samples are analyzed by culture. Tape lifts are analyzed by direct microscopic examination only, and bulk samples can be cultured and microscopically examined.

In general, surface samples are collected to determine whether a surface is supporting fungal growth or if the surface is contaminated from growth site(s) elsewhere. In general, fungal growth is visible and contamination is not. Of course,

TABLE 2.7. Some Reported Fungi Found in House Dust

	Species	Reference
Fungi typically isolated from dry environmental house dust	*Aspergillus penicillioides* *Aspergillus restrictus* *Eurotium amstellodami* *Eurotium chevalieri* *Wallemia sebi*	22
Fungi typically isolated in house dust of homes with conspicuous amplification	*Aspergillus veriscolor* *Penicillium brevicompactum* *Penicillium chrysogenum*	23
House dust as an ecological niche for these fungi	*Aspergillus fumigatus* *Aspergillus versicolor*	24

almost all indoor surfaces are "contaminated" by fungal spores that settle out of the air. This contamination is a form of "background contamination."

The types of fungi typically found in house dust have been published and include those that are commonly found growing in water-damaged materials (see Table 2.7). Thus, it does not suffice to find these fungi in surface contact or settled dust samples to conclude that the surfaces are contaminated above background. One must also document a growth site of the same species of fungi as long as this conclusion is consistent with the chronology of events (e.g., when the leak occurred, actions taken).

The major disadvantage of tape lift methods is that they seldom can provide species information. The major advantage is that a competent laboratory will describe all the morphological structures that they see. Description of these allows the investigator to determine whether vegetative only, or both vegetative and reproductive, growth exists on the surface. Unfortunately, many accredited laboratories report only the genus and a category "hyphae" without association with to any genus. This reporting scheme does not take advantage of the power of tape lift analysis. In addition, some accredited laboratories report quantitative spore counts per unit area, such as spores/mm^2. While this reporting scheme may be useful to determine whether contamination has occurred, it has nothing to do with growth. Similarly, although wipe samples are ideal for culture analysis, some laboratories perform a microscopic examination. Again, only inoculation onto media allows for colonial growth, which provides an abundance of morphological structures for the mycologist to identify the species isolated. It also allows for quantitation. This type of analysis is far superior to a qualitative estimate without species confirmation when analyzed by microscopy.

Table 2.8 summarizes the *optimal* use of surface sampling methods.

2.2.2.7. Laboratory Selection. Few environmental microbiology laboratories provide consistent, competent laboratory analysis for all analyses. One laboratory may provide competent analysis of spore trap air samples; another laboratory may provide competent identification of *Penicillium* species present in culturable

TABLE 2.8. Summary of *Optimal* Use of Surface Samples

Surface Sample Type	Culture or Microscopic Analysis
Wipe	Culture
Tape	Microscopic
Bulk	Culture and microscopic
Surface contact plate	Culture
Surface dust cassette	Culture

air and surface samples. The latter is particularly true in light of the difficulty in differentiating species of *Penicillium*.

Numerous factors are important in selecting the right laboratory for an analysis. The most important factor is the training, education, and experience of the analysts themselves. For example, do not assume that someone with a clinical microbiology background can competently identify environmental fungi. A laboratory that performs various fungal testing must be supervised by a mycologist, typically with an advanced graduate degree. Please refer to Chapter 1 for in-depth discussions on differences among microbiologists, bacteriologists and mycologists. It takes many years of experience at identification to become proficient, even with the appropriate education. Be careful to use laboratories where all the analysts are qualified, not just the laboratory director.

Accurate and consistent laboratory analysis is also a function of a laboratory's standard operating procedures (SOPs). This cannot be overemphasized because of the current lack of standardized analysis for spore trap and culture analysis. This means that spore trap and culture analysis is not done in the same way by all laboratories. More importantly, these differences directly affect the resulting data. Investigators are urged to learn what they need to know in order to evaluate the SOPs for a laboratory under consideration, including information presented in this book (see also Chapters 4 and 5).

Consider the following example in which both laboratories report colony-forming units per gram of dust. The "soil crumb" method and serial dilution method are methods of preparing settled dust for culture analysis. The former tends to result in the isolation of more fungal species without dilution error. Note that "error" means uncertainty regarding a measurement. The latter tends to "swamp out" those fungi at lower concentrations while overemphasizing the concentration of fungi isolated. Without even knowing what sample preparation methods are used, investigators cannot properly evaluate extremely high concentrations samples at high dilution (e.g., 1 : 10,000) or the presence of many different fungi at very low concentrations.

The following is another example of why it is so important to evaluate a laboratory's SOPs. Standard microbiological practice does not allow for combining data from inoculations. In the serial dilution method, most laboratories will inoculate several dilutions, at least one plate per dilution and the results of one chosen dilution reported. It is not standard microbiological and scientific practice to take counts of different fungi from different dilutions and combine all the data into one final result.

Laboratory accreditation, such as the Environmental Microbiology Laboratory Accreditation program (EMLAP) sponsored by the American Industrial Hygiene Association (AIHA), provides some comfort that a laboratory meets certain accreditation criteria. For example, such accreditation establishes a threshold of laboratory training among analysts and requires a laboratory quality assurance/quality control (QA/QC) program. Despite meeting these thresholds, there is significant variability between analysts in the same laboratory and between laboratories. Spore trap analysis of the same samples, called a "round-robin analysis," among AIHA accredited laboratories varied several orders of magnitude in total concentration and the rank order of fungal spores differed (P. Heinsohn, unpublished data). Both are important criteria for interpretation of sample data in ascertaining whether indoor air quality is in fact degraded.

2.2.2.8. Prior Notification. Aspects of the condition of the building and occupant activities prior to a building investigation can directly affect the data. Thus, the following instructions should be given to whatever responsible parties are involved, such as building owner or tenant, to avoid biasing the data. Instructions include the following:

- *Close Doors and Windows the Day (i.e., 24 h) Before Air and Surface Sampling are Conducted.* Air sampling is conducted to determine the *indoor* air quality. Obviously, this means that doors and windows must be closed because four walls, a ceiling, and a floor constitute the typical indoor environment. Air sampling has been done under all manner of conditions of the built environment, such as in the presence of collapsed ceilings and purposefully created holes cut through subflooring exposing an earthen crawlspace. This means that the sample data take into account the effect, large or small, of the air quality in an above-ceiling space or crawlspace, respectively, in these examples. Investigators should not conduct air sampling when the living space or workspace is compromised via open ceilings and crawlspace if the goal is to characterize the air quality with an intact building envelope. Likewise, should the investigator want to know what the air quality is in an attic, then the attic should be physically closed off from the living space or workspace, and samples collected. If windows and doors are not closed prior to sampling, the resulting laboratory data will most likely be negatively biased if a growth site exists that is degrading the indoor air quality, that is, the indoor air quality will more closely resemble the outdoor air than it would otherwise.
- *Do not Vacuum or Dust at Least 24 h Prior to Air or Surface Sampling.* Ostensibly, removal of settled dust from finished flooring materials or horizontal furniture surfaces will likely reduce the fungal burden, thus providing a negative bias. On the other hand, vacuuming and dusting will likely increase airborne concentrations of fungi from resuspending settled spores, thus providing a positive bias.

- *Do not do Any Unusual Cleaning that is not Customary Prior to Air and Surface Sampling.* For example, this includes carpet shampooing, the use of surface disinfectants, deodorizers, and high-efficiency particulate air (HEPA) filtration cleaners. One goal of a baseline investigation is to document the nature and extent of fungal growth and contamination, if any. If surface fungal growth or contamination has been removed or disturbed, then a negative bias is introduced into surface data. Carpet shampooing may actually result in a positive bias, depending on numerous factors including the spore burden prior to shampooing, the process used, efficacy of cleaning, and ventilation following cleaning.
- *Set the Thermostat and Fan Control at Settings Customarily Used for the Time of Year.* Sometimes investigators collect samples in a "fan off" setting, activate the fan, and then collect additional samples with the belief that quantitative and qualitative differences between the two reflects the condition of the system activated. This sampling design is flawed and should not be used. Many paired sample sets would have to be collected to detect a real difference. Furthermore, analytical variability due to variability between analytical methods and between analysts alone may explain differences. Surface sample collection (e.g., swipe, surface contact plate) from *within* the system is the best sampling design to determine whether a ventilation systems or ductwork is contaminated is to collect surface samples. Investigators must also include sampling at control locations in the same building in a nonimpacted area and in a control building using the same surface sampling methods. Much thought should be given to recommending cleaning ductwork or an entire a ventilation system in the absence of unambiguously degraded indoor air quality with the same species of fungi. It is hard to conceive of a ventilation system that does not have background levels of fungi.
- *Presence of Occupants is not Required.* If presence is desirable, engage in activities normally pursued except those excluded above and any other activity that purposefully disturbs settled dust.

2.3. DESTRUCTIVE TESTING INVESTIGATION

Destructive testing is done only when there is a high index of suspicion that water intrusion has occurred. Destructive testing enables a forensic water intrusion investigator to ascertain the cause of water intrusion and recommend a repair. Destructive testing enables a mold investigator to determine the nature and extent of fungal growth that is hidden from view during a baseline investigation.

Destructive testing in search of hidden fungal growth is justified if (1) there is an active leak indicated by elevated moisture content readings taken during a baseline investigation or (2) indoor air quality is unambiguously degraded by fungal species for which growth sites have not been identified.

Destructive testing is usually conducted inside a building. For various reasons, it may be conducted from the exterior of the building. A mold investigator selects areas to be opened according to input from the forensic water intrusion investigator augmented by observation of water stains, water damage, and suspect fungal growth noted in the baseline investigation. Often these areas may be part of the building envelope (e.g., an exterior wall assembly) or within a partition wall (e.g., a plumbing wall).

It is well known that forceful, mechanical disturbance (such as that involved in destructive testing and mold remediation) can generate high concentrations of airborne fungal spores.[21] Thus, destructive testing should be contained so as not to disseminate contamination outside of containment. A protocol should be prepared to maximize personal protection and prevent escape of fungal spores.

Results of a destructive testing survey are used to prepare a scope of work for mold remediation. Careful documentation of destructive testing is essential to interpret the findings. While the same surface sampling methods are used on surfaces within intact rooms, the fact that hidden surfaces are revealed slightly complicates written documentation. It is not sufficient to describe a sample location as "mold on gypsum wallboard." A description such as the following is very informative: "Approximately one square foot of suspect mold growth on the cavity side of gypsum wallboard located behind the bathroom sink cabinet backface. Approximately 64 $in.^2$ of suspect mold growth on the room side of gypsum wallboard behind the bathroom sink cabinet backface. Baseplate to left and right of sink plumbing is water-damaged."

Photographs should be taken as each building material is removed, for example, before the bathroom sink is pulled off the wall, before the room side of gypsum wallboard behind the sink is cut, and the reverse side of the wallboard that was cut out, and so forth.

Destructive testing is obviously intrusive, causes damage, and requires special precautions. These problems make air sampling from wall cavities an appealing alternative. Wall cavity sampling usually involves insertion of tubing through a hole made in gypsum wallboard, use of a high-volume air sampling pump, and collection of a total fungal spore air sample while pounding on the wall above the sample location during the sampling period.

Disadvantages of wall cavity air sampling include the following:

- It is not predictive of hidden growth; elevated concentrations have been found in the absence of growth. The converse has also been found.
- Gypsum dust contamination occurs in the tubing and is subsequently collected on the spore trap trace despite attempts to clean tubing between sample locations.

Such data cannot inform the investigator as to the extent of growth for any building material inside the cavity, which is the reason for destructive testing. Thus, wall air cavity sampling is not recommended as an alternative to destructive testing.

2.4. SAMPLING DURING MOLD REMEDIATION OVERSIGHT AND CLEARANCE

Once the nature, the location(s), and the extent of fungal growth in a building have been determined, an investigator can prepare technical specifications for mold remediation. It is obvious that without this knowledge, mold remediation contractors cannot know what building materials to remediate, how to remediate them, or what constitutes acceptable clearance verifying effective mold remediation. Thus, technical specifications for mold remediation must be building specific to this level of detail.

Unfortunately, building mold investigators will often prepare a generic protocol for mold remediation and not participate in overseeing the mold remediation. They may or may not return as "an independent party" to conduct postremediation sampling at the conclusion of the contractor's efforts. Mold remediation works best when an investigator provides continual oversight through postremediation sampling. In the absence of this, an investigator is forced to rely only on the inspection and sampling that is part of postremediation testing.

Clearance consists of the following:

- Firsthand knowledge of compliance with work practices and procedures in the technical specifications
- Visual inspection for dust
- Surface sampling for fungi
- Air sampling for fungi

The surface and air sampling methods have already been discussed. The same surface sampling methods and analysis are used to verify the absence of fungal growth. Likewise, the same air sampling methods and analysis are used to verify that the air quality within containment is not degraded with respect to the species of fungi originally degrading it and species of growth uncovered by demolition.

2.5. CONCLUSIONS

The importance of planning, designing, and conducting building mold investigations cannot be overstated. Building investigations start with gathering all the relevant data regarding a building's leak history, construction, maintenance, and use, as well as prior investigations by other consultants. Much of the background information is essential in developing a sample design with clear objectives.

While physical inspection focuses on visual identification of all suspect water stains, water damage, and visible fungal growth that might be associated with leaks, it also includes documenting occupant-related sources of moisture, housekeeping, and the presence of non-water-damage reservoirs of fungi. Physical inspection also includes moisture mapping because it can be extremely useful in pinpointing or excluding sources of moisture, or it may elucidate the path of water. "Negative" or "typical" moisture content readings provide some evidence of the absence of a chronic leak.

Proper documentation is essential in providing a factual record for the reliability of all sampling instrumentation used, describing how and where samples were collected relative to architectural details, maintaining the integrity of samples, and identifying the laboratory and analytical methods used. Field notes documenting site conditions provide a written record of a physical inspection. Air and surface sample laboratory data and photographic documentation also form the factual basis on which to base informed decisions regarding the nature and extent of fungal growth and its implications. No one survey tool should be used exclusively to base all decisions.

The reliability of laboratory sample data depends on a sound sampling design, sampling instrumentation, execution of the sampling design, and competent laboratory analysis. Culturable air sampling with species identification must be done in order to detect differences between indoors and outdoors. Collection of replicate air samples using the same methods at the same location allows for assessing variability around an airborne concentration. Laboratory personnel and standard operating procedures (SOPs) greatly affect the quality of sample data.

The key underlying principles of planning, physical inspection, documentation, and sampling discussed in depth in this chapter with respect to baseline investigation of buildings also apply to other types of building investigations, including destructive testing.

REFERENCES

1. California Code of Regulations, Title 8, Subchapter 4, Construction Safety Orders; Article 4, Dusts, Fumes, Mists, Vapors, and Gases; Section 1529, Asbestos.
2. Macher, J. M., *Bioaerosols: Assessment and Control*, American Conf. of Governmental Industrial Hygienists, Cincinnati, OH, 1999.
3. New York City Department of Health, *Guidelines on Assessment and Remediation of Fungi in Indoor Environments*, New York, 2000.
4. Lsitburek, J. and J. Carmody, *Moisture Control Handbook: Principles and Practices for Residential and Small Commercial Buildings*, Wiley, New York, 1994.
5. ASTM, *D4442-92 Test Methods for Direct Moisture Content Measurement of Wood and Wood-Base Materials*, Annual Book of ASTM Standards, Vol. 04.10, Philadelphia, 1992.
6. American Industrial Hygiene Association, *Field Guide for the Determination of Biological Contaminants in Environmental Samples*, AIHA Publications. Fairfax, VA, 1996.
7. Wilcox, W. W., E. E. Botsai, and H. Kubler, *Wood as a Building Material: A Guide for Designers and Builders*, Wiley, New York, 1991.
8. American Industrial Hygiene Association, *Field Guide for the Determination of Biological Contaminants in Environmental Samples*, 2nd ed., AIHA Publications, Fairfax, VA, 2004.

9. Hering, S. V., Inertial and gravitational collectors, in *Air Sampling Instruments for Evaluation of Atmospheric Contaminants*, S. Hering, ed., American Conf. Governmental Industrial Hygienists, Cincinnati, OH, 1989.

10. Chatigny, M. A., J. M. Macher, H. A. Burge, and W. R. Solomon, Sampling airborne microorganisms and aeroallergens, in *Air Sampling Instruments for Evaluation of Atmospheric Contaminants*, S. Hering, ed., American Conf. Governmental Industrial Hygienists, Cincinnati, OH, 1989.

11. Dimmick, R. L. and A. B. Akers, *An Introduction to Experimental Aerobiology*, Wiley-Interscience, New York, 1969.

12. Heinsohn, P. A., *Surrogate Quantitation of Bioaerosols*, Ph.D. dissertation, Environmental Health Sciences, Univ. California at Berkeley, 1994.

13. Aizenberg, V., T. Reponen, S. A. Grinshpun, and K. Willeke, Performance of Air-O-Cell, Burkard, and Button samplers for total enumeration of airborne spores, *Am. Indust. Hyg. Assoc. J.* **61**:855–864 (2000).

14. Tsai, S. M., C. S. Yang, P. Moffett, and A. Puccetti, Comparative studies of collection efficiency of airborne fungal matter using andersen single-stage sampler and Air-O-Cell cassettes, in *Bioaerosols, Fungi and Mycotoxins: Health Effects, Assessment, Prevention and Control*, E. Johanning, ed., Eastern New York Occupational & Environmental Health Center, Albany, NY, 1999, pp. 457–464.

15. Andersen, A. A., New sampler for the collection, sizing, and enumeration of viable airborne particles, *J. Bacteriol.* **76**:471–484 (1958).

16. Brachman, P. S., R. Ehrlich, M. F. Eichenwalk, V. J. Gabelli, T. W. Kethley, S. H. Madin, J. R. Maltman, G. Middlebrook, J. D. Morton, I. H. Silver, and E. K. Wolfe, Standard sampler for assay of airborne microorganisms, *Science* **144**:1295 (1964).

17. Trakumas, S., K. Willeke, T. Reponen, and M. Trunov, *Particle Cut-size Evaluation of Air-O-Cell Sampler*, final report to Zefon International, Univ. of Cincinnati (accessed at http://www.zefon.com/analytical/download/cutsize.pdf; on June 29, 2006), 1998.

18. Cronk, B. and S. A. Olenchock, Industrial workplaces, in *Bioaerosols Handbook*, C. S. Cox and C. M. Wathes, eds., CRC Lewis Publishers, Boca Raton, FL, 1995.

19. Hinds, W. C., *Aerosol Technology: Properties, Behavior, and Measurement of Airborne Particles*, Wiley-Interscience, John Wiley & Sons, Inc. (1982).

20. McDermott, H. J., *Handbook of Ventilation for Contaminant Control*, 2nd ed., Butterworth Publishers, 1985.

21. Rautiala, S. T. Reponen, A. Hyvarinen, A. Nevalainen, T. Husman, A. Vehvilainen, and P. Kalliokoski, Exposure to airborne microbes during the repair of moldy buildings, *Am. Indust. Hyg. Assoc. J.* **57**:279–284 (1996).

22. Samson, R. A., Ecology, detection and identification problems of moulds in indoor environments, in *Bioaerosols, Fungi and Mycotoxins: Health Effects, Assessment, Prevention and Control*, E. Johanning, ed., Eastern New York Occupational and Environmental Health Center, Albany, NY, 1999.

23. Scott, J. A., Heteroduplex DNA Fingerprinting of *Penicillium brevicompactum* from house dust, in *Bioaerosols, Fungi and Mycotoxins: Health Effects, Assessment, Prevention and Control*, E. Johanning, ed., Eastern New York Occupational and Environmental Health Center, Albany, NY, 1999.

24. Gravesen, S., J. C. Frisvad, and R. A. Samson, *Microfungi*, Munksgaard, Copenhagen, Denmark, 1994.

CHAPTER 3

MICROBIOLOGICAL SAMPLING STRATEGIES IN INDOOR ENVIRONMENTS*

PHILIP R. MOREY

3.1. INTRODUCTION

For over a century scientists have carried out sampling for various microorganisms in attempts to associate environmental conditions in buildings with bioaerosol exposures. In the late nineteenth century, Carnelley et al.[1] determined that exposures to both airborne bacteria and fungi were reduced in homes and schools with excellent housekeeping practices (Table 3.1). By the late nineteenth century it was generally accepted that airborne contagion such as *Mycobacterium tuberculosis* could be diluted with large volumes of outdoor air (e.g., $>60\ \mathrm{Ft}^3/\mathrm{min}$ per occupant), thereby reducing the risk of disease.[2] By the midtwentieth century the fundamentals of many microbial sampling techniques used today had been developed[3] because of bioaerosol investigations in occupational environments such as agriculture, fermentation facilities, and pharmaceutical manufacturing. Interest in microbial sampling has continued into the twenty-first century because of health risk in indoor environments associated with exposure to (1) microorganisms and their components such as allergens, endotoxins, and glucans in damp/moldy buildings;[4] (2) *Legionella* aerosols from cooling towers and potable-water systems;[5] and (3) infectious fungi (e.g., *Aspergillus fumigatus*) in medical facilities.[6]

Since 1995 the additional interest in microbial sampling has developed because of moisture failures and microbial growth problems encountered in modern buildings. Many of the construction and finishing materials found in modern buildings (e.g., paper-faced wallboard, particleboards, porous finishes such as carpet that accumulate dust and bioburden) are more susceptible to biodeterioration in the event of a moisture failure as compared to more traditional building materials

*Some of the concepts in this chapter are derived from Section 5.2 of Reference 18, also written by the author.

Sampling and Analysis of Indoor Microorganisms, Edited by Chin S. Yang and Patricia A. Heinsohn

TABLE 3.1. Airborne Microorganisms in One-Room Homes

Cleanliness of Home	Average Number of Fungi and Bacteria[a]
Clean	18
Dirty	41
More dirty	49
Very dirty	93

[a]Number of fungi and bacteria per liter of air. Hesse's sampling method; flowrate approximately 0.33 L/min; total air volume collected was 1–10 L (see Ref. 1).

such as stone, brick, terra cotta, ceramic tile, and pine or hardwood flooring. The intent of this chapter is to provide practical guidance to investigators carrying out microbial sampling in nonindustrial indoor environments.

3.2. SAMPLING STRATEGY

The most important step prior to microbial sampling is development of a plan on how the analytical results of sampling will be interpreted. This is necessary because of the absence of numerical guidelines for microbial agents. Whereas numerical guidelines are well known for many chemical agents [OSHA personal exposure limits (PELs), ACGIH Threshold Limit Values; registered trademark (TLVs)], scientific studies correlating the dose of microbial agents with health effects are unavailable. Reasons for an absence of numerical guidelines[7–9] include (1) the many thousands of fungal and bacterial species, (2) the capacity of microorganisms (whether dead or alive) or their components (e.g., glucans, endotoxins, allergens) to cause disease by various mechanisms (e.g., allergy, infection, toxicity, or irritation), (3) variation in human susceptibility to microbial agents, and (4) an absence of standard sampling and analytical procedures for the development of dose–response relations.

Notwithstanding the difficulties associated with interpretation of sampling results, an appropriate sampling plan begins with a thorough inspection of the building for dampness or water damage and visible signs of microbial growth or contamination. Essential tools for a building inspection include a powerful flashlight, a ruler, a copy of building mechanical/structural plans, moisture and relative humidity meters, and a camera. It is essential during the building inspection to measure and document the location(s) and extent (square feet or meters) of water-damaged and visually moldy materials including the likelihood of possible hidden damage. In focused microbial sampling evaluations such as where legionellosis is a concern, inspection and an understanding of a building's heat rejection (e.g., cooling towers) and potable-water systems is essential in order to select sites for collection of samples possibly containing *Legionella*.

Practical reasons for microbial sampling in indoor nonindustrial environments are numerous and varied. Often the management or occupants of a building wish to know whether interior conditions are "safe." While sampling cannot determine whether a building is safe, a sampling plan may seek to determine whether the general mycological building condition is "normal" or "typical" or similar to conditions in well-maintained, dry buildings or similar to conditions in the outdoor air. By contrast a more focused sampling plan may seek to determine whether specific microbial agents such as *Legionella pneumophila* serogroup 1 or thermotolerant *Aspergillus fumigatus* or *A. terreus* are present in a building component where occupants may have become ill.[10]

Many variables can potentially affect sampling strategies including the following:

- Spatial location
- The time of day
- Seasonal outdoor conditions that affect the indoor environment
- Design, operation, and maintenance of the building including its HVAC system
- Occupant considerations such as complaint and noncomplaint zones and occupant activities

In summary, the investigator should have a thorough understanding of the building condition prior to development of a sampling plan.[11]

3.3. SPATIAL OR LOCATION VARIABLES

Sampling strategies can be affected by the layout of building components or zones.[18] Thus, the objective of sampling may be to determine whether mold spores from a damp attic are being dispersed into occupied zones. Consequently, during the inspection prior to sampling, it is important to determine whether pathways occur that would allow for transport of spores from the attic to occupied zones. In this example, pathways for transport of spores may be hidden from view such as in interstitial or wall cavity passages, especially if an air barrier on the attic floor is absent. Additionally, spore transport can be promoted by negative pressurization in occupied areas. In this example, collection of air samples in the attic and in occupied areas should occur after development of a sampling plan based on a thorough physical inspection.

Another example illustrating the importance of spatial considerations in developing a sampling plan occurs when it is necessary to document that sewage backflow microorganisms that had contaminated one part of a building had been properly removed. An appropriate sampling strategy would be to test surfaces for the presence of pathogenic bacteria such as *Escherichia coli* 0157/H7, *Salmonella*, and *Shigella* in areas physically affected by sewage and also in unaffected control zones or buildings[10] (see Table 8 in Ref. 10; see also Table 3.9 later in this chapter). An absence of differences in concentrations of these Gram-negative bacteria[10] (e.g., bacteria not detected) between sewage impacted and spatially separated control zones suggests that cleanup actions were adequate.

Indoor activities such as vacuum cleaner operation are known to elevate bioaerosol concentrations.[12] Other activities such as walking around and sitting up and down on upholstered furniture can result in a "personal cloud effect"[13] wherein the airborne particle level in the immediate vicinity of the active person is higher than the particle level in other parts of the same room. It may be important, therefore, for the investigator to determine whether personal or area sampling is appropriate. If the intent of the evaluation is to document the personal cloud around the active occupant, then a personal sampling strategy is required.

3.4. TEMPORAL (TIME) VARIABLES

Sampling results in a building are potentially affected by variables including human activity patterns, HVAC system operation, and climatic or seasonal variations such as alternating rainy and dry periods.

For example, sampling for culturable fungi during a prolonged dry season will be less likely to detect water indicator fungi from leak-prone envelope walls than will sampling during a rainy period when the building envelope is leaky and wet.[18] Similarly, sampling for human-shed bacteria in a classroom during unoccupied nighttime hours will yield different results than will sampling in the same classroom with many active students.

Temporal changes in the building envelope affect air sampling results. Solar heating of the building envelope causes air near perimeter walls to warm and move upward. Thus, divergent bioaerosol sampling results can occur in the morning as compared to the afternoon. In this case sampling should be repeated at least several times daily (morning and afternoon) at each testing site. Seasonal changes in moisture levels including condensation potential on the inner surface (closest to occupants) of the envelope wall can affect the ecology of molds on the wall surface and in room air. Thus, during warm, humid seasons condensation and active mold growth can occur on the wall surface especially in depressurized buildings. However, surface and air sampling results obtained in the same areas during the heating season may be different because of the absence of dampness.

In buildings located in cold climates during wintertime conditions, condensation and consequential mold growth can occur on or near inner surfaces of the envelope wall because of factors such as thermal bridging.[14] A sampling strategy should therefore anticipate the changing seasonal pattern of dampness in envelope walls.

A substantial mycology literature exists on seasonal variation in concentrations and kinds of mold spores found in the outdoor air.[15] A more recent study showed that outdoor concentrations of boletes basidiospores varied diurnally by an order of magnitude.[16] Additionally, indoor concentrations of basidiospores followed the same diurnal pattern from a low of about 500 spores/m^3 in the morning and afternoon to a high of about 4000 spores/m^3 in the evening.[16] This study has several practical consequences on sampling strategy including the necessity of collecting comparative indoor and outdoor samples during the same time period. Also, during sampling it is important for the investigator to note the presence of open windows and doors, which will directly affect the infiltration of phylloplane mold spores into the indoor air.

Nosocomial fungal infections (e.g., aspergillosis) have been shown to be temporally associated with aerosolization of renovation and construction dusts.[6,17] Dusts aerosolized during wall or ceiling demolition almost always contain *Aspergillus* spores, some of which have a culturable half life of more than 7 years (*A. fumigatus, A. flavus*; see Table 6.1 in Ref. 18). Aspergillosis epidemics in medical centers are best documented by epidemiology studies.[17] Sampling to document environmental sources of *Aspergillus* spores associated with aspergillosis is likely to be most successful if done concurrently with the epidemiology study rather than after renovation/cleanup activities are completed.

3.5. INDOOR/OUTDOOR COMPARISONS

Because of the absence of interpretive numerical guidelines for bioaerosols, comparative indoor and outdoor sample collection is important in deciding whether the indoor environment is normal and typical. Rank-order comparison of the kinds of fungi found indoors and outdoors[19,20] provides a strategy for determining whether the indoor environment is atypical or suggestive of indoor mold amplification. Table 3.2 provides an example where air sampling was carried out at a building after completion of mold remediation. Sampling results showed that the kinds of molds found indoors and outdoors were similar (Table 3.2). Note that leaf source (phylloplane) fungi such as *Cladosporium cladosporioides* and *Epicoccum nigrum* account for more than half of the identified taxa in both indoor and outdoor samples. Soil fungi (*Penicillium* and *Aspergillus* species; *P. brevicompactum* in Table 3.2) that grow on botanical materials in the soil, as well as on wet construction

TABLE 3.2. Comparison of Indoor and Outdoor Sampling Results at a Building after Mold and Moisture Problems Had Been Remediated

Rank Order (%) of Fungi Types in	
Outdoor Air[a]	Indoor Air[a]
Cladosporium cladosporioides (71)	*Cladosporium cladosporioides* (56)
Epicoccum nigrum (14)	Nonsporulating fungi (16)
Yeasts (5)	*Epicoccum nigrum* (10)
Nonsporulating fungi (4)	*Ulocladium chartarum* (8)
Penicillium brevicompactum (2)	Yeasts (3)
Cladosporium herbarum (1)	*Penicillium brevicompactum* (1)

[a]Ranked types of fungi in percent of total. Minor isolates are not included (numbers in parentheses = % of total).

Source: Adapted from Morey.[10,18]

materials, accounted for only a minor percentage of the culturable propagules found in both the indoor and outdoor air (Table 3.2). The sampling data in Table 3.2 together with a thorough inspection of the building showing an absence of water damage or visible mold growth provides an indication of a normal and typical indoor condition. Comparative indoor and outdoor air sampling for other microbial agents such as endotoxins, β-1,3-D-glucans, and microbial volatile organic compounds (MVOCs) can also be useful in determining whether microbial reservoirs or growth sites are present in the indoor environment.[18,21–23]

The selection of appropriate outdoor sampling sites is essential for achieving meaningful indoor/outdoor bioaerosol comparisons. If possible, outdoor air samples should be collected on the building roof while facing into the wind or at the location of the HVAC system outdoor air inlet.[18,24] If it is not physically possible to access the roof as in most residential evaluations, outdoor air samples should be collected upwind and away from the building in a nondusty location. Outdoor air samples should not be collected by placing the sampler just outside an open window or by placement of the sampler on a porch or patio of the building being evaluated. These outdoor sites just outside the building envelope are readily affected by contaminants sourced in the indoor environment. Additionally, unusual outdoor bioaerosol sources such as lawnmowing activities, composting, and soil excavation should be avoided during the collection of outdoor reference samples.[25]

Comparison of outdoor/indoor sampling results can be confounded by seasonal conditions. For example, during periods when the soil and outdoor vegetation are covered by snow, few, if any, fungi are found in the outdoor air. This makes indoor/outdoor comparisons at that time of the year difficult because during these wintertime conditions, the aerosolization of a few *Penicillium* or *Aspergillus* spores from floor or other interior surfaces is not masked by the normal influx of outdoor phylloplane molds that would occur at other times of the year.[26] By contrast, when crops are harvested, prodigious quantities of outdoor-sourced spores are aerosolized in the atmosphere, also making indoor/outdoor comparisons difficult.

3.6. COMPLAINT AND NONCOMPLAINT ZONES

Occupant complaints in buildings may be environmentally based (dampness, water leaks, visible mold growth, MVOCs, etc.), health-related (e.g., upper respiratory irritation, allergy, infection), or both of these. It should be recognized that results of microbial sampling alone cannot be used to determine whether adverse health outcomes have occurred (microbial PELs or TLVs are unavailable). However, careful inspection of a building for dampness/moisture damage can lead to a microbial sampling strategy, the results of which can provide useful information to a physician or epidemiologist in associating environmental conditions with occupant health complaints.

It is generally accepted that dampness and moisture problems in buildings are associated with occupant respiratory disease.[4,27,28] A prerequisite to developing a sampling strategy useful in understanding occupant complaints is an evaluation of the nature and extent of the dampness/moisture problem ("follow the water"; Ref. 29). Signs of dampness/moisture problems (efflorescence on concrete surfaces; water-stained wallboard, etc.) as well as the proximity of occupants to water-damaged areas[28] should be documented during inspection. It is important to determine whether air is flowing from the water-damaged area to the complaint zone and if visible mold is present in the moisture-affected areas.

A sampling strategy to document the possible association of microbial agents with health (complaint) outcomes depends on appropriate selection of case and control sampling locations. For example, occupants in a building zone with a spray mist humidifier reported a flulike illness. Air sampling for endotoxin was then carried out in the humidifier-affected zone, as well as in an indoor control area (without humidifier) and in the outdoor air. Air sampling results showed that endotoxin levels in the complaint zone were several orders of magnitude greater than that in the indoor and outdoor control areas (Table 3.3). On the basis of this sampling strategy as well as an epidemiology/clinical evaluation of occupants, it was concluded that the flulike illness in the humidifier zone was caused by inhaled endotoxin.[30]

Additional factors to consider with regard to microbial sampling strategy in complaint/noncomplaint areas include the timing of the sampling evaluation as well as the methodology to collect the samples. If microbial sampling is part of a clinical/epidemiologic investigation, it is desirable to carry out all studies simultaneously. Thus, if a possible case of hypersensitivity pneumonitis is being investigated relative to emissions of *Thermoactinomyces vulgaris* from a water sump humidifier,[31] microbial sampling should occur near or at the operating humidifier when the bacterium is growing, and not from the humidifier after it was cleaned.

The methodology used to collect samples in complaint/noncomplaint zones should simulate normal occupant exposure conditions. Collection of air samples from a wall cavity in a complaint zone when wallboard or other material surfaces are being aggressively disturbed (e.g., by drilling or pounding) does not represent normal bioaerosol exposure of room occupants. An appropriate sampling strategy should specify "normal occupant activities and conditions" at the time of air sample collection in both complaint and noncomplaint areas.

TABLE 3.3. Air Sampling for Endotoxin in a Misty Zone of a Building Where a Flulike Illness Occurred

Sample Location	Endotoxin Concentration[a]
Indoor control zone	8
Complaint zone	3200
Outdoor air	23

[a]Data in endotoxin units per cubic meter of air.

Source: Adapted from Morey.[10]

3.7. SOURCE AND AIR SAMPLES

A microbiological sampling strategy can involve collection of source and/or air samples.[18,32,33] Microbial source sampling includes the collection of (1) materials such as pieces of wallboard, fiberboard, or wood; (2) dusts by vacuum cleaner or filter cassette minivacuum;[34] and (3) surface mold spores and structures by cello-tape, contact plate, or swab methodology. Microbial air sampling includes collection of (1) culturable microorganisms by impactor or impinger; (2) mold spores, pollen, and hyphal fragments by spore trap; and (3) microbial particulate by filtration for analysis of endotoxin, glucan, mycotoxins, nucleic acid [polymerase chain reaction (PCR)] composition. Source sampling is most useful in characterizing the microbial composition in building materials or on surfaces. While air sampling has shown only limited direct correlation with health outcomes,[4,18,27] it can be useful in determining whether the air in an indoor environment is microbiologically normal or atypical. Regardless of the type of microbial sampling strategy, the investigator must understand the limitations of the sampling and analytical methods employed.[10,18,32]

The microbial agent of interest often determines the choice of sampling method. Culture-based methods such as collection of bioaerosols by impaction onto an agar plate (culture plate impactor) or analysis of source samples for *Legionella* species and for other Gram-negative bacteria are used where it is important to identify microorganisms to species or subspecies (e.g., *Aspergillus flavus*; *A. terreus*; *Legionella pneumophila* serogroup 1; *Escherichia coli* 0157/H7). Other specific techniques such as PCR can also be used to identify the collected microorganism to the species level, but these techniques cannot distinguish bacteria or fungi that are viable (capable of causing infection) from those that are nonviable or no longer capable of causing infection. It is important therefore for the investigator to know from project onset whether the determination of culturable species or subspecies is important to meet project objectives. For example, in aspergillosis or Legionnaires' disease investigations, establishing the ability of the collected microorganisms to cause infection is important.

The collection of airborne allergens, toxins, or other microbial agents such as glucans and endotoxin is not constrained by loss of culturability, and therefore air sampling for these agents can be accomplished by filtration of particles usually from large volumes ($>1\ m^3$) of air. The collection of culturable bacteria and fungi by filtration methods is constrained by the desiccation of some spores by the large volumes of air that pass over the spores captured in the filter media. In developing a sampling strategy involving filtration methodologies it should be noted that some mold spores are more desiccation-resistant (e.g., *Penicillium, Aspergillus, Eurotium* species) than others (e.g., *Cladosporium, Epicoccum, Stachybotrys* species), and this may affect data interpretation (see Table 6.1 in Ref. 18).

Most air sampling methodologies involve the collection of many samples over short time periods, generally less than 10 min. The ACGIH advises that it is best to collect multiple samples over 3 days at each sample site.[24] This kind of replicate sampling is appropriate primarily for research investigations where statistical methods are used for data evaluation (see Table 5.9 in Ref. 24). For investigations where only a few samples are collected at each site because of time and cost constraints, it is necessary that the investigator appreciate the limitations imposed on data interpretation when less-than-ideal numbers of samples are evaluated.[18,24]

3.8. BULK SAMPLES

Bulk samples can be collected from a variety of solid and liquid materials. For example, pieces of wallboard, fiberboard, or carpet can be analyzed to determine whether hydrophilic (including cellulose degraders) or xerotolerant molds predominate. This analysis may provide an indication of the previous moisture condition in the collected material (e.g., very wet or just damp). Table 3.4 illustrates the results of analysis of visually moldy, moisture-impervious vinyl wall covering from an envelope wall in a depressurized Florida building and of wallcovering from visually nonmoldy interior walls. The presence of mostly xerotolerant *Aspergillus versicolor* and *A. sydowii* in the envelope wall samples suggests a previous damp condition as contrasted with a soaking wet condition where many hydrophilic fungi such as *Acremonium* and yeasts would likely dominate.

Bulk water samples for analysis of legionellas may be collected from cooling tower reservoirs or from potable-water systems, as well as from the water mains entering the building in order to determine whether amplification is occurring in piping or reservoirs.[5,35] It is important for the investigator to work closely with the laboratory when collecting water samples for *Legionella* analysis. For example, sodium thiosulfate should be added to the sterile collection bottles to neutralize chlorine or bromine that may be present in the water sample so as to reduce the biocidal action of chlorine or bromine against culturable *Legionella*.[18] In addition, arrangements must be made to have collected water samples arrive at ambient temperatures at the laboratory by overnight delivery.

When collecting bulk materials for microbial analysis, it is important to obtain samples from representative problem and nonproblem (control) areas. For purposes of data interpretation, it is important to know the normal microbial ecology

TABLE 3.4. Analysis of Moisture-Impervious Vinyl Wallcovering from Envelope and Interior Walls[a]

Types of Wallcovering	Types and Concentrations of Molds
Wallcoverings from envelope walls	*Aspergillus versicolor* and *A. sydowi* dominate; average total concentration 1.7×10^7 CFU/cm^2
Wallcoverings from interior walls	A few *Cladosporium* and *Fusarium* present; concentration 40 CFU/cm^2

[a]Samples removed from wall with disinfected knife; wallcovering ground up in sterile water and dilution plated on malt extract agar.

(background) on material surfaces that are not degraded by water damage or microbial growth.

Bulk material samples should be small enough to be easily transported to the laboratory.[18] Pieces of materials such as wallboard, fiberboard, and carpet are collected in a manner that avoids cross-contamination (disinfect cutting tools between each sample) and then placed in clean, new containers for transport to the laboratory.[18] Aseptic sampling techniques should always be used, although they may not be required if the extent of contamination on the collected sample far outweighs similar contaminants that may be added during sample processing.[18,33] However, it is important to avoid cross-contamination between samples as may occur when an extremely contaminated bulk material is placed in close proximity to a relatively clean sample. Thus, dirty and clean samples should be placed in separate, sealed containers in the same shipping box or in different shipping boxes.[18] In addition, bulk material samples that are collected damp or wet should be packaged at about 5°C (41°F) with desiccant so as to lessen growth during shipment to the laboratory.

Special handling techniques are necessary for the collection of bulk materials and for certain analyses such as endotoxin. Because background endotoxin is present on surfaces of collection containers and instruments, all handling of the bulk material must be carried out with pyrogen-free (or endotoxin-free) containers and methods.[18,36]

Sampling strategies involving collection of liquids, like other materials, must include adequate controls to aid in data interpretation. Table 3.5 provides an example from an investigation where the objective of sample collection was to determine the source(s) of *Legionella pneumophila* serogroup 1 that had caused an outbreak of Legionnaires' disease in a workplace containing water misters.[10] Water samples were collected by standard methods[18,35] from nozzles of misters and from piping and hot-water tanks in the domestic water systems. Sampling also occurred from control locations in the workplace (water mains) and from residences of affected employees (Table 3.5). Analytical results showed that *L. pneumophila* serogroup 1 was present in the workplace water system, but not in well water serving the workplace or in residential water systems of affected

TABLE 3.5. Analysis of Bulk Water Samples for Culturable *Legionella*

Sample Location and Description	Results of Analysis[a]
Hot-water tank and piping from workplace buildings	10–1000 *L. pneumophila* serogroup 1/mL
Water spray nozzle of misters	10–100 *L. pneumophila* serogroup 1/mL
Hot-water tank and piping in occupants' (case) residences; water mains from workplace	Legionellas not detected (LOD = 0.5/mL)

[a]Culture on buffered charcoal yeast extract agar.
Source: Adapted from Morey.[10]

employees (Table 3.5). Decontamination procedures for the workplace water system (chlorination and/or pasteurization) are well known.[5,35]

3.9. DUST SAMPLES

A sampling strategy involving the analysis of settled dust especially from above-floor surfaces[33,37] provides an indication of microbial agents that were likely once airborne. Settled dust is heterogeneous, consisting of particles from people (e.g., skin scales), pets (e.g., Fel d1 allergen), textiles, paper, cooking, and the outdoor air. The composition of settled dust can vary depending on the location in a room or building. Settled dust may be collected by filter cassette minivacuum,[34] adapters fitted into the nozzle of a vacuum cleaner,[38] or by a large precleaned vacuum cleaner.[18,33] As with other kinds of microbial sampling, dust samples should be collected from representative areas (e.g., for floor dusts, collect samples from both high- and low-traffic areas). When collecting dust samples it is important to document both the surface area from which dust was extracted, as well as the time over which the sample was collected. Dust samples should be sieved prior to analysis in the laboratory (sterile sieve) to remove large inert particles (e.g., textile fibers and lint).

Analysis of settled dusts for fungal composition has been used forensically as an indication of a water or dampness problem.[10,34] Table 3.6 provides an example where both air and dust sampling provided an indication of problematic mycological condition. More than 95% of the culturable molds found in the dust samples were *Penicillium, Aspergillus*, and *Eurotium* species (mostly xerotolerant taxa). These sampling data suggested a dampness condition or incomplete cleanup associated with a previous mold remediation (Table 3.6). The composition of fungal taxa in settled dust is typically heterogeneous with a significant percentage of phylloplane fungi (e.g., *Cladosporium, Alternaria, Epicoccum* etc.) being present.[39,40] Dust

TABLE 3.6. Dust and Air Sampling for Molds in a Building that Had Undergone Incomplete Mold Remediation

Sample Description	Sample Analysis
Amalgamated floor dusts collected from many rooms	Dusts contain 10^8 CFU/g mostly (>95%) *Penicillium, Aspergillus*, and *Eurotium* species;<1% phylloplane molds
Air samples indoors by spore trap	*Penicillium/Aspergillus* concentration 11,500 spores/m^3; *Cladosporium* concentration 120 spores/m^3
Air samples outdoors by spore trap	*Penicillium/Aspergillus* concentration 55 spores/m^3; *Cladosporium* concentration 120 spores/m^3

Source: Adapted from Morey.[10]

samples from an indoor environment heavily dominated by *Penicillium/Aspergillus* species (Table 3.6) or cellulose degrading fungi (e.g., *Chaetomium* species) would be considered atypical and forensically indicative of wet or damp conditions present in an indoor environment.[37]

In the example in Table 3.6, *Penicillium/Aspergillus* spores comprised more than 95% of the airborne molds collected by spore trap. However, the relationship between the fungal species in settled dust and the species in indoor air is often less clear than the example in Table 3.6.[41,42] As with air sampling, guideline values relating total concentrations of fungi in dusts with health outcomes have not been established.

Settled dusts may be analyzed for culturable fungi by dilution or direct plating.[18,39] In direct plating a small amount of dust (usually a few milligrams) is sprinkled on the surface of the culture medium. Direct plating is recommended for determination of the actively growing or ecologically important species in the sample.[39,43] Dilution plating provides an indication of the total number of culturable fungi but can overemphasize the presence of those species with the longest half-lives or ability to withstand stress such as repeated mixing with sterile liquids. In addition, the error or deviation in quantitative dilution results increases as the dilution increases.[18] Laboratory data generated by manipulating and combining media and dilutions are erroneous and misleading and should not be relied on.

Dust samples can be collected and analyzed for content of specific allergens such as those from mites, cats, dogs, and cockroaches.[18,44,45] Table 3.7 provides an example where dusts from a complaint office (allergic symptoms), a control office, and soot in a HVAC system outdoor air inlet were analyzed for various allergens. In this example, cat allergen (Fel d1) concentrations were highest in the complaint office. The results in Table 3.7 suggested that pet owners were transporting Fel d1 on clothing into the workplace. The particles containing Fel d1 accumulated in settled dust, especially in the complaint office. A thorough dust removal program (HEPA filter vacuum cleaning), especially for fleecy office materials (upholstered furniture, modular partitions, and carpet), was carried out to lower the Fel d1 in the office environment.

Sampling strategies involving collection of settled dust should include consideration of the type of laboratory analysis most appropriate for evaluation of the microbial composition in the sample. Table 3.8 provides an example where very

TABLE 3.7. Concentration of Allergens in Settled Dust Samples

	Allergens (μg/g)			
Source of Dust	Der f1	Der p1	Fel d1	Can f1
Complaint office	0.5	ND	75	0.5
Control office	0.4	ND	5	0.6
HVAC system soot	ND[a]	ND	ND	ND

[a]ND = less than the detection limit of 0.1 μg/g; increased risk of sensitization among atopic individuals occurs at a level of 0.5–2.0 μg/g of Fel d1 (see Ref. 18).

TABLE 3.8. Analysis of Carpet Dust for *Stachybotrys* Content by Culture and PCR Methodologies

Type of Analysis	*Stachybotrys* Concentration	% *Stachybotrys* among Identified Fungal Taxa
Dilution plating on malt extract agar	1×10^3 CFU/g	Approximately 5% of culturable propagules
PCR,[a] target list of 24 taxa	1×10^7 spore equivalents/g	>95% of spore equivalent taxa

[a]Quantitative PCR assay; see Ref. 18.

different results were obtained by dilution plating of dust for culturable fungi as compared to analysis of the same dust by PCR. The dust sample was obtained from an apartment occupant's vacuum cleaner that had been repeatedly used for housekeeping of carpeted flooring. Previous physical inspection of the building envelope showed that biodeterioration involving extensive *Stachybotrys* growth had occurred on building paper and sheathing in the envelope wall. Water had flowed into the apartment from the envelope, thereby soaking the carpet on numerous occasions. Dilution plating on malt extract agar (MEA) showed that *Stachybotrys* accounted for approximately 5% (or 1×10^3 CFU/g) of culturable molds found in the vacuum cleaner dust. By contrast, PCR analysis of the same dust (24 target fungi including *Cladosporium, Aspergillus*, and *Penicillium* species) showed that *Stachybotrys* accounted for more than 95% of the spore equivalents detected (Table 3.8). Thus, the analytical methods used in this example were important for data interpretation, namely, whether *Stachybotrys* dominated the molds present in the dust sample.

Some microbial investigative strategies include the collection of both air and dust samples (Table 3.6), and perhaps other kinds of surface samples. If the purpose of air sampling is to evaluate occupant exposure to bioaerosols the air samples should be collected first. Dust samples are collected last because disturbance of interior surfaces[12] increases bioaerosol levels in room air.

3.10. SURFACE SAMPLING

Cellotape, sterile swab, or surface contact plate sampling can be used to document the type and/or concentration of microbial agents found on the tested surface.[18] As with other types of microbial sampling, analytical results alone cannot be directly related to health outcomes (absence of microbial TLVs or PELs).

Cellotape sampling together with direct microscopy analysis provides a convenient method for (1) identifying the types of molds that may be present on a surface and (2) documenting whether mold propagules on a surface are characterized as "normal deposition" or as growth, past or present, if the samples are expertly

TABLE 3.9. Surface Sampling for Culturable Bacteria and Air Sampling for Endotoxin at Building Affected by Sewage Backflow

Sample Location	Culturable Bacteria on Surfaces or Airborne Endotoxin
Surface Bacteria[a]	
Surfaces previously covered with sewage	None detected
Control surfaces unaffected by backflow	None detected
Endotoxin (EU/m^3 air)	
Rooms previously affected by backflow	3.1
Rooms unaffected by backflow	0.5
Outdoor air	0.5

[a]*Escherichia coli* 0157/H7, *Salmonella, Shigella, Streptococcus faecalis.*

Source: Adapted from Morey.[10]

analyzed. Cellotape sampling is carried out by applying the adhesive side of a clear tape, sticky side down, onto a material surface and then applying the adhesive side of the tape onto a microscope slide and examination by direct microscopy for spores, hyphae, and fruiting structures.[18,33] The presence of a substantial number fungal fruiting structures (e.g., conidiophores) and a network of hyphae (mycelium) provides clear evidence of mold growth on the sampled surface. Conidiophores indicate reproductive growth, and mycelia indicate vegetative growth. The presence of just a few spores and hyphae fragments on a surface is indicative of a normal condition (normal deposition of spores and hyphal fragments that settle out from air).

Cellotape sampling is useful in detecting the presence of fungal microcolonies (incipient mold growth) on surfaces where growth is not readily visible to the unaided eye. In combination with a thorough inspection of interior surfaces for visible mold growth, cellotape sampling in representative areas aids in estimating the square footage of mold growth, which is critical for implementation of abatement procedures according to governmental and professional society guidelines.[46–48]

A limitation of cellotape sampling is that mold spores can at best be identified only to the genus level and in some cases only to morphological groups (e.g., *Penicillium-Aspergillus*). An obvious limitation of cellotape sampling is that it cannot be used to determine whether mold growth occurs within a porous material such as carpet or upholstery.

Contact plate sampling (surface of culture plate medium is applied to a smooth material) can be used to identify the species of fungi or bacteria that may be present.[33] This kind of sampling has been used to document the extent to which surfaces in operating rooms have been disinfected.[49]

Swab sampling (sterile cotton or gauze applied to a defined area) is used to identify the types of fungi and bacteria or other microbial agents (e.g., β-*N*-acetyl

TABLE 3.10. Analysis of Dust from Porous Surfaces by PCR Where Contact Sampling Is Not Practical

Fungal Taxa in Dust	Spore Equivalents per Milligram
Cladosporium and *Alternaria* species	35,000
Penicillium and *Aspergillus* species	11,000
Memnoniella and *Stachybotrys*	5

hexosaminidase)[50,51] that may be present on surfaces. As with other sampling strategies, it is important to collect control samples for purposes of data interpretation. Thus, if swab sampling is used to demonstrate that pathogenic bacteria from a sewage backflow have been appropriately removed, it is important to collect samples from surfaces previously wetted by sewage, as well as from surfaces unaffected by the backflow. Table 3.9 shows that bacteria typically found in sewage were not detected by sterile gauze sampling of surfaces formerly in contact and unaffected by sewage water. These data suggest that cleanup procedures were effective.

Surface contact sampling by cellotape, contact plate, or sterile swab/gauze methods are not intended to determine whether mold growth occurs within porous materials such as clothing, linens, and upholstery. Table 3.10 provides an example where PCR analysis[18] of dust collected from porous contents was used as a supplement to visual inspection for mold contamination. Visual inspection had shown an absence of mold growth on clothing and linens. A small amount of dust (approximately 1 mg; too small for dilution plating) was extracted by filter cassette microvacuum from the clothing and linen and analyzed by PCR (Table 3.10). Phylloplane molds (e.g., *Cladosporium* and *Alternaria*) were present in greater amounts than were soil fungi (*Penicillium* and *Aspergillus* species) or cellulose degraders (*Stachybotrys* and *Memnoniella*). While there is a little comparative information in the literature to aid in interpretation, the data in Table 3.10 suggest a normal (not contaminated) condition for the tested clothing and linens.

3.11. CULTURE PLATE IMPACTIONS AND LIQUID IMPINGERS

Sampling for culturable bioaerosols may be carried out by impaction (e.g., single-nozzle slit-to-agar, single- or multistage jet, or centrifugal impactors) onto a culture plate inserted into the sampling device or by impingement into a liquid medium.[3,18,32] Culture plate impactors typically have an upper limit of detection of about 1×10^4 culturable particles/m^3 and as such are useful in most bioaerosol evaluations in nonindustrial buildings, where this concentration is seldom exceeded.[18] Impingement samplers, however, are more useful in heavily contaminated environments such as dusty barns or composting operations, where concentrations of culturable particles often exceed 1×10^4/m^3 according to Morey.[52] Impingers have flexibility with regard to choice of laboratory technique used to analyze the collection fluid. Thus, the fluid may be concentrated by filtration to enhance the detection of microorganisms present in trace levels. Also, the collection fluid can be dilution plated simultaneously on different culture media selective for the growth of various kinds of fungi or bacteria.

A necessary requirement for the use of culture plate impactors is the choice of an appropriate medium for the growth of fungi or bacteria being studied. Xerotolerant fungi are often collected on media such as dichloran glycerol (DG-18) or Czapek yeast autolysate agar with 40% sucrose.[18,53] Malt extract agar (MEA; no dextrose or peptone as recommended by Flannigan)[18,54] is a general-purpose culture medium used for the collection of a wide range of fungi, including hydrophilic species. Cornmeal and cellulose agars are used where the primary intent of sampling is to detect cellulose degrading fungi.[18]

Because several different models of culture plate impactors are available, unique sampling strategies can be developed depending on choice of sampling device. Multijet cascade impactors can be used to determine the size characteristics (e.g., respirable or nonrespirable) of culturable particles. A slit-to-agar impactor has the unique ability to detect a sporadic or episodic emission of culturable microbes above background levels. Thus, a slit-to-agar sampler has been used to detect the release of bioaerosols into ventilation air when a disturbance occurs in the air handling unit such as slamming an access door shut.[55] Battery-powered, portable multijet or centrifugal impactors can be used to obtain an integrated sample collected in several nearby locations. The collection of an integrated sample may better simulate general exposure rather than an area sample in a stationary location.

When using culture plate impactors, it is desirable to collect a number of particles on the medium surface that is neither too high nor too low. A loading of about one colony per cm^2 of plate surface area is desirable.[18,24] Overloading the plate (generally more than 60 colonies) or crowding of colonies compromises the ability of the laboratory to identify the microorganisms growing at particle impaction sites. Underloading (<10 colonies) the plate results in collection of too few culturable particles so that the sample may not be representative of the microorganisms present in the air.

3.12. SPORE TRAP SAMPLERS

Spore traps are devices that collect both viable and nonviable spores as well as other morphologically distinct particles (pollens, hyphal fragments, skin scales, plant hairs, textile fibers) by impaction onto a sticky glass or tape surface. Information on sources of commercially available spore traps is found elsewhere.[18,32] The particles collected in a spore trap are examined morphologically by direct optical microscopy with or without staining in order to determine kinds of fungal taxa or pollens that may be present. Mold spores can be identified by morphology at best to the genus or multigenus levels (e.g., *Cladosporium, Alternaria, Penicillium–Aspergillus*). Speciation of collected mold spores cannot be readily achieved by direct microscopic observation of spore trap samples.[9] Species identification is achieved by culture analysis and by other techniques such as PCR methodology.[18] It is important for the investigator to predetermine whether the laboratory chosen to process spore trap samples has analysts skilled in the microscopic identification of mold spores.

Collection of particulate from ~0.05 to ~0.15 m^3 of air is generally adequate to provide a representative number of spores for direct microscopic analysis of spore trap samples.[18] In heavily contaminated environments such as workplaces where mold remediation is underway (see Table 3.6), lesser volumes of sampled air can be collected. In dusty environments nonbiological particles such as soot and gypsum wallboard dust can cover up mold spores on the collection surface, thus making analysis difficult.[18]

Recording or programmable spore trap samplers are available wherein spores are collected continuously over a 24-h or 7-day period. This kind of sampling is useful in identifying the occurrence of an episodic spore release and in identifying diurnal variation in mold taxa levels.

Nonviable mold may account for 90% or more of the spores present in air samples, and as such, spore trap sampling provides an estimate of the total mold burden in the air. In addition, the concentration of some spores such as *Stachybotrys* whose culturability is fragile (e.g., sensitive to desiccation) is best estimated by spore trap as compared to culture air sampling.

Comparison of Tables 3.6 and 3.11 illustrates differences between air sampling for molds by culture plate impactor and spore trap sampling. The sampling results in both tables were atypical in rank-order composition of indoor molds compared to molds in the outdoor air. Note that the culturable *Penicillium* and *Aspergillus* taxa in Table 3.11 are identified to the species level. In the spore trap sample (Table 3.6) *Penicillium* and *Aspergillus* species are combined into the morphological genus *Penicillium–Aspergillus*. However, the spore trap sampling results in Table 3.6, like the culture plate impactor results in Table 3.11, show that the indoor air was atypical or different relative to the kinds of molds found outdoors.

Spore trap sampling is inappropriate when the sampling strategy involves the identification of infective agents. Culture plate impactors, liquid impingers, and

TABLE 3.11. Air Sampling for Culturable Fungal Species in a Building with a Mold Contamination Problem

Rank order (%)[a] of Fungi Types in	
Outdoor Air	Indoor Air
Cladosporium cladosporioides (92)	*Aspergillus versicolor* (24)
Penicillium brevicompactum (2)	*Penicillium corylophilum* (21)
Cladosporium sphaerospermum (1)	*Cladosporium cladosporioides* (14)
Penicillium implicatum (1)	*Penicillium citrinum* (8)
Penicillium sclerotiorum (1)	*Wallemia sebi* (7)

[a]Ranked types of Fungi in percent of total.

Source: Adapted from Morey.[10]

bulk or surface sample culture methods are most useful in the investigation of infectious fungal and bacterial agents.

3.13. SAMPLING BY FILTRATION

Bioaersols can be collected on filters and analyzed by methods including direct microscopy, growth of propagules directly on the filter medium, elution of particles from the filter, and dilution plating.[18] In addition, specialized methods are available for analysis of dust collected on filters for endotoxins, glucans, and mycotoxins, and by PCR.

Personal and area air sampling for hardy fungal and bacterial spores can be carried out using presterilized cassettes with cellulose ester or polycarbonate filters.[56,57] Possible damage (loss of culturability) during sample collection may result from desiccation of sensitive cells and susceptible spores (e.g., *Stachybotrys* spores; vegetative bacterial cells), and this favors the recovery of hardy spores such as *Aspergillus fumigatus*. Personal sampling for culturable propagules by filtration in most appropriate for dusty environments where the concentration of bioaerosols is high (e.g., during mold abatement activities, wood processing, agricultural operations) enough to ensure that an adequate number of culturable propagules are present in the filter wash for dilution plating. Button (trademark) samplers using gelatin filters are reported to be efficient in recovering desiccation sensitive fungi and bacteria.[18]

Filtration is widely used for the collection of airborne endotoxins and glucans. Cassettes, filters, and instruments that touch the filter collection surfaces must be pyrogen-free (surfaces devoid of any environmental sources of endotoxins or glucans). Endotoxins are more readily extracted from polycarbonate, Teflon, and glass fiber than from cellulose ester filters.[58] Sample air volumes adequate for analysis of endotoxins or glucans often exceed 1 m^3 for nonindustrial indoor environments. Filtration sampling for airborne mycotoxins has been successful in dusty agricultural settings where airborne levels of total fungal spores were in the $10^5-10^7/m^3$ range.[59] In one study, using a high-volume sampler (flowrate 0.83 m^3/min; glass fiber filter; total air volume ≤ 50 m^3), several *Fusarium* trichothecene mycotoxins were recovered.[59] However, in nonindustrial indoor environments, detection of specific airborne mycotoxins has been limited to research studies for reasons including limited availability of mycotoxin standards for each of the many mycotoxins that may occur in spores and hyphae.[18] High-volume air filtration sampling has also been used to detect cytotoxins in collected dusts in nonindustrial indoor environments.[60]

Interpretation of data from filtration sampling is hampered by absence of dose–response information not only for culturable and nonculturable fungi and bacteria but also for chemical agents such as glucans and mycotoxins. It is always important therefore, for purposes of data interpretation, to collect background control samples in the outdoor air and where possible in noncomplaint as well as complaint/problem zones.[18] Table 3.3 provides an example where the indoor concentration of

TABLE 3.12. Airborne Endotoxin Upwind and Downwind of Wastewater Treatment Plant[a]

Location	Endotoxin Concentration (EU/m^3)
50 m upwind of plant	14
50 m downwind of plant at entrance to office building	11
Above digester in plant	58

[a]Analysis by *Limulus* assay (see Ref. 61).

endotoxins collected by filter cassette in a complaint zone was at least two orders of magnitude greater than that present at outdoor and indoor control sites. As discussed in an earlier section, this suggested that the flulike illness in the complaint zone was associated with endotoxin exposure.

Table 3.12 provides an additional example of sampling strategy using filter cassettes where the objective was to determine whether airborne endotoxin from a wastewater treatment plant was degrading the air at a nearby office building. Air samples were collected outdoors in the air around the office building, in the air upwind from the wastewater treatment plant, and in the wastewater treatment plant. Endotoxin concentrations in the outdoor air around the office building were similar to those present upwind of the wastewater plant (Table 3.12). The endotoxin concentration in the wastewater plant was only slightly elevated relative to concentrations in the outdoor air at the office building and upwind of the plant. This sampling data suggested that Gram-negative bacterial aerosols from the wastewater treatment plant were not affecting the outdoor air entering the office facility at the time of sampling.

A sampling strategy for β-1,3-D-glucans using filter cassettes is illustrated in Table 3.13. The objective of sampling was to determine the extent of exposure to glucans under the unusual condition where moldy, moisture-impervious wall covering was physically removed from gypsum wallboard (concentration of *Penicillium–Aspergillus* spores $>10^5/m^3$)[34] as might occur during mold remediation activities.

TABLE 3.13. Air Sampling for β-1,3-D-Glucan When Moldy Moisture-Impervious Vinyl Wallcovering Is Removed from Envelope Wall[a]

Sample Location	β-1,3-D-Glucan Concentration (ng/m^3)
Indoor air; wallcovering undisturbed	0.12
Indoor air; panel of wallcovering removed from wall	13,500.0
Outdoor air	0.03

[a]Polycarbonate filter cassette; analysis according to Rylander et al.[62]

Sampling was carried out separately in empty rooms where one sheet of moldy vinyl wall covering was removed and in another room where the wall covering was undisturbed. β-1,3-D-Glucan levels increased by three to four orders of magnitude in rooms where moldy wall covering was peeled back from the wall as compared to indoor and outdoor control locations (Table 3.13; Morey and Rylander, unpublished data).

3.14. CONCLUSIONS

A number of publications contain detailed discussions on microbial sampling strategies in indoor environments.[18,24,54] Listed below are some practical steps that practitioners should consider when developing microbial sampling strategies in nonindustrial indoor environments:

- Sample for the correct microbial agent. For example, in an allergy complaint investigation, do not sample for *Escherichia coli* 0157/H7 or other potentially pathogenic bacteria.
- Never assume that the collection of one or two outdoor samples and one or two indoor air samples fully characterizes indoor air quality.
- Recognize that information in medical mycology books on infections caused by fungi is not directly or even indirectly applicable to environmental sampling where infection is not an issue.
- Do not provide medical/clinical opinion based entirely on the results of environmental sampling.
- Numerical guidelines (TLVs or PELs) for interpreting bioaerosol data do not exist. In other words, there is no scientific basis for relating measurements such as 1000 spores/m^3 or 100,000 CFU/g of dust to adverse health outcomes.
- A sampling strategy for an indoor environment always involves a thorough understanding of the building and its components, and this is achieved by physical inspection. For example, understanding the nature of water damage and the extent of visible mold growth is a prerequisite to development of a sampling strategy in a mycological investigation.
- Never sample for microbial agents without a plan for data interpretation. Remember, there are no TLVs or PELs. The results of laboratory analysis are not self-interpretable.

REFERENCES

1. Carnelley, T., J. S. Haldane, and A. M. Anderson, The carbonic acid, organic matter and micro-organisms in air, more especially in dwellings and schools. *Phil. Trans. Roy. Soc. B* **178**:61–111 (1887).
2. Morey, P. R. and J. E. Woods, Indoor air quality in health care facilities. *Occup. Med.* (State of the Art Reviews) **2**:547–563 (1987).

3. USPHS, *Sampling Microbiological Aerosols*, Public Health Service Monograph 60, PHS Publication 286, Washington, DC, U.S. Public Health Service, 1959.
4. IOM, *Damp Indoor Spaces and Health*, Institute of Medicine, National Academies Press, Washington, DC, 2004.
5. HSE, *The Control of Legionellosis Including Legionnaires' Disease*, Health and Safety Executive, Health and Safety Booklet H.S. (G)70, Sudbury, UK, 1994.
6. Morey, P. R., Fungi in buildings, in *Infection Control during Construction Manual*, 2nd ed., W. Hansen, ed., HCPro, Marblehead, MA, 2004, Chapter 4, pp. 85–112.
7. Storey, E., K. H. Dangman, P. Schenck, R. L. DeBernardo, C. S. Yang, A. Bracker, and M. J. Hodgson, *Guidance for Clinicians on the Recognition and Management of Health Effects Related to Mold Exposure and Moisture Indoors*, Univ. Connecticut Health Center, Farmington, CT, 2004.
8. ACGIH, Data interpretation, in *Bioaerosols Assessment and Control*, American Conference of Governmental Industrial Hygienists, Cincinnati, OH, 1999, Chapter 7, pp. 7-1–7-9.
9. Burge, H. A. and J. A. Otten, Fungi, in *Bioaerosols Assessment and Control*, American Conf. Governmental Industrial Hygienists, Cincinnati, OH, 1999, Chapter 19, pp. 19-1–19-13.
10. Morey, P. R., Microbiological investigations of indoor environments: Interpreting sampling data-selected case studies, in *Microorganisms in Home and Indoor Work Environments*, B. Flannigan, R. A. Samson, and J. D. Miller, eds., Taylor & Francis, London, 2001, pp. 275–284.
11. ACGIH, in *Bioaerosols Assessment and Control*, The building walkthrough, American Conference of Governmental Industrial Hygienists, Cincinnati, OH, 1999, Chapter 4, pp. 4-1–4-11.
12. Hunter, C. A., C. Grant, B. Flannigan, and A. F. Bravery, Mould in buildings: The air spora of domestic dwellings, *Internatl. Biodeteriora. Biodegr.*, **24**:84–101 (1988).
13. Ferro, A. R., R. J. Kopperud, and L. M. Hildemann, Source strengths for indoor human activities that resuspend particulate matter, *Environ. Sci. Technol.* **38**:1759–1764 (2004).
14. Morey, P. R., Control of indoor air pollution, in *Occupational and Environmental Respiratory Disease*, P. Harber, M. B. Schenker, and J. R. Balmes, eds., Mosby, St. Louis, 1996, pp. 981–1003.
15. Mullins, J., Microorganisms in outdoor air, in *Microorganisms in Home and Indoor Work Environments*, B. Flannigan, R. A. Samson, and J. D. Miller, eds., Taylor & Francis, London, 2001, pp. 3–16.
16. Li, D. W., Airborne basidiospores of boletes and their potential to infiltrate a residence in central Connecticut. *Proc. Indoor Air Conf. 2005*, Beijing, 2005, pp. 1450–1454.
17. Arnow, P., R. Anderson, P. Mainous et al., Pulmonary Aspergillosis during hospital renovation, *Am. Rev. Resp. Dis.* **118**:49–53 (1978).
18. AIHA, *Field Guide for the Determination of Biological Contaminants in Environmental Samples*, L.-L. Hung, J. D. Miller, and H. K. Dillon, eds., American Industrial Hygiene Association, Fairfax, VA, 2005.
19. Macher, J., Data analysis, in *Bioaerosols Assessment and Control*, American Conference of Governmental Industrial Hygienists, Cincinnati, OH, 1999, Chapter 13, pp. 13-1–13-16.

20. Morey, P. R., M. C. Hull, and M. Andrew, El Niño water leaks identify rooms with concealed mould growth and degraded indoor air quality, *Internatl. Biodeterior. Biodegr.* **52**:197–202 (2003).

21. Pasanen, A.-L., Evaluation of indoor fungal exposure, *Proc. Healthy Buildings*, International Society of Indoor Air Quality, Helsinki, 2000, Vol. 3, pp. 25–38.

22. Miller, J. D. and J. C. Young, The use of ergosterol to measure exposure to fungal propagules, *Am. Indlist. Hyg. Assoc. J.* **58**:39–43 (1997).

23. Rylander, R., Microbial cell wall constituents in indoor air and their relation to disease, *Indoor Air Suppl.* **4**:59–65 (1998).

24. ACGIH, Developing a sampling plan, in *Bioaerosols Assessment and Control*, American Conf. Governmental Industrial Hygienists, Cincinnati, OH, 1999, pp. 5-1–5-13.

25. Brinkerhoff, D. J. and P. R. Morey, The effect of construction site dust on mold clearance sampling, in *Bioaerosols, Fungi, Bacteria, Mycotoxins and Human Health*, E. Johanning, ed., Fungal Research Group Foundation, Albany, NY, 2005, pp. 440–446.

26. ISIAQ, *Control of Moisture Problems Affecting Biological Indoor Air Quality*, International Society of Indoor Air Quality, Helsinki, 1996.

27. Bonnehag, C. G., G. Bloomquist, F. Gyntelberg et al., Dampness in buildings and health. Nordic interdisciplionary review of the scientific evidence on associations between exposure to "dampness" in buildings and health effects (Nordamp), *Indoor Air* **11**: 72–86 (2001).

28. Hodgson, M. J., P. R. Morey, M. Attfield, W. Sorenson, J. N. Fink, W. W. Rhodes, and G. S. Visvesvara, Pulmonary disease associated with cafeteria flooding, *Arch. Environ. Health* **40**:96–101 (1985).

29. Morey, P. A., Hyvärinen, and T. Meklin, WS17 interpretation of fungal sampling data in building studies, *Workshop Summaries, Healthy Buildings*, 2000, pp. 83–86.

30. Milton, D. K., J. Amsel, C. E. Reed, P. L. Enright, L. R. Brown, G. L. Aughenbaugh, and P. R. Morey, Cross-sectional follow-up of a flu-like respiratory illness among fiberglass manufacturing employees: Endotoxin exposure associated with two distinct sequelae, *Am. J. Indust. Med.* **28**:469–488 (1995).

31. Banaszak, E. F., W. H. Thide, and J. N. Fink, Hypersensitivity pneumonitis due to contamination of an air-conditioner, *N. Engl. J. Med.* **283**:271–276 (1970).

32. Willeke, K. and J. Macher, Air sampling, in *Bioaerosols Assessment and Control*, American Conference of Governmental Industrial Hygienists, Cincinnati, OH, 1999, Chapter 11, pp. 11-1–11-25.

33. Martyny, J. W., K. F. Martinez, and P. R. Morey, Source sampling, in *Bioaerosols Assessment and Control*, American Conf. Governmental Industrial Hygienists, Cincinnati, OH, 1999, Chapter 12, pp. 12-1–12-8.

34. Morey, P. R., Studies on fungi in air-conditioned buildings in a humid climate, in *Fungi and Bacteria in Indoor Air Environments*, E. Johanning and C. S. Yang, eds., Eastern New York Occupational Health Program, Latham, NY, pp. 79–92, 1995.

35. ASTM, *Standard D5952-96, Standard Guide for Inspecting Water Systems for Legionellae and Investigating Possible Outbreaks of Legionellosis (Legionnaires' Disease or Pontiac Fever)*, American Society of Testing and Materials, West Conshohocken, PA, 2002.

36. Milton, D., K. Endotoxin and other bacterial cell-wall components, in *Bioaerosols Assessment and Control*, American Conf. Governmental Industrial Hygienists, Cincinnati, OH, 1999, pp. 23-1–23-14.

37. Pinard, M.-F., P. Widden, T. Debellis, C. Deblois, and C. Mainville, House dust: An efficient and affordable tool to assess microbial contamination in homes, in *Bioaerosols, Fungi, Bacteria, Mycotoxins and Human Health*, E. Johanning, ed., Fungal Research Group Foundation, Albany, NY, 2005, pp. 191–198.

38. Prezant, B., Comparison of PVC cassettes versus vacuum bags for collection and enumeration of culturable fungi in settled dust, in *Bioaerosols, Fungi, Bacteria, Mycotoxins and Human Health*, E. Johanning, ed., Fungal Research Group Foundation, Albany, NY, 2005, pp. 218–225.

39. Dillon, H. K., J. D. Miller, W. G. Sorenson, J. Douwes, and R. J. Jacobs, Review of methods applicable to assessment of mold exposure to children, *Environ. Health Persp.* **107**(Suppl. 3):473–480 (1999).

40. Horner, W. E., A. W. Worthan, and P. R. Morey, Air and dustborne mycoflora in houses free of water damage and fungal growth, *Appl. Environ. Microbiol.* **70**:6394–6400 (2004).

41. Leese, K. E., E. C. Cole, R. M. Hall et al., Measurement of airborne and floor dusts in a nonproblem building, *Am. Indust. Hyg. Assoc. J.* **58**:432–438 (1997).

42. Chew, G. L., C. Rogers, H. A. Burge, M. L. Muilenburg, and D. Gold, Association between dustborne and airborne fungi in residential environments. *Allergy* **58**:13–20 (2003).

43. Miller, J. D., Mycological investigations of indoor environments, in *Microorganisms in Home and Indoor Work Environments*, B. Flannigan, R. A. Samson, and J. D. Miller, eds., Taylor & Francis, London, 2001, pp. 231–246.

44. Pope, A. M., R. Patterson, and H. A. Burge, eds., *Indoor Allergens: Assessing and Controlling Adverse Health Effects*, Institute of Medicine, Washington, DC, 1993.

45. Rose, C. S., Allergens, in *Bioaerosols Assessment and Control*, American Conf. Governmental Industrial Hygienists, Cincinnati, OH, 1999, Chapter 25, pp. 25-1–25-11.

46. Shaughnessy, R. J. and P. R. Morey, Remediation of microbial contamination, in *Bioaerosols Assessment and Control*, American Conf. Governmental Industrial Hygienists, Cincinnati, OH, 1999, Chapter 15, pp. 15-1–15-7.

47. *Guidelines on the Assessment and Remediation of Mold in Indoor Environments*, New York City Department of Health, 2000.

48. OSHA, *A Brief Guide to Mold in the Workplace*, Occupational Safety and Health Administration, U.S. Department of Labor, Washington, DC, 2003.

49. Lowell, J. D. and S. H. Pierson, Ultraviolet irradiation and laminar airflow during total joint replacement, in *Architectural Design and Indoor Microbial Pollution*, R. B. Kundsin, ed., Oxford Univ. Press, Oxford, 1988, pp. 154–173.

50. Reeslev, M. and M. Miller, The mycometer-test: A new rapid method for detection and quantitation of mold in buildings, *Proc. Healthy Buildings Conf.*, 2000, Vol. 1, pp. 589–590.

51. Krause, J. D., Y. Y. Hammad, and L. B. Ball, Application of a fluorometric method for the detection of mold in indoor environments, *Appl. Occup. Environ. Hyg.* **18**:499–503 (2003).

52. Morey, P. R., Practical aspects of sampling for organic dusts and microorganisms, *Am. J. Indust. Med.* **18**:273–278 (1990).

53. Samson, R. A., E. S. Hoekstra, J. C. Frisvad, and O. Filtenborg, *Introduction to Food Borne Fungi*, Centraalbureau voor Schimmelcultures, Baarn, Netherlands, 2000.

54. Flannigan, B., Guidelines for evaluation of airborne microbial contamination of buildings, in *Fungi and Bacteria in Indoor Air Environments*, E. Johanning, and C. S. Yang, eds., Eastern New York Occupational Health Program, Latham, NY, 1995, pp. 123–130.

55. Yoshizawa, S., F. Sugawara, S. Ozawo, Y. Kohsaka, and A. Matsumae, Microbiologial contamination from air conditioning systems in Japanese buildings, *Proc. 4th Internatl. Conf. Indoor Air Quality and Climate*, 1987, Vol. 1, pp. 627–631.

56. Alwis, K. U., J. Mandryk, and A. D., Hocking, Exposure to biohazards in wood dust: Bacteria, fungi, endotoxins, and (1-3)-B-D-glucans, *Appl. Occup. Environ. Hyg.* **14**: 598–608, (1999).

57. Dahlqvist, M., U. Johard, R. Alexandersson, B. Bergström, U. Ekholm, A. Eklund, B. Milosevich, G. Tornling, and U. Ulfvarson, Lung function and precipitating antibodies in low exposed wood trimmers in Sweden, *Am. J. Indust. Med.* **21**:549–559 (1992).

58. Douwes, J., P. Versloot, A. Hollander, D. Heederik, and G. Doekes, Influence of various dust sampling and extraction methods on measurement of airborne endotoxin, *Appl. Environ. Microbiol.*, **61**:1763–1769 (1995).

59. Lappalainen, S., M. Nikalin, S. Berg, P. Parikka, E.-L. Hintikka, and A.-L. Pasanen, *Fusarium* toxins and fungi associated with handling of grain on eight Finnish farms, *Atmos. Environ.* **30**:3059–3065 (1996).

60. Gareis, M., E. Johanning, and R. Dietrich, Mycotoxin cytoxocity screening of field samples, in *Bioaerosols, Fungi and Mycotoxins: Health Effects, Assessment, Prevention and Control*, E. Johanning, ed., Eastern New York Occupational and Environmental Health Center, Albany, NY, 1999, pp. 202–213.

61. Milton, D. K., H. Feldman, D. Neuberg, R. Bruckner, and I. Greaves, Environmental endotoxin measurement: The kinetic *Limulus* assay with resistant-parallel-line estimation, *Environ. Res.* **57**:212–230 (1992).

62. Rylander, R., K. Persson, H. Goto, K. Yuasa, and S. Tanaka, Airborne Beta-1, 3-glucan may be related to symptoms in sick buildings, *Indoor Environ.* **1**:263–267 (1992).

CHAPTER 4

MICROSCOPIC ANALYTICAL METHODS FOR FUNGI

DE-WEI LI, CHIN S. YANG, and F. HARRINGTON

4.1. INTRODUCTION

The presence of indoor fungi and their byproducts may implicate potential health problems, such as allergy, asthma, fungal infections, or detrimental effects derived from mycotoxins.[1–5] Frequently it is necessary to conduct visual inspection or investigation, routine IAQ survey or monitoring and remediation to address indoor-fungi-related water intrusion, indoor air quality, or potential health implications. The presence of indoor fungi often requires remediation by professionals to remove fungi and clean the infested areas. During these processes, taking samples for analysis of indoor fungi is imperative. Several kinds of samples can be taken for indoor fungal investigation. These include sampling the air for airborne fungal spores (spore count) or source sampling by collecting bulk samples, dust samples, tape lifts, and direct swabs. The types and numbers of samples to be taken are determined by the sampling strategies and the purposes of fungal investigations. For most of these samples, a microscopic analysis performed in a laboratory for indoor fungi is inevitable and crucial. The microscopic analysis is to identify fungi present in the samples according to fungal morphological characteristics and to quantify and characterize the fungal structures. Microscopic analysis for indoor fungal investigation is not only for identification of fungal taxa but also to provide additional information, such as fungal structures and quantitative information on the fungi present in the samples. Such information is useful and crucial to field investigators in order to properly evaluate and determine the fungal exposure of occupants and the severity of the fungal infestation.

Presently there are no federal standards for analytical techniques and procedures to identify molds; thus this chapter discusses microscopic methods appropriate for the various sample types (spore traps, bulk, dust, tape lift, swab, and cultures).

Sampling and Analysis of Indoor Microorganisms, Edited by Chin S. Yang and Patricia A. Heinsohn

4.2. PRINCIPLES AND USAGE OF MICROSCOPES

4.2.1. Microscopes

Most introductory microscopy topics, such as (1) anatomy of a microscope, (2) basic microscopy and image analysis, (3) magnification, resolution and contrast, (4) numerical aperture, (5) illumination techniques, (6) depths of fields of focus, and (7) microscope maintenance, are not discussed here but can be found in a number of reference books on microscopes and on the internet at specific manufacturers' Websites, usually with interactive tutorials, illustrations, and in-depth instructions for many of the models. The features specific to that model are explained in detail (see Web resources at the end of this chapter).

Both compound and stereomicroscopes are necessary instruments for fungal analyses. Each one is discussed below. Several brands of both compound and stereomicroscopes are suitable for use. The major manufacturers are Zeiss, Leica, Nikon, Olympus, and smaller companies including Swift, Meiji, Bausch & Lomb, and American Optical. Some of the newer innovations to microscopes include ergonomic designs to decrease eye fatigue from long viewing times. A laboratory's budget determines whether a new or previously used instrument is purchased.

The purpose of the microscopic analytical method is to produce an image in which more details of the specimen can be detected. For compound microscopes, the quality of the image depends on the magnification, the resolving power of the microscope, and the contrast produced in the image.[6] The optical systems within most modern microscopes may be able to produce high-resolution images at high magnifications, but such a capability is worthless without sufficient contrast in the image.[7] Contrast is dependent on the interaction of the specimen with light and the efficiency of the optical system coupled to its ability to reliably record the image with a suitable detector.[8] The contrast of the environmental sample is dependent on several factors, including the proper setting of aperture diaphragms, the degree of optical aberration, the contrast mechanism employed, the type of specimen, the mounting medium, and the characteristics of the detector; the field and condenser aperture diaphragm settings are most critical to the optical system. Figure 4.1

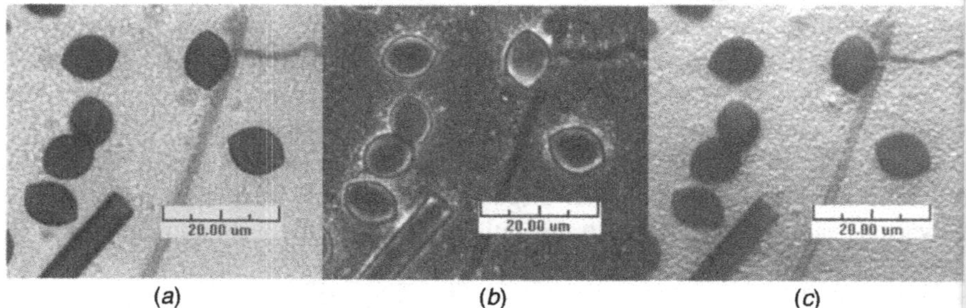

Fig. 4.1. Types of illumination on ascospores of *Chaetomium globosum*: (*a*) brightfield; (*b*) phase contrast; (*c*) differential interference contrast.

illustrates a series of three images captured in transmitted light mode of the same viewing field containing ascospores of *Chaetomium globosum* under differing contrast modes: brightfield, phase contrast, and differential interference contrast (DIC). The three images were taken from the same fungus with the same focus area, but because there is some variation derived from the different contrast modes, the microscopist might arrive at a different conclusion from independent examination of each viewing field. In addition, if the spores are hyaline or colorless, the images would be quite dissimilar and in the brightfield illumination mode, the transparent, unstained spores would be invisible. Brightfield and phase contrast illumination techniques are the prevalent techniques used for fungal analysis. According to the survey conducted by the AIHA taskforce (unpublished data), the majority of the laboratories use brightfield illumination but the colorless or hyaline specimens/spores have to be modified by stains with chemical dyes so that the structures or spores can be noted. These chemical dyes, however, may alter the structures, size, and natural color of the organisms. Not as many laboratories use phase contrast illumination, and a small number of laboratories use DIC illumination. Phase contrast and differential interference microscopes make it possible to see unstained biological material without the addition of chemical dyes. Objectives designed for these microscopes require the assistance of optical detectors to modify events occurring at the objective rear focal plane.[9]

Phase contrast microscopy enhances the phase difference between the specimen and its surroundings with an illumination technique (Fig. 4.1). It requires a special set of objectives with phase rings placed in the back focal plane of the objective. A matching phase ring is placed in the condenser. A phase turret condenser will usually have a brightfield position as well as phase positions to match each phase objective. The phase objective has to match the specified phase position on the turret condenser. The problems with phase contrast microscopy are that the technique suffers from halo artifacts, is restricted to thin specimen preparations, and cannot take advantage of the full condenser and objective apertures.[7]

Differential interference contrast microscopy employs a polarized illumination technique that allows specimens that have a refractive index similar to that of their surroundings to be visually differentiated. DIC achieves this result by placing a Wollaston prism in the front focal plane of the condenser and one in the rear focal plane of the objective. The result creates a three-dimensional (3D) effect of the specimen's surface and produces a monochromatic shadow-cast image that efficiently shows the gradient of optical paths for both high and low spatial frequencies in the specimen. The specimens are brighter (or darker) in the region where the optical paths are enhanced along a reference direction, while regions where the path differences are diminished are in reverse contrast. The primary difference between phase contrast and DIC microscopy is the optical components required for DIC do not mask or obstruct the objective and condenser apertures (as in phase contrast). The result is a dramatic improvement in resolution (particularly along the optical axis), elimination of halo artifacts, and the ability to produce excellent images with relatively thick specimens. In addition, DIC produces an image that can be easily manipulated using digital and video imaging techniques to further enhance contrast.[8] A DIC turret condenser usually has a brightfield illumination position as well as the DIC positions to match each objective. Some models have a DIC turret condenser that includes a brightfield and phase illumination positions. Biological specimens observed with these contrast-enhancing methods often produce images that are quite different in appearance and character (Fig. 4.1). The ability to utilize the microscope at full numerical aperture without the masking effects of phase plates or condenser annuli is the primary advantage of DIC over phase contrast. The other benefit is to improve axial resolution, mainly to produce excellent high-resolution images at large aperture sizes with DIC microscopy. Biological specimens (including fungi) under DIC microscopic techniques have a pseudocolor that is different from the natural color.

Many fungi have displayed fluorescence when H&E (hematoxylin and eosin)-stained samples were examined under a fluorescence microscope,[10] and fluorescence illumination techniques have been used for identification and enumeration of bioaerosols and pathogenic and soil fungi as well.[11–13] Coleman and coworkers reported that quantification of a fungal mass was achieved with a new procedure that detected particle epi-fluorescence.[14] For the assessment of an indoor fungal infestation, a fluorescence staining method incorporating FUN-1 (Molecular Probes, Eugene, OR), Fluorescent Brightener 28 (Sigma Chemical Co., St. Louis MO), and potassium hydroxide was developed to examine fungi in air and physical samples. The results showed that this staining method was most effective in speedily detecting and distinguishing fungal structures with enhanced clarity.[13] Because most fluorescence dyes stain internal organelles or structures, such as DNA in nuclei, the fluorescence may have difficulty emitted from dematiaceous spores, such as *Stachybotrys*. These fluorescence microscopic techniques for identification are still under development and currently not widely used in commercial laboratories.

Because the stereomicroscope provides a 3D image of an object as well as the ability to manipulate the specimen while observing, it is the instrument initially used to determine accurately the location of the fungi on samples of tape, swabs, bulk, and other sources (e.g., dust). The magnification range is normally 10–60×. There can be a single fixed magnification, several discrete magnifications, or a continuous zoom magnification system. Newer versions of stereomicroscopes have become more sophisticated with motor-controlled zooms integrated with digital imaging systems. Besides observing fungal morphological features, especially the reproductive structures, it is also possible to prepare the samples or subsamples for further observation under a compound microscope. It is relatively easy to use a stereomicroscope. Detailed instructions for specific models can be obtained from the manufacturers. Observation starts from low power and proceeds to high power. Light sources may be internal or from an external source. The external illumination provides better observation potential by having the ability to change the angle for a particular field of view. Occasionally, simultaneous lighting from above and below may be necessary to observe certain fungal structures, such as conidial orientation on an environmental source or from a culture.

The morphological features of the reproductive structures (asexual, sexual, or both) are crucial for identifying most fungi to genera or to species, including

Alternaria, *Chaetomium*, *Penicillium*, and *Aspergillus*. The patterns of sporulation and conidial chains are important elements to differentiate among species of *Alternaria*.[15] The orientation of conidial chains and the conidial mass shapes are also important characteristics for identifying the genera of *Aspergillus* and *Penicillium* to species.[16,17] These characteristics are undisturbed for observation under a stereomicroscope, but can be lost in a wet mount preparation observed under a compound microscope.

4.2.2. Microscope Objectives

The compound microscope can be set in a number of ways to greatly increase the contrast images of environmental specimens through objectives and other accessories for specialized applications. Generally, most optical microscopes are equipped with a pair of 10× ocular lenses, or occasionally with 15× eyepieces. Ninety percent of all optical microscopy examinations are conducted using standard achromat or plan achromat objectives, which are inexpensive and easily available.[9] Standard brightfield objectives are the most common and useful types of objectives for examining specimens that are normally colored or have been previously stained (i.e., if the samples or spores were colorless). Objective lenses typically range from 10× to 100× magnification. The American Industrial Hygiene Association (AIHA) Policy Module 2D, Section 2D.3.1.2 indicates: "The laboratory shall have at least one compound microscope having low, high power, and 100× oil immersion objectives." Ideally, all objectives should be parfocal and par-centered to minimize refocusing.

Four objectives, 20×, 40×, 60×, and 100× (oil immersion lens), are commonly used for spore count analysis.[18,19] A 20× lens is used for quickly scanning the whole sample trace of a bioaerosol and to identify large spores. Although some laboratories use both 40× and 60× objectives for detailed spore enumeration and identification,[20–25] the 60× lens is better suited for such analyses because it provides details of fungal structures but does not significantly reduce the viewing area and the depth of observations. The 100× objective (oil immersion lens) is used to observe fine structures on small (<6 μm) spores or conidia. These fine structures may indicate whether they are a specific structure indicating mode of spore development or part of a release mechanism, all enhancing identification. The 100× objective is not used all the time, but it is necessary to have the microscope equipped with it. Keep in mind that the maximum resolving power of the light microscope is about 0.2 μm at 1200× magnification.[26] The smallest fungal spore is approximately 2 μm. A number of research labs have used a 100× objective for their studies without consideration of the disadvantages associated with such an objective lens.[24,27–29] Although a 100× lens provides better resolution, the field depth is significantly reduced. The analyst must constantly adjust the fine focus in order to count and identify all spores. Under a 100× lens it takes a much longer time to analyze a sample, due to reduced viewing area, without actually improving the quality of the data. In addition, immersion oil attracts dust, is difficult to clean from the objectives, and is likely to make the slide samples difficult or impossible to preserve for future examination or for replication and duplication analysis as required by AIHA policy or by a laboratory quality assurance–quality control (QA/QC) policy.

According to the survey of the Environmental Microbiology Proficiency Analytical Testing (EMPAT) task force of AIHA (unpublished data) the majority of accredited environmental labs analyze fungal spores under magnifications ranging from 200× to 600×. Among the surveyed labs, several labs indicated that 1000× was used for their lab examination and analysis.

All compound microscopes used for examinations and analyses should be equipped with an ocular micrometer or microgrid. The ocular micrometer must first be calibrated before it can be used. Each microscope must be individually calibrated at least once a year, because the measurements of the lenses on the microscopes may vary slightly within the same brand or from the same manufacturer. The eyepiece and associated objective lenses are measured against a stage micrometer or a graduated slide. The length of each division of the ocular micrometer must be calculated, recorded, and documented at different magnifications. The ocular micrometer is used not only to measure the size of fungal spores but also to determine the dimensions of sample traces. The micrometer also serves as a guidepost during the analysis of the sample. According to the percentage of the area analyzed, concentrations of airborne fungal spores and structures can be reliably calculated.

Proper alignment and use of the microscope are important for quality examination and analysis. The aperture should be set according to the objective lens being used. For instance, when using a 60× objective lens, the aperture should be closed down to the 0.3 position on a brightfield microscope to achieve the optimal contrast. A wide-open aperture will lead to low contrast and some small colorless spores could be overlooked and undercounted with too bright a background. The condenser normally does not need to be set at a lower position to have sufficient contrast. Phase contrast light microscopy (PCM) is generally not recommended for spore counting because small colorless, round spores and air bubbles are difficult or impossible to differentiate using PCM.

4.2.3. Photomicrographic Accessories

Digital imaging is increasingly applied to image capture for microscopy—an area that demands high resolution, color fidelity, and careful management of limited light conditions.[30] Digital cameras combined with the appropriate software can now offer image quality comparable (but still short of the resolution) to traditional film photography.[30] The powerful software programs allow greater flexibility since images can be captured and stored much more easily than traditional microphotography with 35 mm film. According to Drent,[30] a primary consideration in choosing a digital camera over film is whether the electronic imaging device is of adequate field size and spatial resolution to record the entire viewing field at the optical resolution of the microscope system. The photomicrograph system can either be mounted on the trinocular or the ocular tube; the ocular setup is more economical.

4.3. ASEPTIC TECHNIQUE AND BIOSAFETY

It is important to use aseptic techniques and observe biosafety precautions when working with all microbes. Make sure that the working areas (lab benches, hoods, microscopes, stereomicroscopes, etc.) are cleaned daily with 70% isopropyl alcohol or diluted bleach solutions (5–10% commercial bleach). The analyst should wear and frequently change gloves (latex or nitrile) in order to keep cross-contamination of the samples to a minimum. The tools, such as inoculating needles, forceps, scissors, spatulas, and scalpels with disposable blades, are cleaned for each environmental sample that is prepared. Alcohol wipe pads can be used to clean the tools or can be sterilized in an open flame. Ideally, it would be safest to do the sample preparations in a biosafety cabinet or a HEPA hood. This would protect the analyst from exposures as well as the samples from cross-contamination. Additionally, caution should be taken when culture plates are prepared on laboratory benches. It is necessary to have a HEPA filter installed in the ventilation system in the room. In addition to avoid knocking, shaking, or bumping the culture dish, the culture plate should not be turned upside-down. This will keep the drier conidia from spreading all over other colonies throughout the plate and becoming airborne when the plate is opened.

4.4. SAMPLE PREPARATION FOR SPORE COUNT

At present, several companies manufacture devices or equipment (such as Zefon Air-O-Cell, Burkard, Allergenco, Laro, Cyclex-D, Buck, Rotorod, and industrial hygiene cassettes filled with mixed cellulose ester membrane filters) to collect airborne fungi. These spore traps collect airborne fungal spores (including other fungal structures), as well as other airborne particulates, on a glass slide with a sticky surface or directly onto a filter. The impaction of the airborne materials is generated by an airflow jet. These airborne samples are prepared for direct microscopic examination (spore count), then identified and counted. The data are recorded by hand or directly into a computerized software program.

The information from a spore count is the total number of fungal spores and other fungal elements, whether viable or nonviable, present at that particular moment and at that particular location. The results are presented as fungal structures (or spores)/m^3 of air. This information is crucial for evaluating allergy related indoor air quality and to address relevant health concerns. The two reasons for using a spore count technique are (1) both viable and nonviable fungal spores and fragments are potentially allergenic and (2) some fungi are not able to produce asexual/sexual spores/conidia in culture (e.g., most members of the basidiomycota and all obligate phytopathogenic fungi, such as rust fungi). The major drawback of a spore count is that most fungal spores can be presumptively identified only to groups or genera since the spores are separated from the fruiting structures that are necessary to properly identify them. In most cases, it is possible to identify the spores only to the generic level.

4.5. MATERIALS NEEDED FOR PREPARING SAMPLES

First and foremost, all samples should be prepared in a clean area, preferably in a containment hood to avoid any contamination. The slides prepared from the environmental samples are temporary or semipermanent, and may last for only several days to less than 3 months. The rule of thumb is that a freshly prepared sample will always yield a better image and is much easier to examine for minute fungal structures. In addition, the sooner the slides are analyzed, the possibility of deformation or degradation of the fungal structures from water contained within the spore or fungal structures will be lessened. Crystals may also form in the sample from the combination of stain, mounting medium, and/or the sticky compounds used in the commercial sampling devices, such as Air-O-Cell cassettes.

Basic materials needed for preparing spore count samples include a pair of pointed forceps, filter forceps, small stainless-steel scissors, inoculating/dissecting needles, 25 × 75-mm plain or frosted slides, 22 × 22-mm coverslips, bottles of stock solutions of stains and mounting media (e.g., lactofuchsin, aniline blue, cotton blue), a 15–30-mL drop dispenser bottle for each mounting medium and/or stain, or a 1-mL syringe containing mounting medium and/or staining agent, immersion oil type A, 70% ethanol or isopropyl alcohol, clear nail polish, a fine or extrafine permanent marker, alcohol wipe pads, scalpel with disposable blades, and a cardboard slide container or a slide box.

Some environmental laboratories provide prepared or coated slides for use with Burkard or Allergenco air samplers. Some investigators, who conduct on-site investigations, prepare their own slides. Several coating materials are commonly used, including petrolatum or Vaseline-based high-melting-point wax (VHW) mix, mixed cellulose ester (MCE) cleared with triacetin solution, single- or double-sided clear adhesive tape, silicone in xylene 261, silicone grease 60, and Lubriseal stopcock grease 46. More detailed information about coating materials can be found in the *AIHA Green Book*.[31]

For samples collected on 25 × 75-mm-coated slides, the samples should be numbered and labeled on the uncoated end or frosted area first. Before a slide is prepared, the spore deposit area (known as a "trace") should be outlined with a permanent marker on the uncoated side of the slide. One or two drops of the mounting medium are gently applied to the trace in a manner that avoids disturbing the fungal spore deposit. A 22 × 22-mm coverslip is then placed over the mounting medium on the trace to avoid creating or trapping air bubbles. Let the coverslip settle down naturally to spread the mounting medium over the whole trace. Do not push the coverslip down with a pen or any other instrument. Pushing down the coverslip may lead to splitting the mounting medium and/or stain or smear the sample trace. It is not necessary to fix fungal spores prior to the staining procedure.

For samples collected with Air-O-Cell cassettes or similar devices, make sure that each sample has a matching labeled slide with the appropriate information obtained from the "chain of custody." Cut the sealing tape around the cassette and pry it open. This can be safely done with a stainless-steel weighing spatula. Mark the spore deposition trace on the underside of the slide with a permanent marker. If the sample is to be prepared with collection side up, place a drop of clear nail polish on each of the four corners of the sample slide. With a pair of recently alcohol-wiped forceps, carefully remove the slide from the cassette, turn it over, and place it collection side up on a 25 × 75-mm slide. The nail polish on the corners anchors the sample to the slide. One or two drops of mounting agent can be gently applied onto the trace area. Afterward, a 22 × 22-mm coverslip is placed over the mounting agent and the trace. Although it is possible to prepare a slide sample with the collection side down, it is better to place the collection side up to achieve optimal resolution and clarity. Some laboratories use a simplified preparation method and mount the samples with the collection side down. This method does not require the use of nail polish and a coverslip. The coating material on the sample can anchor itself on a slide. The sample slide serves as the coverslip. This preparation and mounting method greatly reduce the working distance between the sample and the objective lens due to the thickness of the coating. Although it is possible to have a reasonable resolution with a 60× objective, it is impossible to use a 100× lens to observe fine details when such an observation is necessary to properly identify some small spores.

When a sample is collected with a 25-mm mixed cellulose ester (MCE) filter cassette, the filter is removed from the cassette with the filter forceps and mounted on a slide. A clearing agent (such as triacetin) containing a dye is applied to the filter. Again, this process should be conducted in a clean area (e.g., in a hood or a clean Petri dish) to avoid contamination. The clearing process may take approximately 10–30 min. Afterward a mounting agent is applied, a coverslip is gently placed over the sample, and then 25–50% or a fixed area of the filter area is examined. When 37-mm filter cassettes are used for sampling, only a quarter of the filter is gently cut with the alcohol-wiped scalpel blade and mounted on a slide for clearing and staining.

4.6. STAINING AND MOUNTING TECHNIQUES

Depending on which illumination technique will be used, it may be necessary to stain the fungi or fungal structures that are colorless, transparent, or lightly colored, because staining agents improve the contrast of the fungal elements in the samples to be examined. A mounting medium is used to provide a neutral background without altering the color or the chemical composition of the specimen and preserve the specimen from dehydration. In addition, the mounting medium can increase contrast by providing a refractive index higher than that of the glass slide as well as providing a refractive index close to that of the specimen in order to achieve the best transparency. Ideally, the mounting medium should be chemically compatible with the specimen and the stain and mounting medium should not dry out during examination or storage. Although there are many stains and mounting media available for examination of fungi, lactofuchsin, lacto-cotton blue, modified lacto-cotton blue, and lacto-phenol blue are the ones most commonly used by mycologists. Malloch[32] suggests that many fungi are difficult to "wet," even with a wetting agent, and placing a drop or two of 95% ethanol for a few seconds and then, before the alcohol is completely dried out, adding a drop of the required mounting medium will aid the mounting/staining process.

1. *Lactofuchsin (Mounting Medium with Stain).* This is a commonly used stain and mounting medium that stains the cytoplasmic elements a deep pink. Freshly prepared mounts yield the best results. It is especially suitable for microphotography since it increases the contrast to some minute fungal elements, provides excellent clarity, and can be used for semipermanent mounts for most groups of fungi.[33] As the samples age, some fungal structures may lose their natural color (e.g., *Ulocladium* conidia and aeciospores of the rust genus *Gymnosporangium*). To make this medium, dissolve 0.1 g acid fuchsin in 100 mL anhydrous lactic acid (85%).[34] Lactofuchsin is better than cotton blue because the staining process is rapid and cell walls are more noticeable.[35]

2. *Shear's Mounting Medium.* Besides being a useful mounting agent, this medium is also suitable for photomicrography.[36] Since it contains glycerol, it does not dry out as readily as water. To make Shear's medium, use 3 g potassium acetate, 60 mL glycerol, 90 mL ethanol (95%), and 150 mL distilled water.

3. *Poly(vinyl alcohol) in Lactic Acid (PVLG) Mounting Medium and Clearing Solution.* This is an excellent permanent mounting medium that has minimal distortion of fungal elements and sets rapidly. If used as a clearing agent, the mounted specimens require a 2–3-day clearing of spore content before they are studied. To make the mounting medium, use 1.66 g poly(vinyl alcohol) (PVA), 1 mL glycerol, 10 mL lactic acid, and 10 mL distilled water.[35] Mix all liquid ingredients in a dark bottle *Before* adding the PVA. The PVA is added as a powder to the liquid ingredients. The PVA dissolves slowly, and the bottle should be placed in a hot-water bath (70–80°C). The solution will be clear in 4–6 h. This solution is stored in dark bottles and lasts for approximately one year.

4. *Cotton Blue in Lactic Acid (Mounting Medium with Stain).* This is a common medium for both mounting and staining and is considered a stock solution for other mounting media or stains. The components include 0.01 g cotton blue and 100 mL 85% lactic acid. To make this medium, place the lactic acid in a beaker with a magnetic stirrer. Add the cotton blue powder to the solution while it is heating and stirring. Let the solution cool after the cotton blue has dissolved, then filter out the undissolved particles.[33] If a lighter blue color is necessary, the solution can be further diluted with 85% lactic acid.

5. *Lacto-Cotton Blue (Mounting Medium with Stain).* This is a variation of the solution described above and is another popular stain–mounting medium that includes 0.1 g cotton blue, 25 mL 85% lactic acid, 50 mL glycerol, and 25 mL distilled water. As with the previous mounting medium and stain, it can be diluted with 85% lactic acid for lighter color.

6. *Modified Lacto-Cotton Blue (MLCB).* This medium is a modification of lacto–phenol blue.[37] The stain is able to preserve the natural color of some dark-colored spores/conidia and other fungal structures. The staining of hyaline and thin-walled

spores and other structures is subtle, while thick-walled or pigmented spores remain unstained. Sometimes it may be too light for certain spores and the contrast and clarity may not be as good as with lactofuchsin. Sime et al.[37] stated that this stain is able to rehydrate desiccated and deformed spores and allow them to recover their original shapes. The best results can be achieved if several hours of rehydration pass before viewing. Its components include 3 mL cotton blue stock (1.0 g cotton blue and 99 mL 85% lactic acid), 250 mL glycerol, 100 mL 85% lactic acid, and 50 mL deionized water.

7. *Lacto-Phenol Blue (Mounting Medium and Stain).* The use of this particular staining/mounting agent is on the decline because phenol can be irritating and easily absorbed through the skin. Phenol is an irritant and may lead to a number of adverse health effects through long-term skin contact[38] and is considered a possible carcinogen.[33] To make this solution, add 100 mL lactophenol (a mixture of 20 g phenol crystals, 20 g (16 mL) lactic acid, to 40 g (31 mL) glycerol, 20 mL distilled water), 1–5 mL aqueous solution of cotton blue, and up to 20 mL glacial acetic acid.[39]

8. *Water.* When examining fungi, mycologists usually prepare a water mount first for the observation of living material so that the natural shape, size, and color of the structures can be accurately recorded. If fruiting bodies or other structures have dried out, they are first rehydrated in a 10% or diluted potassium hydroxide solutions and then observed under water. Since water quickly evaporates, it should be constantly added to the slide preparation to prevent dehydration of the material. Additionally, a wetting agent may be added to the water (e.g., Triton X100, Tween 80, or Span 60). This solution has the advantage of preventing air bubbles from sticking to many fungal structures.[32] For microscopic analysis in a commercial lab, water is seldom used as a mounting medium.

4.7. PROCEDURES FOR IDENTIFICATION AND QUANTIFICATION OF SPORE TRAPS

Reference data and relevant calculations are as follows:

1. Calibrate the measurements of different magnifications of all objective lenses with a stage micrometer and an ocular micrometer. Attach these measurements to the microscope base for future reference. The magnifications of the microscope should be calibrated at least yearly. The full length of the ocular micrometer under 40×, 60×, and 100× lenses and the diameters of these lenses should be measured also. For instance, the Olympus BH2 microscope has a field of view diameter of 0.39 mm for the 40× objective lens and a 10× ocular lens, and the viewing area is 0.1195 mm^2 at the 400× magnification.
2. Measure the adhesive band for the sample deposit for each sample type or obtain this information directly from the manufacturers. For example,

Burkard samplers collect particles on an adhesive band measuring 14 mm by 2 mm (28 mm^2), Air-O-Cell, 1.055 mm by 14.4 mm (15.19 mm^2), and Allergenco samplers, 14.5 mm by 1.1 mm (16 mm^2), respectively.

3. Use the micrometer measurements and spore trace length to calculate the number of passes necessary to cover the whole trace and/or a portion of it (such as 25%), whether counting transverse or longitudinal passes (or traverses). Both the diameter of the lens and the field of view are used to calculate the number of fields needed to cover the whole trace or a portion of it, if using the random field method.

A number of methods are used to examine spore trap samples, such as random fields, transverse traverses (or passes), and longitudinal traverse methods (Fig. 4.2). These methods offer a systematic but random analysis. Traverse methods include two different approaches: using a whole entire field of view or using a micrometer to guide traverses and the counting process. The primary advantage of the micrometer method is the analysis of a narrower, but more manageable analytical area. The microscopist does not have to turn his/her eyes to scan the entire field of view and can more easily focus on the area within the micrometer. This method is less tiring and less stressful for the eyes.[40] With a micrometer, it is easier to guide traverses and determine whether a spore on the edge or partially on the micrometer should be counted.

The trace area analyzed ranges from ~25% to 100% depending on the laboratory. There are no governmental, professional, or official requirements on the minimal area that should be analyzed to get a reliable result. For a normal spore load, 25% of the covering area is minimal. If unusually high spore loads (>800 spores/sample) with a relatively even spore distribution are present, the area analyzed might be reduced to less than 25%. Extreme caution should be taken when less than 25% of the trace is analyzed. Some labs examine and then analyze their samples with 100% coverage for large spores and only 25% for the remainder of the trace. This approach, however, is not random, and may indicate a bias toward larger spores.

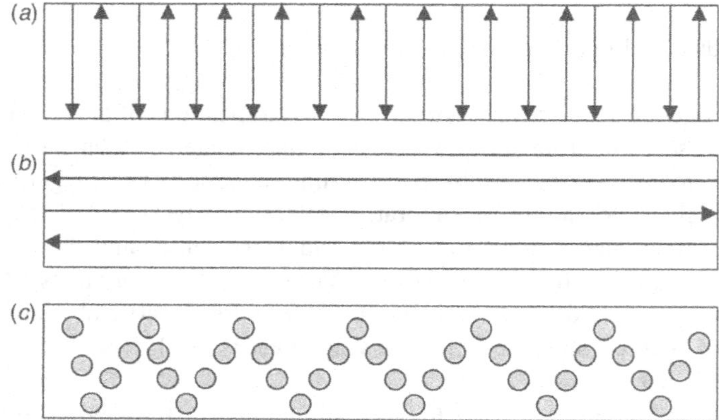

Fig. 4.2. Fungal sample trace reading methods: (*a*) transverse traverse; (*b*) longitudinal traverse; (*c*) random fields.

Käpylä and Penttinen statistically compared the three analytical methods (random field, transverse traverses, and longitudinal traverses) to determine the variations of counts and the sample size necessary to yield valid results.[22] Their results showed that one or two longitudinal traverses may be unreliable because of uneven distribution of the fungal material on the tape sample of a Burkard sampler. Microscope fields or selected transverse traverses were recommended;[22] however, they[22] found that systematic traverses provided more effective results than did randomly spaced ones. Sterling et al. found that 12 transverse traverses usually yielded higher counts than did a single longitudinal traverse.[24] More empirical studies are needed to improve the quality of the analysis and yield more reproducible results.

4.8. TECHNIQUES FOR SPORE COUNT ANALYSIS

The following steps are recommended:

1. Scan the slide. Use a 10× or 20× objective lens to scan the entire trace of the deposit. Determine whether the spore deposit is random or evenly distributed. If clustering and clumping of certain spore/conidial types are observed, note it on the datasheet. If the trace is heavy and thick, rank the load of background particulates on the datasheet. Do not focus the spore counting on a certain section or portion of the adhesive band. Scan for spores over the entire deposit trace. Particularly note the spores of *Chaetomium, Stachybotrys, Memnoniella*, and *Ulocladium. Ulocladium* spores, which are excellent indicators of water-damaged conditions, should be differentiated from those of *Pithomyces*, which is typically of outdoor origin. If any of the spores of these taxa are found during scanning, record the location that they are found on the slide. The process of spore counting must be random and free of bias.
2. Use the ocular micrometer (or graticule) to measure the size of the spores/conidia. Most fungal spores/conidia are >2 μm. If spore-shaped particles are <2 μm, they are probably not fungal spores.
3. Learn to recognize and differentiate spores/conidia, hyphae, and conidiophores (spore-producing structures) of the *Aspergillus–Penicillium* group and *Stachybotrys*. Spores/conidia and conidiophores of these groups are often associated with indoor air contamination.
4. If spores/conidia are in clusters, record their presence. Count individual spores/conidia. Record the number of spores/conidia in each cluster and circle or otherwise mark that number to indicate that it represents a cluster.
5. Generally, a lab assigns each microscope an identification number that is then recorded on each sample's datasheet. Steps 1–5 apply to all analytic methods for spore counting.
6. Transverse traverse method for Air-O-Cell, Burkard, Allergenco samples or similar samplers—if *Chaetomium, Stachybotrys, Memnoniella*, and

Ulocladium spores/conidia are found during scanning, start the first pass from that location. When the first pass is completed, move to the end of the slide and start the second pass.

Use the 60× objective lens for spore identification and counting.

Read >20 passes randomly (or enough to cover a minimum of 25% of the trace) along the entire length of the sample. Count and identify spores/conidia that fall within the ocular micrometer.

In those situations when spores/conidia touch the edge of the micrometer which is guiding the counting pass, always count those that touch the left edge of the micrometer and exclude any that touch the right edge (Fig. 4.3).[40]

7. Random field method—randomly scan across the trace (three to five fields per traverse or pass) in a zigzag pattern at 400× or 600× magnification. The number of fields counted should be equivalent to at least 25% of the band area. Count spores within the entire field of view. Try to randomly scatter fields over the entire trace. For spores present on the perimeter of the field, count only the ones on the left side (Fig. 4.4). If the spores cannot be identified from the partial view, the field can be moved to the left to reveal the entire spore for a more precise identification. Do this only after the identification of the other spores in that field have been processed. Use the total spore count, the number of fields counted, and the air volume to calculate spore density. Spores/m^3 = total spores/number of fields counted × number of fields in the whole trace/air volume in liters × 1000 L.

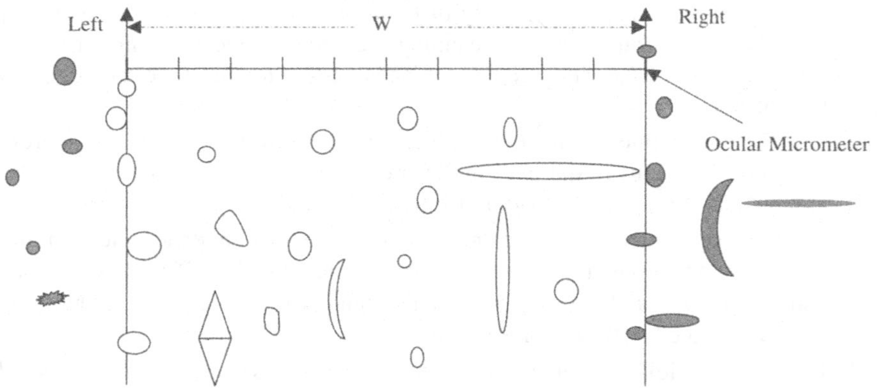

Fig. 4.3. Counting criteria for the traverse method for spore trap slides or samples. Only count the spores touching the left margin of the micrometer pass; ignore the ones touching the right margin. Two vertical solid arrows indicate the direction of a pass guided by a micrometer. *W* is the width of the micrometer/pass. The unshaded spores are counted; the shaded spores are not counted.

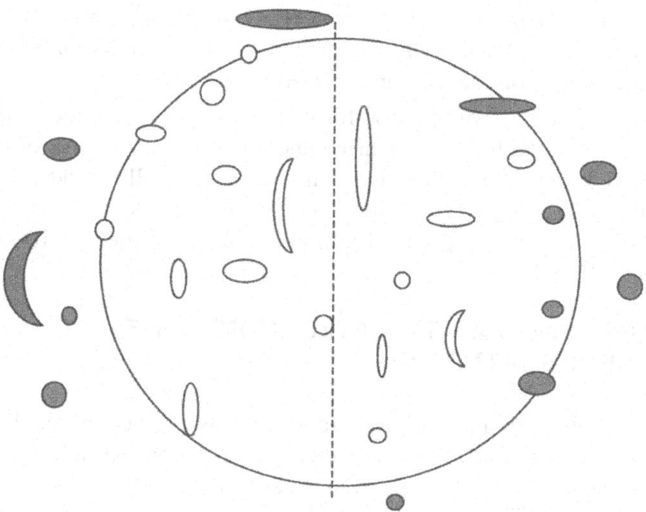

Fig. 4.4. Counting criteria of random field method for spore trap slides or samples. Only spores on the left half of the perimeter of the field of view should be counted; ignore those on the right. The unshaded spores are counted; the shaded spores are not counted.

8. For Cyclex-D or other similar samples:

   Scan the entire circular area with the spore deposit to determine whether the spores are randomly or evenly distributed. Randomly choose one-quarter of the circular area for examination.

   Use a 40× or 60× lens.

   Count the spores/conidia in one quarter of the circular area.

   If the spores/conidia are in clusters, count the individual spores. Record the number of spores in each cluster and circle the number to indicate that it represents a cluster of spores.

9. Filter samples:

   For filter cassettes, use the NIOSH 7400 method as the basis for examination. Using the cleaned filter forceps, extract the filter, cut one-quarter of the filter (use an alcohol-wiped razor blade or scissor), place on a clean slide, then clear with hot acetone vapor (perform in a containment hood or a fume hood), then add the stain/mounting medium lactofuchsin. Scan the slide at 400×. Count and identify all fungal spores and hyphae that fall within the Walton–Beckett graticule or other types of graticule. One hundred fields should be counted. If >200 spores are counted in less than 100 fields, stop. Use the spore counts and field counts for the calculation. Divide the total number of spores counted by the number of fields counted to find an average number of spores per field. Spores/m^3 = average spore count per field/0.00785 mm^2 × 908 mm^2 (for 37 mm filter) or 375 mm^2 (for 25 mm filter)/air volume in liters × 1000 L.

If close-faced sampling is performed, cut a 2-mm-diameter area from the center of the filter for clearing and counting, since most of the particles and spores are concentrated in this area. Follow the counting rules discussed in step 4. Spores/m^3 = average spore count per field/0.000785 mm^2 × 3.14159 mm^2/air volume in liters × 1000 L.

4.9. BACKGROUND PARTICULATES (NONSPORE MISCELLANEOUS MATERIALS)

In addition to fungal spores, airborne samples may contain particulates such as insulation glass fiber fragments, gypsum dust, carbonaceous particles, animal byproducts (e.g., dander, hair), human skin flakes, pollen, and parts of or entire arthropods or insects. These materials are referred to as *background particulates.* The amount of background materials in the sample can influence the laboratory analysis. When the background particulate load is heavy, it will block the light penetrating through the sample as well as the images of the spores by burying the spores within and underneath the background materials. Examination of this type of sample is difficult or even unreadable. The resulting data are expected to have wider variations. Unreadable samples are often reported as overloaded or unreadable. Three factors may be responsible for overloading samples: (1) the sampling period is too long; (2) at the time of sampling, a heavy airborne dust load already exists as a result of various human activities (e.g., demolition, remediation activities); and (3) the areas of concern are poorly maintained or dirty. Laboratories may record the amount of background or debris matter by ranking it into four or more categories: 0—no discernable background matter, 1—low (minimal or no effect on analysis), 2—medium (may reduce the quality of lab analysis), and 3—overloaded (sample is unreadable). An individual laboratory may decide to use other scoring systems. The rank of background materials may help the investigator with future decisions on how to adjust the sampling time according to the site conditions. It is important to note that the background or debris rating has no numeric relationship to the amount of spores counted or missed. When an overloaded sample is indicated, the laboratory may record and report qualitative data, such as taxa of fungal spores or the composition of the background matter.

The presence of gypsum dust (a component of sheetrock) in a sample can be troublesome to the analyst. Gypsum dust reacts with lactofuchsin and forms air bubbles and later crystals in the prepared slide. Some air bubbles may carry spores or resemble fungal spores. In either case, it makes the examination of such samples difficult. The nature of the background particulates, however, may provide useful information. For instance, the presence of pollen may suggest an outdoor origin, poor air filtration, low air ventilation, or any combination of these. Large amounts of skin flakes present may correlate with sampling locations, utilization of the space, maintenance of the buildings, overcrowding, and human activities. Currently, there are no general standards or guidelines on how to report, grade or interpret background particulates. Because of to the variability of background particulate and fiber composition, there are no correlations between background matter on health concerns.

4.10. LIMITS OF DETECTION

The detection limit (DL) depends on the percentage of the sample trace counted and is defined as one spore divided by the percent of the trace analyzed. For example, if 25% of the trace is analyzed, the detection limit of the analysis is $\frac{1}{25}\%$ (or 0.25) = 4.

The quantitation limit (QL) depends on the total air volume (AV in liters) collected and the detection limit or the percentage of the sample analyzed. It can be calculated with the following equation:

$$
\begin{aligned}
\mathrm{QL}(\mathrm{spores/m^3}) &= 1000\,\mathrm{L/AV\ L}/\% \\
&\quad \text{of sample area analyzed or QL} \\
&= \mathrm{DL} \times 1000\,\mathrm{L/AV\ L}.
\end{aligned}
$$

For example, a sample was collected with an airflow rate of 15 L/min for 10 min; 25% of the samples trace was analyzed. The quantitation limit for this sample is $1000/150/0.25 = 27$ spores/m^3.

4.11. DATA PRESENTATION

Although there are currently no standards on how to format the lab reports resulting from the analysis, the following information should be present for enumerating identified fungi: raw count, air volume (airflow rate × sampling time), percentage of sample analyzed (or multiplication factor), fungal spore/conidial concentration (spore/m^3), detection limits, and percentage of particular fungus in the total fungal population. Clients should be encouraged to provide information on the chain of custody listing collection date, time, and location. This information is useful for the interpretation of lab results. For AIHA-accredited labs, AIHA laboratory accreditation policy should be followed to present the data and information in their lab reports.

4.12. VARIATION OF REPLICATIONS AND DUPLICATIONS

The accepted variations of replications and duplications will be determined in the future with more empirical research. AIHA laboratory accreditation policy has required the accredited labs to determine the accepted ranges of the variations for quality control purpose. Generally speaking, the higher the percentage of the trace is analyzed, the smaller the variation is expected.

4.13. SAMPLE PREPARATION FOR DIRECT EXAMINATION OR FROM CULTURES

The key issue here is how to preserve diagnostic features of fungi and the integrity of fungal structures, such as conidiophores, phialides, pycnidia, and ascomata.

4.13.1. Bulk Samples

In addition to the basic materials used to examine the samples of spore traps, a roll of transparent tape (19.0 mm × 32.9 m), and a dropper bottle of 70% ethanol is needed to process bulk, dust, tape, and swab samples. Do not use either packing tape (too wide and too sticky to handle) or frosted tape (not clear and will reduce the resolution and clarity of the images) for the microscopic examination. Remember to use aseptic techniques and clean all instruments between every sample.

The following procedure is recommended:

1. Directly examine the entire bulk sample macroscopically (with naked eyes) or under a dissecting microscope to locate fungal colonies and evaluate the degree of fungal infestation. Discoloration of the sample can be an indication of fungi. However, early infestations may not yield obvious discoloration. When there are no obvious areas of discoloration, an examination of the material should be conducted under a dissecting microscope. It is easy to miss the colonies at early stages of development because of their light color and small size. When an area with the potential of fungal colonies is located, a tape lift sample as a subsample should be made from that location.
2. Using a pair of scissors cut a piece of clear tape (~2 × 2 cm or 0.8 × 0.8 in.) and then hold the tape with the forceps. Place the adhesive side of the tape over the area to be examined, using the handle of the forceps (or the blunt end of a pen) to firmly and evenly press the area. Peel off the tape from the sample.
3. While holding the tape in the air over a paper towel, rinse with a few drops of 70% ethanol or isopropyl alcohol.
4. Put the tape on a slide with adhesive side up.
5. Stain it with a drop of mounting medium with stain. Coat both sides of the tape with the medium by picking it up with the forceps, then putting it down again. Again, make sure that the adhesive side of the tape is face up.
6. Gently place a coverslip over the tape.
7. Examine this wet mount (i.e., the sample side) under a compound microscope.
8. Scan the entire tape with a 20×, 40×, or 60× lens for fungal structures. Use a higher magnification to observe fine structures.

If fungal structures are not present on the tape, additional subsamples or tape lifts should be made to avoid assuming a false negative. Each area of discoloration needs to be examined to determine whether more than one species of mold is present. For each different species/taxon, a tape lift should be taken to verify the identification.

For bulk carpet samples when the pile is deep, tape lifts are not suitable. Tape lifts are also of no value when the fungi are growing on the reverse side of the carpet. It is also difficult to make contact with fungi when they are at an early stage of growth. In these circumstances, it is necessary to pull out fibers from the carpet or backing for microscopic examination.

4.13.2. Bulk Dust Samples

The following steps are recommended:

1. Shake the dust cassette by hand or place on a vortex instrument to produce a more homogenous mixture.
2. Leave the cassette on a bench to allow the dust to settle.
3. Open the cassette (cut the tape with a weighing spatula; pry apart).
4. Subsample the dust by grabbing a portion using filter forceps or with a micro-spatula, then place it on a slide.
5. Add a drop of 70% of alcohol or isopropyl alcohol to wet and fix the sample, then spread the dust with an inoculating needle to form a thin layer that will allow light to penetrate sufficiently. Allow the alcohol to evaporate.
6. Stain the dust subsample with a drop of staining–mounting agent.
7. Place a 22 × 22-mm coverslip over the sample; view with a microscope.
8. Scan the subsample with a 20×, 40×, or 60× lens. Use a higher magnification to observe fine structures.

When only a small amount of dust is in the cassette, a wet (sterile distilled water) inoculating needle can be rubbed against the cassette wall to collect any particulates. If still not enough dust is on the cassette wall, it is necessary to remove the filter from the cassette and transfer it onto a slide. Record this situation for future reference. If the dust sample contains gypsum, the prepared sample should be observed without any delay, because the staining agent will react with gypsum and generate air bubbles. These bubbles will obscure and distort spores and other fungal elements, thus making observation more difficult.

4.13.3. Swab Samples

Taking a swab sample for the direct microscopic examination described in this section is seldom recommended because intact fungal structures are disturbed and separated during sampling and preparation of slides. However, this sampling–analytical method can sometimes provide quick and useful information when circumstances do not allow other sampling and testing. Swabs can be cotton, rayon, or other synthetic fibers and may or may not be commercially prepared (i.e., presterilized in a tube with a buffer solution). For swab tips made with nonfibrous (e.g., spongy) material, it may be necessary to cut or pull off some of this substance and mount it on a slide. Sometimes, the investigator takes a nonsterilized cotton swab, samples the area, and places the sample in a plastic bag. It should be recorded on the datasheet that the swab was not previously sterilized. When a swab sample does not have any obvious discoloration, it should be examined under a stereomicroscope. An empty Petri dish can be employed to support the sample to avoid contamination during the observation. Remember to use aseptic techniques and clean all instruments between samples. The following procedure is recommended:

1. Remove the swab from the tube and thoroughly examine for areas of discoloration.
2. At the points of discoloration, pull off several fibers with pointed forceps.
3. Place the fibers on a slide with the aid of an inoculating needle.
4. Fix the sample with a drop or two of 70% of alcohol and allow the alcohol to evaporate.
5. Stain it with a drop or two of the staining–mounting medium.
6. Put a coverslip over it and examine under a compound microscope to locate and observe fungal structures.
7. If fungal structures are absent, additional fibers from different areas should be pulled off for further examination.
8. If no discoloration is found on the tip of the swab, fibers should be pulled off from a minimum of four areas, covering most of the swab surface. Since many fungi or fungal elements are naturally transparent or lightly colored, it may be more difficult to locate these colonies.

4.13.4. Tape Lift Samples

Although field investigators may prepare tape samples using different methods, it is important that only clear adhesive tape (i.e., sticky cellophane tapes) be used to collect samples. Most tape samples are fixed or taped to a glass slide or taped within a plastic bag for shipping to the environmental laboratory. Commercially prepared kits have become common and popular since 2002. Becasue of the composition of the tape adhesive, it is important that the tape sample be examined as soon as it is prepared. The staining–mounting agent reacts with the sticky coating material on the tape, and any delay will lead to reduced resolution of images. The fungal structures on the tape will become blurry with time. Remember to use aseptic techniques and clean all instruments between every sample. The protocol for preparing tape samples for microscopic examination is similar or identical to that of the bulk samples described earlier.

If fungal structures are not present on the initial subsample, the remaining tape should be scrutinized and additional subsamples should be prepared. If the tape sample has large dust particles on it, gently remove them with forceps or with an inoculating needle. Otherwise, the large particles may prevent examination under a higher-power lens because the thickness of the sample is greater than the working distance of the higher-power lens. If gypsum particles are present on the tape, it would be another reason to examine the sample as soon as possible.

Avoid mounting a tape sample with the adhesive side down and without using a coverslip. Such a preparation will have a reduced resolution. Tape samples with large fungal bodies or structures may interfere with the quality of analysis since a minute fungal characteristic (e.g., annellation on the tip of a conidiophore) would be overlooked and may lead to a misidentification. It is better to extract the larger fungal bodies, then make a squash mount to examine them and examine the other mycelia/hyphae on another slide.

4.13.5. Culture Samples

Detailed information on how to examine and then analyze culture samples is covered in a Chapter 5. Only the procedure for preparing the sample from cultures is described here. At the end of the incubation period, actively growing fungal colonies need to be identified by preparing wet mounts. Two common methods used to select fungal structures are by tape lift or with a dissecting needle.

A tape lift is often used to avoid damaging the reproductive fungal structures:

1. Examine the colony to determine whether it is slow-, medium-, or fast-growing. For a fast-growing colony, the tape lift should be made at the colony's edge since the center of the colony contains the oldest reproductive structures. In the center of fast-growing species such as some *Penicillium* spp. and some *Aspergillus* spp., there are excessive numbers of conidia, so much so that finding intact conidiophores would be unlikely.[16] If the sample tape lift contains conidia exclusively or very few reproductive structures, another tape lift should be prepared from the edge or from a younger area of the colony. For a slower-growing colony, a tape lift can be made from the center.
2. Use scissors and forceps to cut a small piece of 3 × 3-mm clear tape.
3. Open the Petri dish just enough to allow the forceps with a small piece of tape attached to enter the plate, to minimize the chance for spores/conidia becoming airborne. Press the tape gently on a specific fungal colony with the forceps (Fig. 4.5). The tips of the forceps should be nongrooved. The grooves on the tips easily trap some fungal structures and lead to cross-contamination.
4. Lift the tape, remove from the plate, and fix it with a drop of 70% alcohol.
5. Place the tape in a drop of staining agent on the labeled glass slide. Make sure that both sides are coated with the staining agent and that the adhesive side is facing upward.
6. Carefully place a coverslip over the sample, then examine under a compound microscope to observe the fungal structures for identification.
7. After use, the forceps should be thoroughly rinsed and wiped with 70% isopropyl alcohol to ensure that there are no residual fungal elements remaining on the tips of the forceps. Blot the forcep tips on a clean paper towel.

Fig. 4.5. Making a tape lift from a fungal colony.

It is difficult to collect the larger fungal structures such as ascomata, acervuli, and pycnidia with tape lifts. If a tape lift fails to pick up these structures, a pair of forceps or an inoculating needle can be used to extract them from the Petri dish. The forceps can be used to pick up a fruiting body from directly the medium. Alternatively, an inoculating/dissecting needle can be inserted beneath the fungal body and gently lifted with a bit of the culture medium. Some mycologists use a needle to prepare mitosporic fungi also. For some analysts, it may be harder to manipulate a needle than a small piece of tape. When using a needle, first separate a tiny area of colony with mature fungal structures by cutting the edge of the area with the tip of the needle.[16] Then, pick up the tiny piece from the culture and place it on a slide sideways to fix and stain it. If the piece is too big, it may be difficult to find intact conidiophores with conidia for identification. The advantage of this method is that the wet mount is prepared without a layer of tape and provides better resolution of the fungal structures. The disadvantage is that it is harder to pick up some fungal elements without resultant damage. The advantage of the tape lift is that it is easy to pick up intact fungal structures; however, the tape will reduce the resolution of the fungal structures and the reaction of the staining agent and tape coating will blur the fungal image, especially if the preparation is not observed immediately.

For descriptive information of common fungi and fungal spores from microscopic analysis, please see the Appendix.

4.14. EVALUATION OF FUNGAL INFESTATION

Mastering microscopy alone does not guarantee a quality microscopic analysis because the analysis involves the combination of adept microscopic abilities, well-prepared slides, and a strong background with experience in fungal taxonomy, biology, and ecology. The latter is the most important factor for consistent, reliable analyses. The goal of the analysis is to identify and characterize the fungi or fungal spores from the samples so that the results will aid the field investigators to evaluate the level of indoor fungal contamination. Fungal identification and characterization is not an easy task. According to Hawksworth, there are 1.5 million fungi but only 100,000 fungi have been described.[41] Records from a laboratory indicate that there are over 600 species of indoor fungi identified from the samples (unpublished data). Introductory mycology textbooks such as *The Fifth Kingdom*[42] and *Introductory Mycology*[43] are important starter references in any laboratory that conducts fungal identifications. A collection of mycological journals and advanced references, such as monographs of various important fungal groups or genera, are a must if a laboratory and its analysts are expected to identify fungi and their spores from a wide range of background and conditions. New and uncommon fungal species have been reported from indoor samples.[44–46]

Key morphological features such as mycelia, conidia, conidiophores, ascomata, or pycnidia are necessary to determine whether fungal growth is present. The presence of mycelia alone indicates that a fungus has grown and colonized the sample but is at only an early stage of development or not in reproductive mode. The presence of reproductive structures and spores/conidia indicates that the sample is well established and has completed one lifecycle either asexually (conidiophores or pycnidia) or sexually (ascomata or basidiomata). These fungi are at a mature stage. When examining bulk samples, it is relatively easier to evaluate fungal contamination, since the presence of fungal colonies can be obtained by observing the discoloration or fungal growth of the samples. Discoloration by itself, however, does not indicate the presence of fungal growth since nonorganic factors can lead to discoloration (e.g., weathering, oxidation, or chemical reactions of nails or other metal materials), as well as the activities of other biological agents. If visible colonies are present on the bulk samples and mycelia and reproductive structures are identified from samples examined microscopically, then a fungal infestation is confirmed. If only a few isolated spores or hyphal fragments are present and other fungal structures are absent, the presence of fungal growth or infestation will not be indicated. When spores/conidia belonging to a mixture of several different taxa, in particular those of phylloplane fungi, are found together and other fungal structures are absent, the result does not indicate fungal contamination. These spores probably resulted from the settling of airborne spores with an outdoor origin. When small numbers of spores/conidia belonging to a single genus or species are found, this may also indicate an outdoor source or may be the result of a fungi-infested area in a different indoor location. Otherwise other fungal structures would be present on the sample. When large numbers of spores of one to many fungal genera or species are present (e.g., *Aspergillus*/*Penicillium*-like) but with no other fungal structures, this may indicate the presence of a fungal infestation because large numbers of spores are the result of amplification. The absence of mycelia and reproductive structures is probably due to the massive production of billions of fungal spores in layers, which prevents an investigator from efficiently collecting other fungal structures underneath the spore masses. For bulk samples, when in doubt, confirmation of the presence of conidiophores can be obtained by rubbing and/or removing the conidia from the colony to expose the reproductive structures and mycelia, then preparing another tape lift.

Caution should be taken whenever spores of *Chaetomium*, *Memnoniella*, *Stachybotrys*, or *Ulocladium* are detected. These fungi and their spores are excellent indicators of water-damaged environments, usually on a long-term basis of months or even years. The detection of a single spore or low numbers from one of these genera may indicate a fungal infestation or contamination. Investigators should review and evaluate for any missed signals or hidden mold growth and contamination. The spore may also have originated from an outdoor source or from contaminated materials. It may also be due to the settlement of airborne spores that originated from a different area indoors.

The presence of obligate plant pathogens, such as powdery mildew, smut, and rust, does not indicate fungal contamination on building materials or furniture. These fungi need living plants to develop and complete their lifecycles[47] and are most likely from an outdoor source. On occasion, powdery mildew may develop on houseplants, such as miniature roses and chrysanthemums, but does not signal water-related damage in the building.

4.15. TRAINING OF MICROSCOPY ANALYSTS

Miller et al. indicated that the skill of analysts in quantifying and identifying fungal structures, especially for small colorless spores, is exceptionally important.[48] A basic knowledge of mycology is one of the prerequisites for all analysts. New analysts should be trained in mycology until they have enough experience and background to perform analyses. The training should cover general mycology and familiarity with the most common indoor fungi. The minimal educational requirements of analysts for spore traps and other samples should be a B.S. in biology or in a related field. Analysts in training should first pass competency tests before they can analyze clients' samples. The competency test should include (1) correct identification of spores/conidia most commonly and occasionally found on indoor samples, (2) ability to differentiate major fungal groups on the basis of key characteristics, and (3) for spore trap analyses, the ability to enumerate spores from whole samples and portions of the samples following their laboratory's standard operating procedures. Samples of airborne fungal spores at low, moderate, and high concentrations should be used for proficiency training with the spore trap analyses. For other types of samples, samples collected from different building materials should also be used for the test. Novice analysts should be supervised with 100% of their work checked by a supervising or senior analyst for at least 4 months for quality control before being allowed to work independently. Records for training should be documented and filed. Continuous training to improve and challenge the knowledge and the quality of the analyses should be planned at regular intervals. Replications and duplications should be carried out routinely for not only quality control but also proficiency improvement of microscopic analysis.

4.16. QUALITY ASSURANCE/QUALITY CONTROL PROCEDURES

Proper QA/QC is important to ensure the quality of the microscopic analyses. There are many aspects and areas where QA/QC will play a role in a microbiology laboratory. However, the present section will deal only with the issues and steps related to the microscopic analysis. As stated previously, no standard protocols are available for laboratory analyses and no numeric standards or guidelines exist for interpreting laboratory results of airborne fungal structures. It is unlikely that such standards, thresholds, and guidelines will be developed in the near future. Since standards and guidelines are not available, the current approach to the interpretation of the results relies on comparisons of results from indoors versus outdoors, complaint versus noncomplaint areas, or both:

1. A set of reference slides should be prepared and maintained as part of the laboratory quality assurance program. The reference slide collection should include the most common and less common airborne fungal spores/conidia. These slides need to be verified by relevant authorities of fungal groups. In addition, a reference culture collection should be maintained. These cultures can be commercially purchased (e.g., ATCC) or obtained from authorities of each group.
2. Establish a comprehensive mycological reference collection. The monographs of various fungal groups are especially helpful for any fungal analysis. The reference slides and literature should be accessible to all analysts.
3. Important and indicator spores, such as those of *Stachybotrys* and *Memnoniella*, must always be confirmed by a senior mycologist. An analyst must be able to consistently identify spores/conidia of indicator fungi, such as *Chaetomium*, *Memnoniella*, *Stachybotrys*, and *Ulocladium*, as well as differentiate them from similarly shaped and colored spores. Needless to say, the labs should be able to identify these fungi correctly with confidence.
4. It is important to differentiate hyphal fragments, mycelial fragments, and fungal hairs, such as those of *Chaetomium* species, from other organic and nonorganic fibers. The presence and number of hyphal fragments and mycelial fragments is significant for evaluating the degree of fungal infestation. Under certain conditions, such as remediation, this is indicative of active growth. Make sure that all analysts are competent to differentiate these structures through thorough and proper training. Do not lump hyphal fragments with other fungal categories.
5. Whenever a spore/conidium or other fungal element is unfamiliar to an analyst, consultation with colleagues, reference slide collections, fungal

identification references (keys, monographs, etc.), and with the senior mycologist or analyst may be necessary before categorizing it as an "unknown." If the "unknown" is the dominant spore type, the laboratory should indicate this in the report or inform the investigator for consideration to recollect the sample.

6. Quality assurance and quality control regarding the results from microscopic fungal analyses should be qualitative, quantitative, and semiquantitative. The qualitative aspect is the correct or proper identification of fungal spores/conidia and structures. Because of the difficulties of quantifying fungal elements from bulk, tape, dust, and swab samples reliably, a number of laboratories report their results in a descriptive or semiquantitative manner, with ranks such as 0, I, II, III, IV, V, or trace, few, many, numerous, and massive. Consistency of the characterization, however, is important. A set of reference slides covering the range of categories should be regularly consulted. Quantitative QA/QC may involve intra- and interlaboratory testing. Ten percent of all the analyzed slides are selected for replicate and duplicate analyses to assess the accuracy of intralaboratory analyses and to examine the analytical variations of a laboratory. Any significant discrepancy in results should be investigated to determine the cause. Such discrepancy should be documented, including an explanation for the discrepancy, with a solution and appropriate corrective action. A minimum of 20 spore count samples should be exchanged and analyzed among participating laboratories for round-robin analysis every 4 months to comply with the requirement of the AIHA EMLAP. Intra- and interlaboratory results are collected and analyzed statistically for precision, accuracy, and deviation. It is necessary to document any significant discrepancy in results and to convene a discussion among the laboratories to determine the reason and recommended corrective action, then address the appropriate solution or corrective action for the problem.
7. Generally this is not a problem in most modern biological laboratory settings, but all textured surfaces found in carpets, furniture, or other surfaces are not allowed.
8. All laboratories and all analysts engaged in fungal microscopic analyses should participate in a proficiency testing program (e.g., EMPAT).

4.17. WEB RESOURCES

Leica Microsystems for microscopes: http://www.leica-microsystems.com/home/index4.html.

Microscope World large selections of microscopes: http://www.microscopeworld.com

Nikon USA microscopes: http://www.nikonusa.com/template.php?cat=5&grp=22.

Nikon's MicroscopyU: an educational forum for all aspects of optical microscopy http://www.microscopyu.com/.

Olympus Microscopy Resource Center, "designed to provide an Internet-based educational forum on all aspects of optical microscopy, photomicrography, and digital imaging": http://www.olympusmicro.com/.

Swift microsopes: http://www.swift-optics.com/products/scientific.

Zeiss: http://www.zeiss.com/.

REFERENCES

1. Anyanwu, E. C., A. W. Campbell, and J. E. Ehiri, Mycotoxins and antifungal drug interactions: Implications in the treatment of illnesses due to indoor chronic toxigenic mold exposures, *Sci. World J.* **12**:167–177 (2004).
2. Crameri, R., M. Weichel, A. G. Glaser, S. Fluckiger, and C. Rhyner, Fungal allergies: A yet unsolved problem, *Chem. Immunol. Allergy* **91**:121–133 (2006).
3. Portnoy, J. M., K. Kwak, P. Dowling, T. VanOsdol, and C. Barnes, Health effects of indoor fungi, *Ann. Allergy Asthma Immunol.* **94**:313–319 (2005).
4. Portnoy, J. M., K. Kennedy, and C. Barnes, Sampling for indoor fungi: What the clinician needs to know, *Curr. Opin. Otolaryngol. Head Neck Surg.* **13**:165–170 (2005).
5. Laumbach, R. J. and H. M. Kipen, Bioaerosols and sick building syndrome: Particles, inflammation, and allergy, *Curr. Opin. Allergy Clin. Immunol.* **5**:135–139 (2005).
6. Slayter, E. M., Optical microscope, in *The Encyclopedia of Microscopy and Microtechniques*, P. Gray, ed., Van Nostrand Reinhold, New York, 1973, pp. 382–389.
7. Murphy, D. B., R. Hoffman, K. R. Spring, and M. W. Davidson, *Specimen Contrast in Optical Microscopy* (accessed at www.microscopyu.com/articles/formulas/specimencontrast.html on Dec. 12, 2005).
8. Murphy, D. B., J. Hinsch, E. D. Salmon, K. R. Spring, H. E. Keller, M. Abramowitz, and M. W. Davidson, *Differential Interference Contrast, Fundamental Concepts*, Florida State Univ., 2005 (http://micro.magnet.fsu.edu/primer/techniques/dic/dicintro.html).
9. Spring, K. R. and M. W. Davidson, Specialized microscope objectives (accessed at http://www.microscopyu.com/articles/optics/objectivespecial.html on Dec. 12, 2005).
10. Elston, D. M., Fluorescence of fungi in superficial and deep fungal infections, *BMC Microbiol.* **1**:21 (2001).
11. Cheng, Y. S., E. B. Barr, B. J. Fan, P. J. Hargis, D. J. Rader, T. J. O' Hern, J. R. Torczynski, G. C. Tisone, B. L. Preppernau, S. A. Young, and R. J. Radloff, Detection of bioaerosols using multiwavelength UV fluorescence spectroscopy, *Aerosol Sci. Technol.* **30**:186–201 (1999).
12. Jensen, C., H. Neumeister, and G. Lysek, Fluorescence microscopy for the observation of nematophagous fungi inside soil, *Mycologist* **12**:107–111 (1998).
13. Franck, M. M., R. L. Brigmon, P. C. McKinsey, M. A. Heitkamp, and C. B. Fliermans, *An Improved Method for Direct Fungi Identification and Enumeration*, Research Report for U.S. Department of Energy under Contract DE-AC09-96SR18500 (accessed http://sti.srs.gov/fulltext/ms2000310/ms2000310.html on July 14, 2005).
14. Coleman, T., J. V. Madassery, G. S. Kobayashi, M. H. Nahm, and J. R. Little, New fluorescence assay for the quantitation of fungi. *J. Clin. Microbiol.* **27**:2003–2007 (1989).
15. Simmons, E. G., *Alternaria* themes and variations (226–235) classification of citrus pathogens, *Mycotaxon* **70**:263–323 (1999).

16. Pitt, J. J., A Laboratory Guide to Common *Penicillium* Species, North Ryde, Australia, 2000.
17. Klich, M. A., *Identification of Common Aspergillus Species*, Centralbureau voor Schimmelcultures, Utrecht, Netherlands, 2002.
18. Aylor, D. E., Vertical variation of aerial concentration of *Venturia inaequalis* ascopsores in an apple orchard, *Phytopathology* **85**:175–181 (1995).
19. Muilenberg, M. L., Aeroallergen assessment by microscopy and culture, *Immunol. Allergy Clin. N. Am.* **9**:245–268 (1989).
20. Calderon, C., J. Lacey, H. A. McCartney, and I. Rosas, Seasonal and diurnal variation of airborne basidiomycete spore concentrations in Mexico City, *Grana* **34**:260–268 (1995).
21. Li, D. W. and B. Kendrick, Functional and causal relationships between indoor and outdoor airborne fungi, *Can. J. Bot.* **74**:194–209 (1996).
22. Käpylä, M. and A. Penttinen, An evaluation of the microscopic counting methods of the tape in Hirst-Burkard pollen and spore trap, *Grana* **20**:131–141 (1981).
23. Oh, J. W., H. B. Lee, H. R. Lee, B. Y. Pyun, Y. M. Ahn, K. E. Kim, S. Y. Lee, and S. I. Lee, Aerobiological study of pollen and mold in Seoul, Korea. *Allergol. Internat.* **47**:263–270 (1998).
24. Sterling, M., C. Rogers, and E. Levetin, An evaluation of two methods used for microscopic analysis of airborne fungal spore concentrations from the Burkard Spore Trap, *Aerobiologia* **15**:9–18 (1999).
25. Tsai, S. M., C. S. Yang, P. Moffett, and A. Puccetti, A comparative study of collection efficiency of airborne fungal matter using Andersen single-stage N6 impactor and the Air-O-Cell cassettes, *Proc. 8th Internat. Conf. Indoor Air Quality and Climate*, Construction Research Communications Ltd., London, 1999, pp. 776–781.
26. Morris, K. J., Modern microscopic methods of bioaerosol analysis, in *Bioaerosols Handbook*, C. S. Cox and C. M. Wathes, eds., Lewis Publishers, Boca Raton, FL, 1995, pp. 285–316.
27. Gilliam, M. S., Periodicity of spore release in *Marasmius rotula, Mich. Bot.* **14**:83–90 (1975).
28. Hestbjerg, H. and H. Dissing, Studies on the concentration of *Ramularia beticola* conidia in the air above sugar beet fields in Denmark. *J. Phytopathol.* **143**:269–273 (1995).
29. Vismer, H. F., A. Cadman, A. P. S. Terblanche, and J. F. Dames, Airspora concentrations in the Vaal Triangle: Monitoring and potential health effects. 2, Fungal spores. *S. Afr. J. Sci.* **91**:408–411 (1995).
30. Drent, P., *Digital Imaging—New Opportunities for Microscopy*, Nikon, 2005 (accessed at http://www.microscopyu.com/articles/digitalimaging/drentdigital.html on Dec. 11, 2005).
31. Li, D. W. and C. S. Yang, Spore counting, in *AIHA Green Book* (in press).
32. Malloch, D., *Moulds, under the Microscope, Mounting Media,* Univ. Toronto, 1997 (accessed at www.botany.utoronto.ca/ResearchLabs/MallochLab/Malloch/Moulds/Examination.html on Dec. 18, 2005).
33. Kirk, P. M., P. F. Cannon, J. C. David, and J. A. Stapler, eds., *Dictionary of the Fungi*, 9th ed., CABI Publishing, Wallingford, CT, 2001.
34. Carmichael, J. W., Lacto-fuchsin: A new medium for mounting fungi, *Mycologia* **47**:611 (1955).

35. Dhingra, O. D. and J. B. Sinclair, *Basic Plant Pathology Methods*, CRC Press, Boca Raton, FL, 1985.
36. Samson, R. A., E. S. Hoekstra, and J. C. Frisvad, *Introduction to Food- and Airborne Fungi*, 7th ed., CBS, Utrecht, Netherlands, 2004.
37. Sime, A. D., L. L. Abbott, and S. P. Abbott, A mounting medium for use in indoor air quality spore-trap analyses, *Mycologia* **94**:1087–1088 (2002).
38. Sigma-Aldrich, *Material Safety Data Sheet P4161 Phenol ACS Reagent*, Sigma Chemical Co., St. Louis, MO, 2000, p. 6.
39. Maneval, W. E., Lacto-phenol preparations, *Stain Technol.* **11**:9 (1936).
40. Lacey, J. and J. Venette, Outdoor air sampling techniques, in *Bioaerosols Handbook*, C. S. Cox and C. M. Wathes, eds., Lewis Publishers, Boca Raton, FL, 1995, pp. 407–471.
41. Hawksworth, D., Fungal diversity and its implication for genetic resource collections, *Stud. Mycol.* **50**:9–18 (2004).
42. Kendrick, B., *The Fifth Kingdom* 3rd ed., Focus Publication, R. Pullins Co., Newburyport, MA, 2000.
43. Alexopoulos, C. J., C. W. Mims, and M. Blackwell, *Introductory Mycology*, 4th ed., Wiley, New York, 1996.
44. Morgan-Jones, G. and B. J. Jacobsen, Notes on hyphomycetes. LVIII. Some dematiaceous taxa, including two undescribed species of *Cladosporium*, associated with biodeterioration of carpet, plaster and wallpaper, *Mycotaxon* **32**:223–236 (1988).
45. Li, D.-W. and C. S. Yang, Notes on indoor fungi I: New records and noteworthy fungi from indoor environments. *Mycotaxon* **89**:473–488 (2004).
46. Fernando, A. A., S. E. Anagnost, S. R. Morey, S. Zhou, C. M. Catranis, and C. J. K. Wang, Noteworthy microfungi from air samples, *Mycotaxon* **92**:323–338 (2005).
47. Agrios, G., *Plant Pathology*, Academic Press, San Diego, 1997.
48. Miller, J. D., R. Dale, and J. White, Exposure measures for studies of mold and dampness and respiratory health, in *Bioareosols, Fungi and Mycotoxins: Health Effects, Assessment, Prevention and Control*, E. Johanning, ed., Mount Sinai School of Medicine, New York, 2001.

CHAPTER 5

CULTURE-BASED ANALYTICAL METHODS FOR INVESTIGATION OF INDOOR FUNGI

FLORENCE Q. WU

This chapter focuses on the isolation and identification of culturable fungi from indoor environments. Topics include the pros and cons of using culture-based analytical methods for fungal investigations, the factors that may influence the results, culturable sampling considerations, fungal isolation and identification techniques, data interpretation, and quality control measures. Bacterial isolation and identification require different techniques and are not covered in this chapter. The purpose of this chapter is to provide a framework for culturable fungal analysis based on sound mycology.

5.1. ADVANTAGES AND LIMITATIONS OF CULTURE-BASED ANALYTICAL METHODS

Using spore-trapping samplers to collect fungal spores followed by direct microscopy in the laboratory to identify and enumerate the spores is frequently done in indoor air quality (IAQ) surveys. This practice has its practical merits, but it is not without limitations. One problem is the direct examination of the samples by microscopy, which results in uncertainty in the identification of the fungal spores. Only a small number of fungal spore types can be identified to generic level with confidence. Species identification based on spore morphology is usually unachievable. Even the most significant genera such as *Aspergillus* and *Penicillium* cannot be differentiated on spore trap samples. This problem hinders the accurate assessment of fungal contaminants in IAQ projects. Culture-based methods offer a convenient and practical solution to this problem, but generally require at least a seven-day incubation period.

Sampling and Analysis of Indoor Microorganisms, Edited by Chin S. Yang and Patricia A. Heinsohn

The study and naming of filamentous fungi (including molds) rely mostly on the morphological characteristics of the colony, hyphae, sporulating mechanisms and structures, and spores. These characteristics are best observed by obtaining pure cultures of the organisms. The fungi that are capable of growing on indoor building materials (mostly dead organic materials) are primarily saprophytic, which means that they can usually be cultured on artificial media. Almost all common indoor fungi can be grown on media and produce sporulating colonies within a reasonable timeframe (~7–14 days). In addition, the spores of many indoor fungal species can remain viable and culturable for a very long time (e.g., years),[1,2] which makes recovery from the environment a relatively easy task. Culture based analytical methods take advantage of these facts and allow for more detailed analysis.

Obtaining cultures of filamentous fungi is essential for identifying the fungi to species. Most filamentous species are discovered, described, and referenced on the basis of the full body of the representative culture (called *type specimen*). Identification is achieved by comparing the unknown organism with the type specimen or the original/reliable description of the type specimen. This approach is time-honored and scientifically sound. Indoor air quality surveys often require that species identification is known, which is possible only through use of a culture-based method performed by a quality laboratory.

The term *species identification* should not be confused with *speciation*. These two terms are not interchangeable. *Species identification* refers to a human activity of determining the identity of an organism to species level that has been discovered, described, and given a scientific name. *Speciation* refers to the formation of new species and is an evolutionary process that includes gene mutation, genetic exchange, geographic isolation, and natural selection.

Culture-based methods can provide quantitative data on viable and culturable fungi from nearly all sample types (spore trap air and tape lift surface samples are exceptions). Because the viability of a microorganism determines its infectivity and growth potential, the use of culturable methods is required in certain environments, such as in public buildings or in the critical areas of hospitals and in cleanrooms. In this case, the sampling and analytical protocols can be designed to target specific groups, for example, human pathogens. Because only live pathogens can cause infectious disease, culture methods are ideal for isolating and identifying potential pathogens.

Culture-based analysis is one of the most economical ways to identify molds to species level, especially when compared to existing molecular methods. If desired, the culture can be preserved and archived for confirmative identification or other use.

As do all other analytical methods, culture-based analysis has its limitations. One of the most important limitations is its dependence on spore viability and culture conditions. Nonviable or nonculturable fungi are not detected by this method, although these fungi can be allergenic and/or irritating. The concentrations of culturable samples, reported in colony-forming units (CFUs), are often substantially lower than the total spore counts reported by direct microscopic examination.

Additionally, some spores travel in groups or clusters, and each cluster has the potential of forming a colony. Therefore, the actual number of spores forming a colony is unknown, and the number of colonies may not reflect the total number of spores that were actually collected in the sample.

Moreover, because of the need to incubate and grow the cultures, analysis usually takes 1–2 weeks to complete. Depending on the media used, faster-growing fungi may overgrow and obscure slower-growing colonies, which may result in false negatives. Nevertheless, culture-based methods are still the most commonly used methods to complement direct microscopy and to provide valuable information that cannot be obtained otherwise.

5.2. FACTORS INFLUENCING THE RESULTS OF CULTURE-BASED ANALYSIS

5.2.1. Ecological Considerations

Understanding the basic ecology of those fungi that can grow in the indoor environments is helpful in assessing the indoor environment, locating growth sites, identifying reservoirs, and interpreting the sample results reported by the laboratory. The air exchange with the outdoors may be relatively continuous in a mechanically ventilated building. If there is no indoor fungal growth and contamination, the composition of the indoor airborne fungal spores may resemble that of the outdoors, although spores may be present in lower concentrations indoors. Outdoor airborne spore composition, rank order, species diversity, and relative frequency and abundance vary with geographic location, vegetation, season, weather, time of day, and human activities.

There are generally three sources for indoor spores: influx from outdoors, spore accumulation over time, and indoor fungal growth. Surface and carpet dust often harbor fungal spores. Obviously, spores originating from indoor mold growth are of the most concern to IAQ and microbial investigators.

Essentially all fungal spores originate outdoors. Fungal spores or hyphal fragments generated outdoors can be transported into the indoor environment by human activities or into the building structure in various ways.[3] Except the fungi that are parasitic to indoor plants, most fungi that are capable of growing indoors are saprophytic, that is, capable of utilizing dead organic materials as food sources. Some fungi are accustomed to living in human environments. For example, those wood-rotting fungi capable of utilizing copper can overcome the fungicidal treatment of lumber, causing extensive damage to wood-frame buildings. *Chaetomium* and *Stachybotrys* species are often found flourishing on water-damaged walls built of gypsum sheetrock. Framing lumber that has not been kiln-dried may be colonized by various filamentous fungi, including *Alternaria, Aspergillus, Aureobasidium pullulans, Chaetomium, Cladosporium, Penicillium, Phoma*, and *Ulocladium* species.[4,5]

Although typical building materials (wood, gypsum sheetrock, tile, carpet, glass, etc.) do not necessarily support fungal growth under dry conditions, any organic material is a potential nutrient source for fungi in damp conditions. Fungi have been isolated from damp or water-damaged building and finishing materials and building contents. These include gypsum sheetrock, wood, wallpaper, carpet, cardboard, spoiled foods, and clothing. Some fungi can use only simple carbohydrates such as starch and hemicellulose as food, while others are capable of breaking down more complicated compound such as cellulose and lignin, which can cause structural damage to the substrate the fungi are growing on.

The growth requirements of fungi include sufficient moisture, a nutrient source, and an optimum temperature range, which is 25°C ± 5°C for mesophilic fungi, 37°C for temperature-tolerant fungi and/or human/animal pathogens, and 50°C for thermophilic fungi. In a typical indoor environment, nutrients and water are the factors typically limiting fungal growth. It is now known that about 60 species are commonly isolated from indoor substrates.[6]

5.2.2. Viability of Fungal Spores

Most of the microbial IAQ sampling methods collect fungal spores and hyphae. Spores are minute propagative units (propagules) that function as seeds, usually with some food reserve, but not containing a preformed embryo. The fungal spores that are important in IAQ investigations are generally between 2.0 and 10 μm in length. The mechanism by which spores are produced is the basis for classifying fungi. This is why identification using culture analysis is the primary method of choice. Spore types that are important in IAQ investigations include ascospores, basidiospores, conidia, and sporangiospores. The major functions of fungal spores are maintaining the populations, spreading genetic variability, and extending their distribution range. Spores are dispersed mostly by air, but can also be dispersed by water, arthropods, larger animals, plant seeds or propagating materials, and human activities. Because fungi have the ability to produce massive numbers of spores, fungal spores are among the most common microbial particles in outdoor and indoor environments.

Many factors such as water, nutrition, temperature, light, and pH can determine the viability of a fungal spore, as well as influence the process of a fungal spore developing into a sporulating colony, as illustrated in Figure 5.1. Understanding of these factors may help both laboratory and investigators design a suitable culture method to maximize recovery of fungi.

Fungal spores can remain viable for a short period, for weeks or as long as many years, depending on the fungal species, spore types, and storage conditions.[1] It appears that under the same conditions, chlamydospores and resting spores survive longer than do other types of spores; ascospores, longer than basidiospores; and smaller-sized conidia (e.g., of *Aspergillus* or *Penicillium*), longer than dark-colored, larger spores (e.g., *Stachybotrys*, *Ulocladium*), although this kind of generalization is often challenged. Most spore longevity data are derived from experimental records, and those laboratory conditions may not reflect field conditions. These data should be used only as an estimation of spore longevity. Indoor fungi are mostly in an open environment that has many factors affecting the survivability of fungi, such as desiccation, temperature, chemicals, and radiation. The common belief is that the number of viable fungal spores reported as CFUs

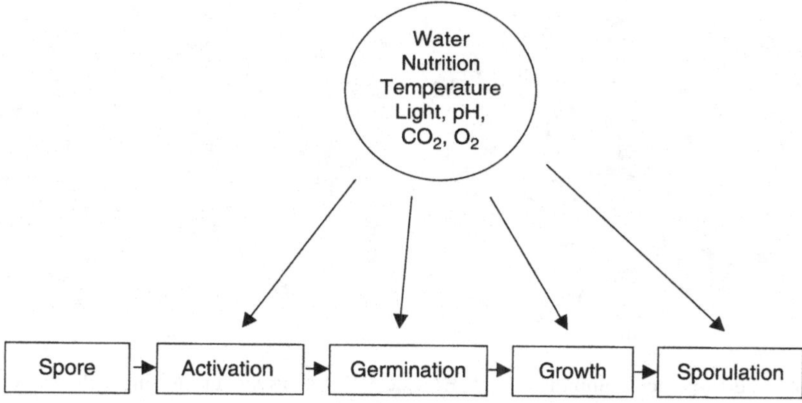

Fig. 5.1. To develop into a sporulating colony, a viable spore must undergo a process consisting of spore activation, germination, growth, and sporulation. Many factors such as water, nutrition, temperature, light, and pH can influence the process.

represents only about 1% of total spores.[6] Because there are so many variables, it is not realistic to estimate the survival rate of fungal spores in an indoor environment.

5.2.3. Selection of Culture Media

The media for culturing fungi can be divided into two functional types: primary recovery media and standard identification media. A recovery medium receives and revives the sampled spores (or occasionally hyphal fragments) and allows the fungal propagules to germinate and grow into a colony on the medium. Ideal recovery media maximize the diversity of fungi isolated from environmental samples. However, no medium allows competing fungi to grow at similar rates, thereby maximizing the recovery of all viable species present on the plate. Different media have different nutrients and water activity. All commonly used media are selective. For example, malt extract agar (MEA) has high sugar and water content, which favors hydrophilic and fast-growing species. Dichloran Glycerol (DG-18) can limit rapid growth, permitting more diversity, but it does not support the growth of some hydrophilic fungi because of its low a_w content (Fig. 5.2).

Standard identification media are those cited in authoritative literature for confirmative identification of fungal species. Mycologists working on different groups of fungi may employ different media. In IAQ surveys, confirmative identification of *Aspergillus* and *Penicillium* species is often based on colonies developed on standard identification media, for example, MEA, to observe subtle but defining characteristics.

5.2.4. Concentration Variations

The concentrations of airborne fungal spores, both viable and nonviable, are influenced by a variety of factors. These factors include air movement, building

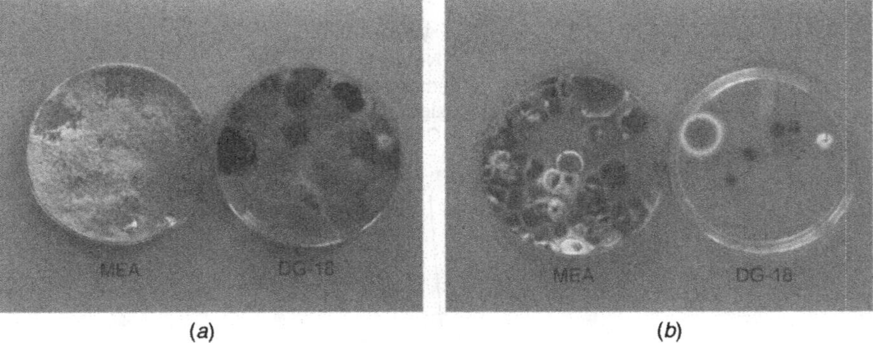

Fig. 5.2. Performance comparison of MEA and DG-18 in recovering fungal diversity. The plates shown in each photograph are cultures from the same concentration of spore suspension: (*a*) a *Trichoderma* colony on the MEA medium inhibited the growth of all other fungi—these fungi were recovered on the DG-18 medium; (*b*) *Stachybotrys* species were recovered on the MEA medium, but not on the DG-18 medium.

ventilation rate, temporal and spatial variability, building maintenance, housekeeping, activities in the sampling area, and sampling time and sampler performance. These factors affect the composition of airborne spore populations. Designing a proper sampling strategy should take these factors into consideration.

5.2.5. Sampler Performance, Sampling Time, and Culture Preparation

There are variations in performance among the culturable samplers that are commercially available, such as the inertial impactors including the Andersen or similar samplers, Burkard viable sampler, the Surface Air System (SAS) samplers, the Reuter Centrifugal Samplers (RCS), the AGI-30 impinger, or filter cassettes. Investigators should be familiar with the proper usage and performance of the air sampler that they choose to use.[6] Sampling time is another crucial factor affecting the results. Efforts should be made to maximize sampling time, which increases the representativeness of the airborne population recovered and statistical reliability of the sample, while avoiding the desiccation of the fungal spores and overloading the sample medium.

For bulk, dust, and swab samples, how the samples are plated and cultured directly affects the results. Common methods include direct plating or spore extraction and serial dilution. Depending on the species present in the sample, a culture plate suitable for fungal identification should ideally have 30–60 colonies, even though a mycologist can work with a plate with more colonies. Manually preparing culture plates is labor-intensive, especially when multiple serial dilutions have to be made and plated, but it is a reliable method and commonly used in mycology laboratories and reported in the literature for many years for samples with a varied fungal community, such as dust samples. Because fungal colonies can grow quite large compared to bacterial colonies, the accuracy of using an automatic spiral plating machine, which was originally designed for bacterial plating, for estimating fungal concentration in the sample has no relevant scientific support and the result is highly questionable.

Direct plating involves placing a part of or the entire sample directly on the culture medium. Using this method, a swab or wipe sample is streaked or wiped across the surface of the culture medium. Dust samples may be sprinkled directly on the culture medium or after weighing. Cassette filters can be put on media directly. This method is useful when dust on a filter is too little to be weighed. Direct plating is best used when only qualitative results are desired, or only a limited number of species is suspected.

Dilution plating of a sample is accomplished by placing the sample in a known volume of a sterile diluent [such as peptone water (0.1% v/v) or sterile distilled water with Tween 80 (0.1% v/v)] to wash microorganisms off the sample matrix and suspend spores and other propagules in the diluent. Various quantities or aliquots of the suspension are pipetted and spread over the surface of agar. It is often necessary to make serial dilutions out of the spore suspension and subsequently plate the dilutions. Dilution plating usually results in multiple plates with evenly distributed colonies from which the optimum plates can be selected for further identification and analysis. This method is best used when the sample is suspected to contain a high number of viable propagules and/or quantitative data are desired. The disadvantages of dilution plating are that spores with low numbers might be lost in the process. Another disadvantage is that the fact that bias increased with the dilution factor.

5.2.6. Experience of the Analyst

Identification of filamentous fungi is based on culture morphology and the reproductive process, which dictates that the analyst's training and experience are very important for obtaining accurate results. A competent analyst should have proper training in fungal morphology and classification, understand how genera and species are delineated, know what distinguishing traits to look for, and have sufficient hands-on experience with the groups of fungi to be identified. Previous experience with mycology, including the identification of soil, food, and plant pathogenic fungi, is useful, while experience with bacterial identification is not directly relevant to fungal identification, due to the different approaches used for these microorganisms.

Because of the richness in the variety of fungal species, fungus identification can be a daunting task even for an experienced fungal analyst. Misidentification of fungi is not uncommon. Misidentification is worse than no identification because it can lead to misinterpretation of the results. Different fungal species have different ecological niches and may have different implications for IAQ investigations. Savvy investigators often check the qualifications and experience of the laboratory and its staff who perform the identification, not just the overall laboratory accreditation.

5.3. CULTURABLE SAMPLING CONSIDERATIONS

There are many references published by technical and professional communities as well as manufacturers of sampling devices on the collection efficacy and proper usage of various sampling methods.[6] Readers are encouraged to research such information of their chosen samplers. Therefore, this section discusses only the sampling considerations that are required to ensure good sample quality for reliable analytical results. As a general rule, precautions should always be taken to protect the viability of sampled microorganisms and the integrity of the samples during sampling, storage, and shipping.

5.3.1. Air Samples

The general mechanism of air sampling is to draw air through a sampling device that captures and retains the microorganisms according to its design characteristics, then allows the sampled air to exit the device. The airborne particles are typically collected by the mechanisms of impaction, interception, and diffusion. Examples of sieve impaction devices include the Andersen N-6 and similar impaction samplers. Impaction and trapping into a liquid, or impingement, is the mechanism by which particles are collected in an AGI-30 impinger. Filters take advantage of all three mechanisms of collection to trap particles. The most common method for culturable fungi is sieve impaction onto a culture medium.

The following factors should be considered:

1. A representative amount of air should be sampled to obtain results that are scientifically and statistically reasonable. This means that the limit of quantitation should be met, which is dependent on the sampler selected, the sampling time, and the specific laboratory SOP for the analytical method.
2. Choice of collection medium and incubation temperature should be made prior to the sampling and based on the type of fungi suspected in the environment to be sampled (Table 5.1).
3. Similar to sampling for direct microscopy, outdoor and nonaffected area reference samples should also be collected. In addition, blanks should be collected during each sampling campaign. Field blanks are those opened and placed in the sampler in the environment to be sampled but without activating the pump to pull air. Sealed blanks are those original laboratory-supplied plates to serve as quality control of the media supply as well as handling and shipping.

5.3.2. Surface Samples

Surface samples provide information on settled culturable fungal spores or fungal colonies on a surface. A surface swab sample collects fungal spores or structures from hard, smooth surfaces. A contact (e.g., Rodac) plate can be used to collect fungal spores and structures from even and nonporous surfaces. A tape lift sample from a surface is not appropriate for culture-based analysis, although it is possible

TABLE 5.1. Recommended Culture Media and Incubation Conditions for Recovery and Identification of Indoor Filamentous Fungi

Purpose	Recommended Media[a]	Incubation Temperature
Recovery and identification of hydrophilic to moderately hydrophilic fungi	Malt extract agar (MEA)[11,12,15]	25°C
Recovery and identification of xerophilic fungi	Dichloran glycerol (DG-18)[33]	25°C
Recovery and identification of cellulytic fungi (*Chaetomium*, *Stachybotrys*, *Ulocladium*, etc.)	Cornmeal agar (CMA),[7,8] cellulose agar[6,34]	25°C
Recovery and identification of potentially pathogenic *Aspergillus* species	MEA; Czapek yeast agar (CYA)[11]	37°C
Confirmative identification of *Penicillium* species	CYA, MEA, 25% glycerol nitrate agar (G25N)[12]	25°C, 37°C, 5°C
Recovery and identification of *Cryptococcus neoformans*[b]	Birdseed agar (*Guizotia abyssinica*–creatinine agar, GACA)[15]	25–30°C
Recovery and identification of *Histoplasma capsulatum*[b]	Yeast extract phosphate (YEP) agar[34]	25°C
Recovery of yeast or yeast-like fungi	MEA, CMA, etc.;[15] identification often requires physiological tests	25°C or 37°C

[a]See the reference given for identification method and/or the medium recipe.
[b]Direct culturing of *Cryptococcus neoformans* and *Histoplasma capsulatum* is not always reliable because of interference of environmental organisms; extra procedures may be needed. (*Note*: *Cryptococcus neoformans* is a biosafety level II/III agent; *Histoplasma capsulatum* is a biosafety level III agent.)

to culture the spores or hyphae imprinted on the tape provided that they are viable and in sufficient quantity and the sticky substance on the tape is not inhibitory to the fungal structures.

When collecting surface samples for culture-based analysis, sterile supplies should be used to avoid contamination during sampling and shipping. If the surface material appears to be colonized by multiple fungi because of variations in coloration or texture, then each distinct fungal type should be included in the sample to ensure proper and adequate identification, and to maximize the utility of the sample.

5.3.3. Bulk Samples

A bulk sample is a three-dimensional piece of material, for example, wallpaper, carpet, duct lining, or wood. Fungal propagules in the bulk sample can be extracted using sterile water. The resultant suspension can be plated directly or undergo serial dilution followed by plating. A bulk sample can also be broken into small pieces and cultured directly onto a medium plate. Whenever wood decay is of concern, swab sampling may not be the best sampling method. Bulk samples are necessary, if it is possible to take such samples. It is recommended that the investigator inform the laboratory so that special procedures can be taken to avoid isolating only surface fungi.

5.3.4. Dust Samples

Dust samples provide information on culturable fungal spores settled in carpet dust or on other surfaces. The presence of these spores may be due to the accumulation of airborne spores over time or to the effects of excess moisture in the sampled material. Dust samples are usually collected using an air sampling pump to collect either a visible quantity of dust or within a defined surface area into a filter cassette. Because some fast-growing and widely propagating fungi are common in dust samples, it is best to use both MEA and DG-18 to culture dust samples for more reliable results. A dust sample can be immersed in sterile water for suspension or dilution plating or plated directly on culture media.

5.4. SAMPLE PREPARATION METHODS

The general principle for fungal recovery is to provide fungal spores with the nutrients and optimal conditions that they require for germination, growth, and sporulation. In an environmental sample, there is often a mixed population of fungal species. The choice of recovery and isolation methods should be based on the ecology of the targeted group (e.g., mesophiles or thermopiles), the need for selective isolation (e.g., for pathogens), and the objective of the sampling (e.g., investigator's hypothesis). Investigators are encouraged to communicate with the laboratory to determine the proper medium to use for air sampling or for plating of bulk, surface, and dust samples. The media used and the incubation conditions are critical factors that determine the diversity of species that are reported. Obviously, the more media and incubation conditions that are employed to culture a sample, the better the chance of recovering more species is. However, in most investigations it is often unrealistic to use more than two media and two temperature conditions. The most commonly used isolation scenario is 2% malt extract agar (MEA) with incubation at 25°C or room temperature.

For more comprehensive results, the simultaneous use of MEA and DG-18 should be considered in IAQ surveys for two main reasons: (1) MEA recovers a broad range of hydrophilic and mesophilic fungi, while DG-18 is preferred for recovering xerophilic as well as moderately xerophilic fungi—the combination of these two media provides a reasonably diverse recovery of fungi from environmental samples; and (2) a few common indoor fungi can grow rapidly on MEA that they may mask, inhibit, interfere with, or entirely suppress the germination and growth of other fungi. Clever and experienced laboratories and mycologists may be able to overcome such interference and count and identify as many colonies as they possibly can. This may significantly reduce the number of species recovered. For example, colonies of *Botrytis, Fusarium, Mucor, Rhizopus, or Trichoderma* tend to overgrow and mask slower-growing fungi on MEA. However, these species have limited growth on DG-18, allowing the recovery of other fungi in the sample. Other media can also be used to select specific fungi. For example, cellulose agar (CA) is usually used to demonstrate the cellulolytic capability of fungi in the laboratory and has been used to enhance the recovery of *Chaetomium, Stachybotrys,* and other cellulolytic species. However, colonies on CA are often diffuse, making it difficult to accurately count the number of colonies. Cornmeal agar (CMA) is used to isolate hydrophilic fungi and identify *Stachybotrys* to species.[7,8] If fungal pathogens are of concern, then the laboratory should be requested to incubate the sample at 35–37°C, which is near body temperature. An antibacterial compound (e.g., chloramphenicol at 0.1 g/L) is often added to fungal isolation media to avoid interference from competing bacterial growth.

Samples with solid medium (such as Andersen plates and contact plates) do not require additional preparation or processing other than labeling and incubation on arrival at the laboratory. All other samples require preparation and processing in the laboratory by direct plating, extraction–dilution plating, or both. Fungal isolation techniques aimed at recovering fungal biodiversity may be referenced in developing protocols for recovering indoor fungi, for example, those described for saprobic soil fungi[9] and for microfungi.[10]

An aliquot of a liquid impinger sample can be plated directly onto an agar plate using a spreader, or used to make serial dilutions before plating. Conversely, samples with low counts may be concentrated by filtration prior to culturing. In most cases, the typical 1 : 10 serial dilution method in microbiology laboratory manuals can be applied in mycology analysis. The advantages of suspension/dilution plating are that the method allows thorough washing of the sample, even distribution of colonies on the agar, and quantification of the sample.

The details of the analytical procedures vary among laboratories. The laboratory can precheck the sample before plating. For example, if it is observed that both *Stachybotrys* and *Penicillium* are major components in the original sample, then the use of cornmeal agar as well as multiple dilutions may be needed to adequately analyze that sample.

Positive controls (culturing of a fungal isolate) and negative controls (the media and reagent water used in sample preparation) should be set up and incubated alongside the samples. Another aspect of culture preparation is laboratory contaminant control. For example, *Chrysonilia sitophila* is a notorious contaminant that grows rapidly and can extend outside the Petri dish. If this species is observed on a sample, the sample should be kept in isolation. The incubator where the species is found growing should be thoroughly cleaned with 10% (v/v) bleach.

Normally a minimum of 7 days' incubation time is necessary to allow fungi to sporulate and to exhibit characteristic morphology, according to major mycology identification references for *Aspergillus*[11] and *Penicillium*.[12] The identification of

Aspergillus and *Penicillium* species after exactly 7 days of incubation is more reliable because this matches the conditions described in the authoritative literature. For most common indoor fungi that produce conidia, 7 days are sufficient for sporulation. However, for slow-growing and/or ascospores-producing fungi (e.g., *Eurotium*), a longer incubation time is often necessary. Most basidiomycetes cannot form fruiting structures on culture medium within a reasonable timeframe, and their culture-based identification is much more complex.

5.5. IDENTIFICATION OF FILAMENTOUS FUNGI

Identification of fungi frequently requires that microscopic mounts of the colony be made and examined. Generally, there are two ways of picking out fungal structures from its colony to make wet mounts: the needle method and the tape method. The method of choice depends on fungal structure, colony texture, the analyst's experience, and preference. Sometimes, both methods have to be used to make a satisfactory slide preparation.

The Needle Method. An inoculating needle or a sterile disposable toothpick is used to pick out a small portion of the colony that includes sporulating structures. A stereomicroscope is used to guide the search for the appropriate section of the colony where fruiting structures and spores are available. The portion of colony on the needle is floated onto a slide with the aid of a drop of 70% ethanol. A pair of the needles is used to tease apart the web of fungal aggregates so that details of the fungal structures can be observed under the compound microscope. When most of the alcohol has evaporated, a drop of lactic acid (for observing spore ornamentation using phase contrast) or lactofuchsin stain, or another mycological stain (for observing hyphae and conidiophores using brightfield) is added to the sample on the glass slide. The sample is covered with a glass coverslip for microscopic examination. The needle method is particularly useful for identifying fungi that produce large numbers of spores, such as *Aspergillus* and *Penicillium* species. For species forming more complex fruiting structures, for example, *Chaetomium, Eurotium*, and *Phoma*, the structures can be flattened by gently tapping the cover slide with a pen.

The Tape Method. A piece of clear adhesive tape is used to remove sporulating structures. An observation slide is prepared from the adhesive tape for microscopic examination. For details of such preparation, please see the Chapter 4. If carefully done, the tape method does not distort the conidiophore structure, and is best used to observe sporulating structures. It works well with many filamentous fungi, including *Acremonium*, *Alternaria*, *Curvularia*, *Phialophora*, and *Stachybotrys*.

Discussion. The spore-producing structure and mechanism are of particular importance in fungal identification. A clear understanding of relevant mycology terms (ascus, blastoconidia, phialoconidia, poroconidia, etc.) and accurate observation of spore production (e.g., solitary, in chains, sympodial) are essential for identification. It is also important to know what characteristics are important and what characteristics are more variable. These skills and knowledge can be obtained through proper mycology training and experience.

Important literature sources for fungal identification include Refs. 7 and 13–27.

Additional time, as well as extra steps may be needed for the identification of *Aspergillus* and *Penicillium* to species. Identification is based primarily on the culturing, isolation, and identification schemes described by Klich[11] and Pitt,[12] respectively.

5.6. DATA REPORTING

Fungal Identification. Genus or species information should be clearly and correctly presented in the analytical report. When more than two media are used, the results should be reported separately to provide more detailed information. It is not standard microbiological practice to combine individual results of the same sample inoculated onto different media into one result. It is also not standard microbiological practice to combine results from different media inoculated with different serial dilutions of the same sample into one result. The laboratory report should include the results of only one dilution in an extraction and serial dilution method. There should be no "manipulation or normalization" of data by combining results from different media and different dilutions. Normally the dilution plate with reasonably scattered and individually identifiable colonies and of diverse species is to be analyzed. Colonies on plates from different dilutions can be used as a reference to aid the identification.

Most true yeast and the yeast form of some filamentous fungi are normally difficult to identify solely on the basis of morphology. Biochemical tests and reaction databases are often needed. Unless specifically requested by the investigator, it is acceptable to report them as yeast colonies without further differentiating into genus or species. The term *nonsporulating colony* refers to any colony that does not produce a reproductive structure and is not in a yeast form. Whenever clamp connections are present, the colony should be identified as a basidiomycete. However, many basidiomycete fungi do not form clamp connections.

Quantification and Reporting Unit. If the volume of air sample is provided to the laboratory, and the serial dilution method is used for sample preparation, quantification is possible. For sieve impaction air samples, CFU/m^3 = number of colonies/total air volume in cubic meter (sampling rate in m^3 per minute × sampling time in minutes). Note that the unit of flowrate of most air sampling pumps is liters per minute (L/min). The conversion factor is as follows: 10^3 L = 1 m^3. If the result is reported and understood as colony-forming units (CFUs), then positive-hole correction is unwarranted. For dust and bulk samples, CFU/g = (number of colonies × dilution factor)/weight (g) of dust used in analysis. If only a portion of the dust sample is analyzed, then the total dust weight should be reported in addition to the analyzed dust weight to facilitate the interpretation of the data.

Detection Limit (DL) and Quantitative Limit (QL). *Detection limit* in mycology refers to the smallest unit (i.e., colony) of an analyte (i.e., fungus) that can be visually detected by the analyst. For culture-based analyses, the sampled fungal spore must form a colony to be visually detected, thus, the DL is one colony. *Quantitative limits* refer to both the lowest and highest concentrations of the fungi that can be detected and quantified in the sample as defined by the sampling, processing, and analysis parameters. For air samples collected with an Andersen N-6 sampler, the lowest QL is $\text{CFU/m}^3 = 1$ CFU/sampled air volume in m^3. For other sample types, such as dust and bulk samples, where serial dilutions are made, the lowest QL in CFU/g = 1 CFU × dilution factor/weight used. For air samples collected with impaction samplers, such as Andersen and SAS samplers, the highest QL may also be a factor because of the maximum loading limit imposed by the samplers.[28] For an Andersen N-6 single-stage sampler, the maximum number of colonies that can be recorded is 400. The highest QL of an air sample collected with an Andersen N-6 single-stage impactor is $\text{CFU/m}^3 = 400$ CFU/sampled air volume in m^3. For an Andersen two-stage sampler, the maximum colony count is 200 colonies per stage.

Media Used and Incubation Temperature. The laboratory report should include the media used and incubation temperatures along with the taxonomic and numerical data to give the investigator a better understanding of the data. The actual number of colonies of each taxa identified should be given. Other relevant information such as sample type, dilution factor, and incubation time may also need to be presented on the analytical report.

5.7. DATA INTERPRETATION

Interpretation of culturable analysis is often a complex task. The common practice is to compare the airborne concentrations outdoors with those indoors. The indoor concentration may also be compared to a reference area (unaffected area). The relative abundance (rank order), presence of indicator species and/or species of concern, and concentration are factors that can be applied to understanding the results of culture-based analysis.[6,20] Information on the ecology, pathogenicity, and toxigenic properties of the reported fungi may be helpful to the investigator in data interpretation and report writing, and should be available from reputable laboratories. Such information is available in this book as well as in Hung et al.[6] The following is not intended to provide comprehensive data interpretation. Instead, it attempts to address concerns specific to the nature of culturable analysis.

False Positive and Negative. In culturable analysis, false positives and negatives are unlikely if appropriate quality control measures were in place during sampling and analysis. Investigators should follow appropriate sampling protocols in this book as well as in reputable publications, such as the new edition of AIHA's *Field Guide for the Determination of Biological Contaminants in Environmental Samples*.[6] A false negative may result from collecting the sample in the wrong location, using the wrong sampling method, insufficient sample volume, the use of inappropriate media and/or incubation conditions, limitations in the sample handling techniques, storing and transporting samples in extreme conditions, misidentification, or others. If the sampling protocol is not specifically designed to culture particular fungi, it is not prudent to conclude that the fungus is absent solely on the basis of limited air samples, especially when one is concerned with pathogens in healthcare facilities or cleanrooms. If a laboratory has a sound quality assurance and quality control program and carries it out according to the program, the chance of laboratory false positives and negatives is minimal.

Competition and Interference Among Fungi. The presence of fast growing and aggressive fungi often results in underestimation of both the diversity and the abundance of the sampled fungal population. For example, if *Fusarium*, *Mucor*, *Rhizopus*, or *Trichoderma* are reported on MEA, chances are that the growth of other fungi (e.g., *Aspergillus versicolor* or *Eurotium*) were suppressed or masked, and were not detected even if viable spores were present. This further highlights the importance of using a more limiting medium along with the primary medium.

Source of Contamination. The ability to provide species information is an important advantage of culture analysis and should be used properly. Data of culturable air samples accompanied by surface samples and careful visual inspections may provide a better chance of identifying the source of growth and/or contamination in relation to the air quality. Indoor growth of *Aspergillus*, *Penicillium*, *Stachybotrys*, and other indicator fungi of water-damaged environments warrants a risk assessment, which usually requires further investigation. Knowing what may be present in an air or dust sample of a normal building will also be helpful.[29]

Pathogenic Fungal Species. Data on a culture-based analytical report confirm both the presence and viability of the fungal spores that were detected. While detection of airborne viable spores does not necessarily confirm indoor growth, the presence of potential pathogens at even low concentrations may cause concern in a building occupied by people with diverse health conditions. Examples of pathogenic fungi include certain *Aspergillus* species (e.g., *A. fumigatus*, *A. flavus*, *A. terreus*), *Cryptococcus neoformans*, *Fusarium verticillioides*, *Histoplasma capsulatum*, *Trichophyton* spp.[30,31] Some indoor fungi may have multiple health implications on humans, including *Aspergillus versicolor*, *Stachybotrys chartarum*.[32]

5.8. LABORATORY QUALITY ASSURANCE

To ensure the quality of data, the laboratory must have a quality assurance–quality control (QA/QC) program and should comply with QA/QC policies of the laboratory as well as those of accreditation agencies. In addition to general requirements, quality control of culture analysis should have at least the following specific components: (1) quality control of regent water, media, and other consumables used for plating; (2) calibration and regular check of instrumentation such as analytical balances and pipettes; (3) calibration and maintenance of biosafety cabinet, incubator, refrigerator, and autoclave; (4) daily temperature check of incubator and refrigerator; (5) use of HEPA-filtered workstations during analysis to avoid contamination of the laboratory; (6) regular check of the airborne spores in the laboratory; (7) positive and negative controls performed with each batch of culture or sample preparation; (8) maintenance and use of reference culture collections whose identification has been confirmed by a reliable source; (9) qualification and training of analysts; and (10) interanalyst, intraanalyst, and intralaboratory quality control as well as proficiency testing.

Among the quality assurance measures mentioned above, the importance of having a competent analyst perform fungal identification should be emphasized because of the nature of the morphology-based culture analysis. In addition, the use of a reference culture collection of correctly identified fungi that are relevant to IAQ investigations is beneficial for training of the analyst and for quality assurance.

REFERENCES

1. Sussman, A. S., Longevity and survivability of fungi, in *The Fungi*, Vol. III, G. Ainsworth and A. S. Sussman, eds., Academic Press, New York, 1968, pp. 447–485.
2. Wilson, S. C. et al., Culturability and toxicity of sick building syndrome-related fungi over time, *J. Occup. Environ. Hyg.* **1**:500–504 (2004).
3. Airaksinen, M., J. Kurnitski, P. Pasanen, and O. Seppanen, Fungal spore transport through a building structure, *Indoor Air* **14**:92–104 (2003).
4. Kang, S.-M. and J. J. Morrell, Fungal colonization of Douglas-fir sapwood lumber, *Mycologia* **92**:609–615 (2000).
5. Yang, C. S. and E. Johanning, Airborne fungi and mycotoxins, in *Manual of Environmental Microbiology*, C. J. Hurst, ed., American Society for Microbiology, Washington, DC, 1997.
6. Hung, L.-L., J. D. Miller, and H. K. Dillion, eds., *Field Guide for the Determination of Biological Contaminants in Environmental Samples*, 2nd ed., American Industrial Hygiene Association, Fairfax, VA, 2005.
7. Jong, S. C. and E. E. Davis, Contribution to the knowledge of *Stachybotrys* and *Memnoniella* in culture, *Mycotaxon*, **3**(3):409–485 (1976).
8. Miller, J. D., T. G. Rand, and B. B. Jarvis, *Stachybotrys chartarum*: Cause of human disease or media darling? *Med. Mycol.* **41**:271–291 (2003).
9. Bills, G. F., M. Christensen, M. Powell, and G. Thorn, Saprobic soil fungi, in *Biodiversity of Fungi: Inventory and Monitoring Methods*, G. M. Mueller, G. F. Bills, and M. S. Foster, eds., Elsevier Academic Press, Burlington, MA, 2004, pp. 271–302.
10. Cannon, P. F. and B. C. Sutton, Microfungi on wood and plant debris, in *Biodiversity of Fungi: Inventory and Monitoring Methods*, G. M. Mueller, G. F. Bills, and M. S. Foster, eds., Elsevier Academic Press, Burlington, MA, 2004, pp. 217–239.

11. Klich, M. A., *Identification of Common Aspergillus Species*, Centraalbureau voor Schimmelcultures, Utrecht, Netherlands, 2002.
12. Pitt, J. I., *A Laboratory Guide to Common Penicillium Species*, 3rd ed., Food Science Australia, 2000.
13. Barnett, H. L. and B. B. Hunter, *Illustrated Genera of Imperfect Fungi*, 4th ed., American Phytopathological Society Press, St. Paul, MN, 1998.
14. Domsch, K. and W. Gams, *Compendium of Soil Fungi*, Vol. 1, IHW-Verlag, 1993.
15. de Hoog, G., J. Guarro, J. Gené, and M. J. Figueras, *Atlas of Clinical Fungi*, 2nd ed., Centraalbureau voor Schimmelcultures, Utrecht, Netherlands, and Uverversitat Rovira I Virgili, Reus, Spain, 2000.
16. Ellis, M. B., *Dematiaceous Hyphomycetes*, Commonwealth Mycological Institute, Kew, Surrey, UK, 1971.
17. Ellis, M. B., *More Dematiaceous Hyphomycetes*, Commonwealth Mycological Institute, Kew, Surrey, UK, 1976.
18. Ellis, M. B. and J. P. Ellis, *Microfungi on Land Plants: An Identification Handbook*, Richmond Publishing, Slough, UK, 1997.
19. Ellis, M. B. and J. P. Ellis, *Microfungi on Miscellaneous Substrates: An Identification Handbook*, Richmond Publishing, Slough, UK, 1998.
20. Flannigan, B., R. A. Samson, and J. D. Miller, eds., *Microorganisms in Home and Indoor Work Environments: Diversity, Health Impacts, Investigation and Control*, CRC Press, Boca Raton, FL, 2001.
21. Hanlin, R. T., *Illustrated Genera of Ascomycetes*, Vols. 1 and 2, American Phytopathological Society Press, St. Paul, MN, 1997.
22. Kiffer, E. and M. Morelet, *The Deuteromycetes, Mitosporic Fungi: Classification and Generic Keys*, Science Publishers, Enfield, NH, 2000.
23. Larone, D. H., *Medically Important Fungi, a Guide to Identification*, 3rd ed., ASM Press, Washington, DC, 1995.
24. McGinnis, M. R., R. F. D'Amato, and G. A. Land, *Pictorical Handbook of Medically Important Fungi and Aerobic Actinomycetes*, Praeger Scientific, New York, 1982.
25. Samson, R. A., E. S. Hoekstra, J. C. Frisvad, and O. Filtenborg, eds., *Introduction to Food- and Airborne Fungi*, 6th ed., Centraalbureau voor Schimmelcultures, Utrecht, Netherlands, 2000.
26. Samson, R. A. and J. I. Pitt, eds., *Intergration of Modern Taxonomic Methods for Penicillium and Aspergillus Classification*, Hardwood Academic Publishers, Amsterdam, Netherlands, 2000.
27. Wang, C. J. K. and R. A. Zabel, *Identification Manual for Fungi from Utility Poles in the Eastern United States*, American Type Culture Collection, Rockville, MD, 1990.
28. Buttner, M. P. and L. D. Stetzenbach, Monitoring of fungal spores in an experimental indoor environment to evaluate sampling methods and the effects of human activity on air sampling, *Appl. Environ. Microbiol.* **59**:219–226 (1993).
29. Horner, W. E., A. G. Worthan, and P. R. Morey, Air- and dustborne mycoflora in houses free of water damage and fungal growth, *Appl. Environ. Microbiol.* **70**:6394–6400 (2004).
30. Howard, D. H., ed., *Pathogenic Fungi in Humans and Animals*, 2nd ed., Marcel Dekker, New York, 2003.

31. Rippon, J., *Medical Mycology*, Saunders, Philadelphia, 1988.
32. Kuhn, D. M. and M. A. Ghannoum, Indoor mold, toxigenic fungi, and *Stachybotrys chartarum*: Infectious disease perspective, *Clin. Microbiol. Rev.*, **16**(1):144–172 (2003).
33. Hocking, A. D. and J. I. Pitt, Dichloran glycerol medium for enumeration of xerophilic fungi from low-moisture foods, *Appl. Environ. Microbiol.* **39**:488–492 (1980).
34. Atlas, R. M., *Handbook of Microbiological Media*, 3rd ed., CRC Press, Boca Raton, FL, 2004.

CHAPTER 6

AIRBORNE BACTERIA IN INDOOR ENVIRONMENTS

LINDA D. STETZENBACH

6.1. INTRODUCTION

Bacteria occur naturally in the environment. When environmental and nutritional conditions are favorable, bacteria will colonize and grow on virtually any surface and in most liquids. Bacteria become airborne as a result of dispersal from a site of colonization and growth. The presence of bacteria in the air and on surfaces indoors may result in human health concerns for building occupants. It is a complex task to assess the populations and concentrations of airborne bacteria that may impact the indoor environment and the health of building occupants. This chapter discusses factors that affect bacterial presence and growth in indoor environments, lists selected adverse health effects that may result from exposure of building occupants, and details some analytical methods for monitoring airborne and surface-associated bacteria indoors.

6.2. BACTERIAL AEROSOLS

Bacterial aerosols can include cells, cellular fragments, and byproducts of metabolism present as particulate, liquid, or volatile organic compounds.[1] A variety of activities can result in the generation of bacterial aerosols indoors, including operation of heating–ventilation–air-conditioning (HVAC) systems, hot-water systems, and water spray devices (e.g., showerheads, humidifiers).[2] Talking and coughing generate aerosols with concentrations of 10^4–10^5 droplets/m^3 of air.[3] Manufacturing practices and biofermentation procedures generate additional bacterial aerosols in industrial workplace environments,[4] medical[5,6] and dental procedures,[7,8] and patient care activities[9] generate aerosols in healthcare facilities.

Sampling and Analysis of Indoor Microorganisms, Edited by Chin S. Yang and Patricia A. Heinsohn

6.3. SELECTED BACTERIA-ASSOCIATED HEALTH EFFECTS

Numerous Gram-positive bacteria are present in air, and bacteria shed from humans are seldom harmful to others,[10] but high ratios of airborne bacteria indoors compared to levels found outdoors have been used as indicators of high occupancy rate, poor ventilation, or inadequate building maintenance.[11] The bacterial pathogens, bacterial byproducts, and environmental bacteria associated with water-damaged building materials and associated fungi are different from common human-source bacteria.[12]

Bacterial pathogens are capable of causing severe disease following inhalation exposure, and this is the primary route of exposure to bacterial aerosols for building occupants. However, ingestion and dermal contact with airborne bacterial cells also results in exposure. Approximately 10 m^3 of air per day are inhaled by the average human[13] with the larger particles lodged in the upper respiratory tract (nose and nasopharynx) and particles <6 μm in diameter being transported to the lung. The 1–2-μm particles have the greatest retention in the alveoli of the lungs.

Infectious *Mycobacterium tuberculosis*,[14] nontubercal mycobacteria,[15] and *M. leprae*[16] are associated with severe illness following inhalation exposure. The airborne transmission of *M. tuberculosis* from infected people occurs during coughing, talking, speaking, laughing, and sneezing.[17] Nosocomial *M. tuberculosis* infections have been reported in healthcare facilities resulting from aerosols generated by the operation of suctioning instruments during treatment, manipulation of tuberculosis ulcers, drainage of necrotic tissue, and exposure during autopsy.[6] Inflammatory responses in mouse lung have been reported from laboratory experiments with *M. terrae* that was isolated from a water-damaged building,[18] and respiratory infection in immunocompromised patients has been linked to exposure to the *M. avium* complex, which is commonly isolated from soil, water, and the air.[19]

Inhalation exposure to *Legionella pneumophila*, a vegetative Gram-negative bacterium that is ubiquitous in freshwater environments worldwide,[20,21] can result in severe pneumonia termed Legionnaires' disease. This disease was first reported following the outbreak of respiratory illness among American Legion conventioneers in Philadelphia in 1976.[22] An estimated 10,000–15,000 cases of Legionnaires' disease per year are reported in the United States.[23] However, *L. pneumophila* is not the only *Legionella* bacterium that is pathogenic, and Legionnaires' disease is not the only result from inhalation exposure. Nineteen other *Legionella* species are also human pathogens,[24] and the disease Pontiac fever, a mild flulike syndrome, is attributed to the inhalation of *Legionella* that cannot replicate in host tissue.[25,26]

Mechanical air-conditioning systems with water spray components, cooling towers, and aerosol-generating water devices (e.g., showerheads, faucets, whirlpool tubs, decorative fountains, humidifiers) are all associated with *Legionella* exposure, but no human–human transmission has been documented.[25] Other airborne Gram-negative bacteria including *Escherichia coli*, *Pantoea* (*Enterobacter*) *agglomerans*, *Pseudomonas* spp., *Alcaligenes faecalis*, and *Acinetobacter* spp. are associated with human exposure and resulting disease. These and other Gram-negative bacteria are commonly isolated in cow barns, pig houses, and poultry barns,[27] wastewater treatment plants,[28–31] recycling facilities,[32] and other industrial facilities.[33] These bacteria can cause a variety of health problems, especially to the young, elderly, and immunocompromised persons.

Endotoxin is a component of the Gram-negative bacterial cell wall that is released during active cellular growth and after cell lysis.[34] Endotoxins are biologically active materials that affect cellular and humoral mediation systems,[34] and these materials can affect many organ systems.[35] Endotoxin exposure is a health concern in occupational settings such as cotton processing and agricultural facilities with high concentrations of organic dusts,[36] industrial environments with water spray components (e.g., machining fluids),[37] and in indoor environments with humidified mechanical air-conditioning systems.[38,39] In residential settings, the presence of dogs, moisture, and settled dust increases the likelihood of occupant exposure to airborne endotoxin.[40]

A report by the Institute of Medicine[12] lists environmental bacterial genera isolated from studies of moldy building materials including *Acinetobacter*, *Bacillus*, *Flavobacterium*, *Nocardia*, *Streptomyces*, and *Thermomonospora*, and it states that additional studies are needed to elucidate connections between exposure and health effects.

Staphylococcus spp. and *Micrococcus* spp. are non-endospore-forming bacteria that are commonly dispersed into the air from the skin, oral and nasal surfaces, and hair of humans.[41] These organisms are associated with increases in nosocomial infections in healthcare settings,[42] including in-home care and assisted-living environments.

Although they resemble fungi in their production of mycelia and dry spores, Actinomycetes are filamentous non-endospore-forming Gram-positive bacteria.[43] Actinomycetes are atypical in nonproblem buildings, but they are associated water-damaged buildings and illness of occupants.[44] They are also found in association with numerous fungi that colonize building materials and furnishings following water intrusion with concentrations of $\geq 10^4$ CFU/m^3 of actinomycetes reported during repair of damaged structures.[45,46]

Infectious and toxigenic biological agents released in aerosols indoors as an act of bioterrorism may result in severe illness.[47,48] The Centers for Disease Control and Prevention (CDC) stated that attacks with biological agents are most likely to be covert with diagnosis of illness due to the occurrence terrorism only after infectious disease has spread via person-to-person contact.[49] Therefore, investigation of contaminated indoor environments may be delayed, thereby complicating the monitoring of suspected indoor environments.

6.4. ANALYSIS OF SAMPLES FOR AIRBORNE BACTERIA

The concentrations of airborne bacteria are determined by volumetric air sampling,[11] but air sampling alone does not verify that an area is free of biological contamination.[50] Airborne cells settle onto surfaces and may not be present in the air at the time of sampling, although they are reaerosolized during routine activity.[51,52]

Therefore, surface sampling in conjunction with air sampling assists in determining areas of contamination, identifying the source of biocontamination, and determining the effectiveness of remediation.[50,53] Discussion of air and surface sampling is presented in other chapters of this book. The following material focuses on analysis of surface and air samples for bacteria. Agar-based impaction sampling methods are limited to culture analysis. Culture methods require that the bacterial cells be viable and culturable at the incubation conditions and on the growth medium selected. This can be problematic as no agar formulation can culture all bacterial cells.[53] For example, a series of media amended with antibiotics and growth factors are needed for the isolation and identification of *Legionella pneumophila*.[25] *Mycobacterium* species are also difficult as many of them require extended incubation time. Resuscitation may be needed for stressed organisms.[54]

All agar impaction have upper and lower limits of quantitation as no dilution or concentration of the sample is possible. For example, the Andersen single-stage impactor sampler operated at the manufacturer's recommended flowrate of 28.3 L/min operated at a sampling time of 2 min results in a lower limit of detection (LDL) of 18 colony-forming units (CFUs) per cubic meter of air. The upper limit of quantitation for the Andersen single-stage impactor sampler under these conditions is 10^4 CFU/m^3 of air. Altering the flowrate or sampling time may reduce the LDL or increase the upper limit of quantitation of impactor samplers, but particle bounce,[55] insufficient embedding of cells into the collection medium, and increased injury to bacterial cells[56] can be a problem when flowrates are altered. In situations where the aerosol concentration is high, overloading of the agar surface by impaction of too many cells can result in colony masking.[57] In addition, agar collection methods that utilize multiple-jet impaction (e.g., Andersen impactor samplers, SAS Bio Samplers) for collection of cells have manufacturer's recommended positive hole corrections to account for the likelihood that multiple cells may pass through the same collection hole and settle onto the same place on the agar surface.[53,58] The correction factor increases as the number of colonies collected on the agar surfaces increases, thereby magnifying uncertainty in the measurement.[59]

The same concerns for culturability and incubation conditions described above for agar-based collection apply to air samples collected by impingement into liquid. However, because the bacterial aerosol is collected into liquid, analysis can also be performed using analytical chemistry and/or molecular biology. A variety of microscopy techniques, chemical methods, spectrometry, and flow cytometry can be used to identify and enumerate airborne cells and their components.[60] Many of these are used for atmospheric aerosols and are not practical for routine indoor air quality monitoring.

Light scattering spectrometry is applicable for monitoring of bacterial aerosols indoors, and this method results in rapid counting of bacterial cells in liquid.[61] Chemical assays for muramic acid, peptidoglycan, and fatty acids are also applicable and are used to detect markers of bacterial cells and endotoxin.[62,63] *Limulus* amebocyte lysate assays have been widely used for endotoxin analysis.[64] The data are expressed as endotoxin units (EU) or nanograms of endotoxin per cubic meter of air with one EU equivalent to 0.10 ng of a reference standard,[65] but variations in extraction efficiency and the protocols for analysis result in differences in the results reported.[66] Immunoassay and radioimmunoassay methods are used to detect specific antigens. Fluorescence immunoassay incorporates a fluorescently labeled antibody and radioimmunoassay utilizes an antibody linked with a radioactive label. The availability of specific and sensitive antibodies to the bacteria of interest limit the use of immunoassay. Therefore, these techniques are being considered for monitoring of biothreat agents[67] rather than used for monitoring airborne bacteria in water-damaged buildings. Flow cytometry incorporates the use of a target-specific fluorescent probe and a laser beam.[67] Lange et al.[68] utilized this technique when monitoring airborne bacteria in swine confinement buildings, finding that the results were comparable to microscopic analysis, and Zollinger et al.[69] used flow cytometry to monitor airborne bacteria in indoor prefermentation grape processing. Polymerase chain reaction (PCR) amplification and quantitative PCR (QPCR) provides enhanced detection and rapid identification of specific airborne and surface-associated bacteria.[50,70–72] These molecular techniques require primers and probes that target unique gene sequences and specific amplification conditions for detection of the microorganism of interest, and confirmation of their specificity and sensitivity in laboratory studies. Internal controls and rigid quality control are also necessary to minimize reporting of false negatives due to environmental interference caused by other microbial DNA, background particulate, and other confounders.

6.5. SUMMARY

Exposure to bacterial aerosols may cause adverse human health effects prompting the desire to monitor indoor water-damaged environments. Monitoring for bacteria requires selection of appropriate analysis methods to determine the concentration and population of cells and their components. Interpretation of the data remains difficult as insufficient data are available to formulate an association of bacterial aerosols with occupant health, and studies directed to dampness and bacterial exposure indoors are needed.[12]

REFERENCES

1. Cox, C. S. and C. M. Wathes, Bioaerosols in the environment, in *Bioaerosols Handbook*, C. S. Cox and C. M. Wathes, eds., Lewis Publishers, Boca Raton, FL, 1995, pp. 11–14.
2. Stetzenbach, L. D., Airborne bacteria, in *Topley & Wilson's Microbiology & Microbial Infections*, 10th ed., S. P. Borriello, P. R. Murray, and G. Funke, eds., HodderArnold Publishers and ASM Press, Washington, DC, 2005, pp. 185–194.
3. Papineni R. S. and F. S. Rosenthal, The size distribution of droplets in the exhaled breath of health human subjects, *J. Aerosol Med.* **10**:105–116 (1997).
4. Abrams, L., N. Seixas, R. Robins, H. Burge, M. Muilenberg, and A. Franzblau, Characterization of metalworking fluid exposure indices for a study of acute respiratory effects. *Appl. Occup. Environ. Hyg.* **15**:492–502 (2000).

5. Demers, R. R., Bacterial/viral filtration: Let the breather beware, *Chest* **120**:1377–1389, (2001).
6. Nottrebart, H. S., Nosocomial infections acquired by hospital employees. *Infect. Control.* **1**:257–259 (1980).
7. Depaola, L. G., D. Mangan, S. E. Mills, W. Costerton, J. Barbeau, B. Shearer, and J. Bartlett, A review of the science regarding dental unit waterlines, *J. Am. Dental Assoc.* **133**:1199–1206 (2002).
8. Leggat, P. A. and U. Kedjarune, Bacterial aerosols in the dental clinic: A review, *Internatl. Dentistry J.* **51**:39–44 (2001).
9. Shiomori, T., H. Miyamoto, and K. Makishima, Significance of airborne transmission of methicillin-resistant *Staphylococcus aureus* in an otolaryngology-head and neck surgery unit, *Arch. Otolaryngol. Head Neck Surg.* **127**:644–648 (2001).
10. Burge, H. A., J. M. Macher, D. K. Milton, and H. A. Ammann, Data evaluation, in *Bioaerosols: Assessment and Control*, American Conf. Governmental Hygienists, Cincinnati, OH, 1999, pp. 14-1–14-11.
11. Macher, J., ed., *Bioaerosols: Assessment and Control*, American Conf. Governmental Hygienists, Cincinnati, OH, 1999.
12. Institute of Medicine (IOM) of the National Academies, Committee on Damp Indoor Spaces and Health, Board on Health Promotion and Disease Prevention, *Damp Indoor Spaces and Health*, National Academy of Science, Washington, DC, 2004, pp. 29–89.
13. Lynch, J. M. and N. J. Poole, Aerosol dispersal and the development of microbial communities, in J. M. Lynch and N. J. Poole, eds., *Microbial Ecology: A Conceptual Approach*, Wiley, New York, 1979, pp. 140–170.
14. Kaufmann, S. H. E. and J. D. A. van Embden, Tuberculosis: A neglected disease strikes back, *Trends Microbiol.* **1**:2–5 (1993).
15. Contreras, M. A., O. T. Cheung, D. E. Sanders, and R. S. Goldstein, Pulmonary infection with nontuberculosis mycobacteria, *Am. Rev. Resp. Dis.* **137**:149–152 (1988).
16. Bryceson, A. and R. E. Pflatzgraft, *Leprosy*, Churchill Livingstone, London, (1990).
17. Rubin, J., Mycobacterial disinfection and control, in *Disinfection, Sterilization, and Preservation*, 4th ed., S. S. Block, ed., Lea & Febiger, Philadelphia, 1991, pp. 377–384.
18. Jussila, J., H. Komulainen, K. Huttunen, M. Reponen, E. Iivanainen, P. Torkko, V. M. Kosma, J. Pelkonen, and M. R. Hirvonen, *Mycobacterium terrae* isolated from indoor air of a moisture-damaged building induces sustained biphasis inflammatory response in mouse lungs, *Environ. Health Perspect.* **110**:1119–1125 (2002).
19. Wolinsky, E., Nontuberculosis mycobacteria and associated diseases, *Am. Rev. Resp. Dis.* **119**:107–159 (1979).
20. Fliermans, C. B., W. B. Cherry, L. H. Orrison, S. J. Smith, D. L. Tison, and D. H. Pope, Ecological distribution of *Legionella pneumophila*, *Appl. Environ. Microbiol.* **41**:9–16 (1981).
21. Steinert, M., U. Hentschel, and J. Hacker, *Legionella pneumophila*: An aquatic microbe goes astray, *FEMS Microbiol. Rev.* **26**:149–162 (2002).
22. Fraser, D. W., T. F. Rsai, W. Orenstein, W. E. Parkin, H. J. Beecham, R. G. Sarrar, H. Harris, G. F. Mallison, S. M. Martin, J. E. McDade, C. C. Shepard, P. S. Brachman, and The Field Investigation Team, Legionnaires' disease: Description of an epidemic of pneumonia, *New Engl. J. Med.* **297**:1189–1197 (1977).

23. Marston, B. J., J. F. Plouffe, R. F. Breiman, T. M. File, R. F. Denson, M. Moyenuddin, W. L. Thacker, K. H. Wong, S. Skelton, D. Hackman, S. J. Salstrom, J. M. Barbaree, and the Community-based Pneumonia Incidence Study Group, Preliminary findings of a community-based pneumonia incidence study, in *Legionella: Current Status and Emerging Perspectives*, J. M. Barbaree, R. F. Breiman, and A. P. Dufour, eds., American Society for Microbiology, Washington, DC, 1993.

24. Muder, R. R. and V. L. Yu, Infection due to *Legionella* species other than *L. pneumophila, Clin. Infect. Dis.* **35**:990–998 (2002).

25. Fields, B. S., *Legionella* and legionnaires' disease, *in Manual of Environmental Microbiology*, 2nd ed., C. J. Hurst, R. L. Crawford, G. Knudsen, M. McInerney, and L. D. Stetzenbach, eds., ASM Press, Washington, DC, 2002, pp. 860–870.

26. Kaufman, A. F., J. E. McDade, C. M. Patton, J. V. Bennet, P. Skaliy, J. C. Feeley, D. C. Anderson, M. E. Potter, V. F. Newhouse, M. B. Gregg, and P. S. Brachman, Pontiac fever: Isolation of the etiologic agent (*Legionella pneumophila*) and demonstration of its mode of transmission, *Am. J. Epidemiol.* **111**:337–339 (1981).

27. Zucker, B. A., S. Trojan, and W. Muller, Airborne gram-negative bacterial flora in animal houses, *J. Vet. Med. Infect. Dis. Vet. Publ. Health.* **47**:37–46 (2000).

28. Brandi, G., M. Sisti, and G. Amagliani, Evaluation of the environmental impact of microbial aerosols generated by wastewater treatment plants utilizing different aeration systems, *J. Appl. Microbiol.* **88**:845–852 (2000).

29. Brenner, K. P., P. V. Scarpino, and C. S. Clark, Animal viruses, coliphage, and bacteria in aerosols and wastewater at a spray irrigation site, *Appl. Environ. Microbiol.* **54**:409–415 (1988).

30. Crawford, G. V. and P. H. Jones, Sampling and differentiation techniques for airborne organisms emitted from wastewater, *Water Res.* **13**:393–399 (1979).

31. Kenline, P. A. and P. V. Scarpino, Bacterial air pollution from sewage treatment plants, *Am. Indust. Hyg. Assoc. J.* **33**:346–352 (1972).

32. Reinthaler, F. F., D. Haas, G. Feierl, R. Schlacher, F. O. Pichler-Semmelrock, M. Kock, G. Wust, O. Feenstra, and E. Marth, Comparative investigations of airborne culturable microorganisms in selected waste treatment facilities and in neighbouring residential areas, *Zentralbl. Hyg. Umweltmed.* **202**:1–17 (1999).

33. Dutkiewicz, J., C. Skorska, J. Milanowski, B. Mackiewicz, E. Krysinska-Traczyk, E. Dutkiewicz, A. Matuszyk, J. Sitkowska, and M. Golec, Response of herb processing workers to work-related airborne allergens, *Ann. Agric. Environ. Med.* **8**: 275–283 (2001).

34. Bradley, S. G., Cellular and molecular mechanisms of action of bacterial endotoxins. *Ann. Rev. Microbiol.* **33**:67–04 (1979).

35. Hewett, J. A. and R. A. Roth, Hepatic and extrahepatic pathobiology of bacterial lipopolysaccharides, *Pharm. Rev.* **45**:381–411 (1993).

36. Buchan, R. M., P. Riijal, D. Sandfort, and T. Keefe, Evaluation of airborne dust and endotoxin in corn storage and processing facilities in Colorado, *Int. J. Occup. Med. Environ. Health* **15**:57–64 (2002).

37. Gordon, T., Acute respiratory effects of endotoxin-contaminated machining fluid aerosols in guinea pigs, *Fund. Am. Appl. Toxicol.* **19**:117–123 (1992).

38. Dutkiewicz, J., L. Jablonski, and S. A. Olenchock, Occupational biohazards: A review. *Am. J. Indust. Med.* **14**:605–623 (1988).

39. Flaherty, D. K., F. H. Deck, J. Cooper, K. Bishop, P. A. Winzenburger, L. R. Smith, L. Bynum, and W. B. Witmer, Bacterial endotoxin isolated from a water spray air humidification system as a putative agent of occupational-related lung disease, *Infect. Immunity* **43**:206–212 (1984).

40. Park, J. H., D. L. Spiegelman, D. R. Gold, H. A. Burge, and D. K. Milton, Predictors of airborne endotoxin in the home, *Environ. Health Perspect.* **109**:859–864 (2001).

41. Favero, M. S., J. R. Puleo, J. H. Marshall, and G. S. Oxborrow, Comparative levels and types of microbial contamination detected in industrial clean rooms, *Appl. Microbiol.* **14**:539–551 (1966).

42. Schall, K. P., Medical and microbiological problems arising from airborne infection in hospitals, *J. Hosp. Infect.* **18**:451–459 (1991).

43. Reponen, T. A., S. V. Gazenko, S. A. Grinshpun, K. Willeke, and E. C. Cole, Characteristics of airborne actinomycete spores, *Appl. Environ. Microbiol.* **64**:3807–3812 (1998).

44. Hyvarinen, A., T. Meklin, A. Vepsalainen, and A. Nevalainen, Fungi and actinobacteria in moisture-damaged building materials—concentrations and diversity, *Internatl. Biodeter. Biodegrad.* **49**:27–37 (2002).

45. Nevalainen, A., A.-L. Pasanen, M. Niininen, T. Reponen, M. J. Jantuen, and P. Kalliokoski, The indoor air quality in Finnish homes with mold problems, *Environ. Internatl.* **17**:299–302 (1991).

46. Rautiala, S., T. Reponen, A. Hyvarinen, A. Nevalainen, T. Husman, A. Vehcilainen, and O. Kalliokoski, Exposure to airborne microbes during repair of moldy buildings, *Am. Indust. Hyg. Assoc. J.* **57**:279–284 (1996).

47. Sheeran, T. J., Bioterrorism, in *Encyclopedia of Environmental Microbiology*, G. Bitton, ed., Wiley, New York, 2002, pp. 771–782.

48. Wiener, S. L., Biological warfare defense, in *Biological Warfare Modern Offense and Defense*, R. A. Zilinskas, ed., Lynne Rienner, Boulder, CO, 2000, pp. 119–146.

49. Centers for Disease Control and Prevention (CDC), *Biological and Chemical Terrorism: Strategic Plan for Preparedness and Response*, U.S. Department of Health and Human Services, Centers for Disease Control and Prevention, Atlanta, 2000.

50. Higgins, J. A., M. Cooper, L. Schroeder-Tucker, S. Black, D. Miller, J. S. Karns, E. Manthey, T. Breeze, and M. L. Perdue, Field investigation of *Bacillus anthracis* contamination of U.S. Department of Agriculture and other Washington, D.C. buildings during the anthrax attack of October 2001, *Appl. Environ. Microbiol.* **69**:593–599 (2003).

51. Buttner, M. P. and L. D. Stetzenbach, Monitoring of fungal spores in an experimental indoor environment to evaluate sampling methods and the effects of human activity on air sampling, *Appl. Environ. Microbiol.* **59**:219–226 (1993).

52. Weis, C. P., A. J. Intrepido, A. K. Miller, P. G. Cowin, M. A. Durno, J. S. Gebhardt, and R. Bull, Secondary aerosolization of viable *Bacillus anthracis* spores in contaminated US Senate office, *J. Am. Med. Assoc.* **288**:2853–2858 (2002).

53. Buttner, M. P., K. Willeke, and S. A. Grinshpun, Sampling and analysis of airborne microorganisms, in *Manual of Environmental Microbiology*, 2nd ed., C. J. Hurst, R. L. Crawford, G. Knudsen, M. McInerney, and L. D. Stetzenbach, eds., ASM Press, Washington, DC, 2002, pp. 814–826.

54. Crozier-Dodson, B. A. and D. Y. Fung, Comparison of recovery of airborne microorganisms in a dairy cattle facility using selective agar and thin agar layer resuscitation media, *J. Food Protect.* **65**:1488–1492 (2002).

55. Grinshpun, S. A., K. Willeke, V. Ulevicius, A. Juozaitis, S. Terzieva, J. Donnelly, G. N. Stelma, and K. P. Brenner, Effect of impaction, bounce and reaerosolization on the collection efficiency of impingers, *Aerosol Sci. Technol.* **26**:326–342 (1997).

56. Stewart, S. L., S. A. Grinshpun, K. Willeke, S. Terzieva, V. Ulevicius, and J. Donnelly, Effect of impact stress on microbial recovery on an agar surface, *Appl. Environ. Microbiol.* **61**:1232–1239 (1995).

57. Chang, C.-W, Y.-H. Hwang, S. A. Grinshpun, J. M. Macher, and K. Willeke, Evaluation of counting error due to colony masking in bioaerosol sampling, *Appl. Environ. Microbiol.* **60**:3732–3738 (1994).

58. Macher, J. M., Positive-hole correction of multiple-jet impactors for collection viable microorganisms, *Am. Indust. Hyg. Assoc. J.* **50**:561–568 (1989).

59. Andersen, A. A., New sampler for the collection, sizing, and enumeration of viable airborne particles, *J. Bacteriol.* **76**:471–484 (1958).

60. Spurny, K. R., Chemical analysis of bioaerosols, in *Bioaerosols Handbook*, C. S. Cox and C. M. Wathes, eds., Lewis Publishers, Boca Raton, FL, 1995, pp. 317–334.

61. Mainelis, G., R. L. Gorny, K. Willeke, and T. Reponen, Rapid counting of liquid-borne microorganisms by light scattering spectrometry, *Ann. Agric. Environ. Med.* **12**:141–148 (2005).

62. Laitinen, S., J. Kangas, K. Husman, and P. Susitaival, Evaluation of exposure to airborne bacterial endotoxins and peptidoglycans in selected work environments, *Ann. Agric. Environ. Med.* **8**:213–219 (2001).

63. Liu, L. J., M. Krahmer, A. Fox, C. E. Feigley, A. Featherstone, A. Saraf, and L. Larsson, Investigation of the concentration of bacteria and their cell envelope components in indoor air in two elementary schools, *J. Air Waste Manage.* **50**:1957–1967 (2000).

64. Olenchock, S. A., Airborne endotoxin, in *Manual of Environmental Microbiology*, 2nd ed., C. J. Hurst, R. L. Crawford, G. Knudsen, M. McInerney, and L. D. Stetzenbach, eds., ASM Press, Washington, DC, 2002, pp. 853–859.

65. Milton, D. K., Endotoxin, in *Bioaerosols*, H. S. Burge, ed., CRC Press, Boca Raton, FL, 1995, pp. 77–86.

66. Reynolds, S. J., P. S. Thorne, K. J. Donham, E. A. Croteau, K. M. Kelly, D. Lewis, M. Whitmer, D. J. Heederick, J. Douwes, I. Connaughton, S. Koch, P. Malmberg, B. M. Larsson, and D. K. Milton, Comparison of endotoxin assays using agriculture dusts, *Am. Indust. Hyg. Assoc. J.* **63**:430–438 (2002).

67. Andreotti, P. E., G. V. Ludwig, A. H. Peruski, J. J. Tuite, S. S. Morse, and L. F. Peruski, Jr., Immunoassay of infectious agents, *BioTechniques* **35**:850–859 (2003).

68. Lange, J. L., P. S. Thorne, and N. Lynch, Application of flow cytometry and fluorescent in situ hybridization for assessment of exposures to airborne bacteria, *Appl. Environ. Microbiol.* **63**:1557–1563 (1997).

69. Zollinger, M., W. Krebs, and H. Brandl, Bioaerosol formation during grape stemming and crushing, *Sci. Total Environ.* (in press).

70. Alvarez, A. J., M. P. Buttner, G. A. Toranzos, E. A. Dvorsky, A. Toro, T. B. Heikes, L. E. Mertikas, and L. D. Stetzenbach, The use of solid-phase polymerase chain reaction for the enhanced detection of airborne microorganisms, *Appl. Environ. Microbiol.* **60**:374–376 (1994).

71. Buttner, M. P., P. Cruz-Perez, and L. D. Stetzenbach, Enhanced detection of surface-associated bacteria in indoor environments using quantitative PCR, *Appl. Environ. Microbiol.* **67**:2564–2570 (2001).

72. Pascual, L., S. Perez-Luz, A. Amo, C. Moreno, D. Apraiz, and V. Catalan, Detection of *Legionella pneumophila* in bioaerosols by polymerase chain reaction, *Can. J. Microbiol.* **47**:41–346 (2001).

CHAPTER 7

GENETICS-BASED ANALYTICAL METHODS FOR BACTERIA AND FUNGI IN THE INDOOR ENVIRONMENT

RICHARD A. HAUGLAND and STEPHEN J. VESPER

7.1. INTRODUCTION

Since the mid-1980s, advances in high-throughput sequencing technologies have led to a veritable explosion in the generation of nucleic acid sequence information.[1] While these advances are illustrated most prominently by the successful sequencing of the human genome, they have also factored heavily in our current knowledge of nucleic acid sequences from a variety of microorganisms. Concurrent with this growth in sequencing activity has been the development of a variety of techniques for the amplification and manipulation of nucleic acids. The combination of these technologies is currently supporting a gradual shift away from the use of traditional culture-based analyses for the detection and characterization of microorganisms and toward the use of new analytical methods based on genetic composition, namely, nucleic acids.[2,3] This shift is evident in a number of fields, including the microbiological analyses of indoor environments.

Genetic analysis methods for bacterial and fungal microorganisms are presently becoming increasingly widespread in their applications, not only in research settings but also in clinical and environmental testing laboratories. Advantages of these methods over culture and phenotypic analyses can include higher speed, sensitivity, and accuracy in the detection and identification of microorganisms as well as better resolution between similar organisms and an ability to detect noncultivatable or fastidious organisms. Through their potential for automation, these methods also offer the potential for less reliance on analyst training and decreased labor investments. As will be discussed in more detail below, current limitations of many of these genetic analysis methods can include ongoing uncertainty of the extent to which

Sampling and Analysis of Indoor Microorganisms, Edited by Chin S. Yang and Patricia A. Heinsohn

available sequence information circumscribes the genetic variability of different target microbial groups, technical and quality control issues, and higher costs related to expenditures for instruments and reagents. It is reasonable to expect, however, that each of these limitations will decrease in the future, making genetic microbial testing methods an increasingly attractive option for many clinical and environmental applications.

Section 7.2 of this chapter provides an overview (with references for additional reading) of currently available genetics-based analytical techniques that may be useful for investigations of bacterial and fungal microorganisms in the indoor environment. These techniques are grouped into four general categories: (1) *in vitro* nucleic acid amplification, (2) hybridization probes, (3) nucleic acid sequencing, and (4) microbial strain typing. Section 7.3 provides example applications of techniques within each of these categories for the study of indoor microbiology. Section 7.4 provides a synopsis of quality assurance issues and presently accepted quality control measures for laboratories performing genetic analysis methods, focusing primarily on methods involving nucleic acid amplification techniques. Other current strengths and limitations of these methods and their outlook for the future are also discussed in Section 7.5.

7.2. GENETICS-BASED ANALYTICAL TECHNIQUES

7.2.1. *In Vitro* Nucleic Acid Amplification

The purpose of *in vitro* nucleic acid amplification techniques is to increase the quantities of total, or more commonly, specific nucleic acid segments present in small or complex samples to sufficiently high levels to allow their detection by optical, physical, and other methods and/or to allow their further manipulation. Another category of techniques meets the objective of specific nucleic acid detection through the amplification of probe signals or other indirect markers of target sequences rather than the target sequences, themselves.[4] Detection and manipulation of specific nucleic acid sequences has been integral to many advances in biological research and testing in the present era. A salient example of this is the important role that nucleic acid amplification techniques have played in the development of modern nucleic acid sequencing methods.[5,6] Nucleic acid amplification also forms the basis for the majority of available genetic methods for detecting and characterizing microorganisms that are of interest in the indoor environment.

The earliest widely accepted technique for nucleic acid amplification was the polymerase chain reaction (PCR),[7] which was first practically demonstrated in 1985.[8] Despite the emergence of numerous alternative amplification methods over the years, the elegance and simplicity of this basic technique, combined with its flexibility in accommodating numerous modifications for different purposes, and a steady progression of readily available and easy-to-use instrumentation, have contributed to the fact that PCR remains the most widely used technique for nucleic acid amplification to this day.

PCR uses a thermostable DNA polymerase to amplify a segment of target DNA defined at each end by specific oligonucleotide primers. PCR typically consists of three basic steps: (1) the reaction mixture containing polymerase, oligonucleotide primers, deoxynucleotide triphosphates (building blocks of DNA), magnesium (a necessary cofactor for polymerase activity), buffer constituents, and sample containing target DNA is heated to separate or denature the two strands of the DNA in the sample; (2) the reaction temperature is lowered, allowing the oligonucleotide primers to anneal to each of the separated single strands of the target DNA segment; (3) the reaction temperature is raised, allowing the polymerase to extend the primers to make complementary copies of both strands of the target DNA segment. The primer annealing and extension steps are often performed at the same temperature. Each repetition of these three steps is referred to as a *thermal cycle* that theoretically results in a doubling of the amount of target DNA present in the reaction. The number of thermal cycles required for a PCR application may be dependent on the amount of target DNA present at the start of the reaction and the number of amplified product copies—also referred to as *amplicons*—desired for post-PCR detection or manipulation. Generally 30–50 thermal cycles are sufficient to detect 100 or fewer target sequence copies in a reaction, and detection of a single target sequence is possible.[9,10]

Taq DNA polymerase, which is isolated from the thermophilic bacterium *Thermus aquaticus*, was the first enzyme used for the amplification of DNA in PCR procedures. More recently a number of modifications of this enzyme, as well as other DNA polymerases with similar properties, have become available. Important examples of these modified enzymes are the "hot start" DNA polymerases.[11,12] These enzymes are inactive until a specified temperature is reached—usually greater than the annealing temperatures of the primers. Use of hot-start enzymes reduces the production of nonspecific amplification products by preventing the elongation of primers that have nonspecifically annealed to the DNA sample during preparation of the reaction mixture or prior to thermal cycling.

Oligonucleotide primers are essential components of PCR as well as most other nucleic acid amplification techniques. These are normally synthetically prepared, single-strand sequences of ~16–27 nucleotides that are constructed to be homologous to nucleotide sequences flanking the DNA segment targeted for amplification. Primer design often requires some prior knowledge of the sequence of the target DNA region. This knowledge is ususally obtained from the sequencing of cloned DNA fragments or products from nucleic acid amplifications using primers that are highly conserved among different organisms (i.e., broad-range PCR—see below). In a general sense, primer selection plays an important role in the efficiency of amplification and hence the sensitivity of a method. A number of computer programs are available to aid in the determination of optimal annealing temperatures and avoidance of intra- and intermolecular dimers and hairpin loop formations that can decrease amplification efficiency. For detection of specific groups of microorganisms, such as species or genera, knowledge of the target sequence in as many different isolates representing these groups as possible is important for obtaining optimal test method sensitivity. Similarly, corresponding sequence information from related organisms that are not intended to be detected is important for the selection of PCR primers with the desired levels of specificity.[13]

As mentioned above, one factor contributing to the longevity and widespread use of PCR in research and testing laboratories is its ability to be modified for different applications. Some of the more widely used modifications of this technique have included reverse transcriptase PCR (RT-PCR), nested PCR, multiplex PCR, broad-range PCR, quantitative-competitive PCR, and genomic fingerprinting PCR (Table 7.1). One of the most recent and perhaps significant modifications of this technique is real-time PCR.

Real-time PCR is so named because it detects and reports amplification of target nucleic acids as it occurs in the reaction.[2] This feature provides important advantages over the basic PCR technique by reducing analysis time and labor as well as potential artifacts from contamination. A unique feature of real-time PCR is the use of target sequence-homologous oligonucleotide probes labeled with two fluorescent dyes that report amplification based on fluorescence resonance energy transfer (FRET) mechanisms.[14]

Several different probe and dye configurations have been developed for achieving FRET-based reporting in the reactions, including hydrolysis probes,[15] molecular beacons,[16] scorpion probes,[17] and adjacent dual-hybridization probes.[18] In each case, the level of reported fluorescence is directly related to the amount of amplified DNA. An alternative mechanism for detecting amplification is the use of intercalating dyes—most commonly SYBR Green I, which fluoresce in proportion to the amount of double-stranded DNA products that have been generated by amplification.[19] This approach eliminates the need for designing internal hybridization probes but can suffer from lower specificity since nonspecific amplification products such as primer–dimers are also reported. The fluorescence generated by each of the abovementioned mechanisms is detected by a fluorimeter that is coupled with the

TABLE 7.1. Modifications of the Basic PCR Technique for Different Applications

Modified Technique	Application	Reference
Reverse transcriptase PCR	Amplification of RNA sequences	129
Nested PCR	Two rounds of amplification with different primers sets for increased sensitivity and specificity	130
Multiplex PCR	Simultaneous amplification of multiple target sequences in a single reaction	131
Broad-range PCR	Detection and/or sequence analysis of new organisms using phylogenetically conserved primer sequences	132, 133
Quantitative–competitive PCR	Quantification of target sequences	134
Genomic fingerprinting PCR	Genetic typing of microbial strains	See Section 7.2.4

thermal cycling instrument to provide highly sensitive monitoring of amplicon production as the reaction proceeds.

Another significant benefit of real-time PCR is its amenability to quantitative analyses of target sequences. Because of this widely used feature, a common synonym of real-time PCR analysis is quantitative PCR or QPCR. The number of thermal cycles required for the fluorimeter to initially detect a significant increase in fluorescence above background levels is inversely correlated with the log of the initial target sequence concentration in the reactions over a dynamic range of up to seven log units. Real-time data collection software calculates this cycle number for each reaction as a noninteger value by interpolation of the significantly above-background fluorescence value on the fluorescence growth curve of the reaction and reports it as the cycle threshold (CT) value. Standard curves of CT values from different known quantities of target sequences can be prepared and used for the quantification of target sequences in unknown samples.[19]

Other more recently developed target sequence amplification techniques, including the ligase chain reaction,[20] strand displacement amplification,[21,22] rolling-circle amplification[23,24] and several forms of transcription-based amplification,[25–27] have been reviewed in detail elsewhere.[4,28] Some of these techniques offer distinct advantages compared to PCR such as isothermal cycling, which lowers instrumentation costs. Others have also been applied to whole-genome amplification.[29] While these techniques are beginning to make significant inroads in clinical research applications, they have not yet been widely used for indoor environmental research or testing. Similarly, several innovative signal amplification methods, including branched DNA amplification,[30] QB replicase,[31,32] cleavage-based signal amplification,[33] and cycling probe technology,[34] have been developed but have not yet been widely employed in environmental studies.

7.2.2. Hybridization Probes

The earliest applications of genetic information for microbial identification and characterization were by means of nucleic acid hybridization probes. While currently augmented or supplanted by nucleic acid amplification techniques in many instances, direct hybridization with oligonucleotide, gene length, and whole-genome probes continues to be an important tool. In clinical laboratories this technique is widely used as a confirmatory tool for the identification of cultured microbial isolates, and a number of kits are commercially available for important bacterial and fungal pathogens.[35] Another extensively used application of direct hybridization analysis has been the identification or characterization of microorganisms *in situ* within clinical and environmental samples using fluorescent labeled probes referred to as *FISH* (Fluorescence in situ hybridization).[36] FISH techniques typically target sequences within ribosomal RNA molecules because of their taxonomic and phylogenetic information and also their naturally high abundance in microorganisms.

Hybridization of target nucleic acid sequences with probes can be performed either in solution, in conjunction with various hybrid capture mechanisms, or with the targets directly affixed to a solid support. The former technique facilitates the speed of analyses due to the relatively rapid kinetics of nucleic acid hybridization in solution, whereas the latter is amenable to simultaneous testing of multiple samples. A method for attaching various probes to differentially labeled fluorescent particles, used in conjunction with flow cytometric analysis, has also been developed that combines the more favorable aspects of both of these techniques.[37] The use of surface-localized probes has more recently entered into a new era with the advent of microarray production technologies.[38] These technologies allow the localization of up to tens of thousands of probes on a single support and have formed the basis for major recent advances in gene discovery and gene expression studies in selected eukaryotic and prokaryotic organisms. Additional applications of this technology are being directed toward the identification of microbial pathogens in clinical samples[39] as well as microbial community analyses.[40] Technologies of this type may be useful in indoor microbial analyses in the future.

7.2.3. Nucleic Acid Sequencing

Despite certain limitations, nucleic acid sequence analysis is generally accepted as the most rigorous of all genetics-based microbial identification techniques and serves as the underlying basis for most other methods. Improved standardization in the form of commercially available amplification and sequencing kits such as the MicroSeq 500 16S rDNA kit for bacteria and the MicroSeq D2 LSU rDNA kit for fungi (Applied Biosystems, Foster City, CA) have lead to the increasing use of nucleic acid sequencing as a diagnostic test for pathogen identification or confirmation in larger clinical laboratories. These tests are directed predominantly at sequences of the ribosomal RNA genes for which the largest reference databases from known organisms are currently available through organizations such as the National Center for Biotechnology Information (http://www.ncbi.nlm.nih.gov), the Ribosomal Database Project (http://rdp.cme.msu.edu/index.jsp), and various commercial sources. The small subunit 16S ribosomal RNA gene is most commonly used for bacterial identification; however, for fungi, sequencing of the 28S large subunit RNA gene has moved forward because of its higher content of variable sequence regions that can be used for discrimination of different species. As with nucleic acid amplification and hybridization-probe-based tests, limitations of this technique exist in dealing with intraspecific nucleotide sequence variability that may not be represented by existing sequence reference databases and also with the lack of sequence variability in some instances between closely related species. Sequencing of other, more variable genes has been used in some instances to obtain greater levels of taxonomic resolution. Another limitation of this technique stems from the lack of generally accepted standards for determining the percentage of sequence identity within a single gene that constitute a species.[41] Despite these limitations, several studies have demonstrated a high success rate by this technique in correctly identifying fungal (particularly yeasts) and bacterial isolates from clinical sources, including a number of isolates that have been refractory to conventional identification methods. Another advantage of sequence analysis is that its ongoing use should continue to fortify existing sequence reference databases,[42] which will presumably further increase the success rate of future identifications made by this technique.

7.2.4. Microbial Strain Typing

Strain typing is used for the discrimination or recognition of similar or different cultured microbial isolates within a given genus or species. Analyses falling within this category have proved to be a valuable tool in a variety of arenas, including medicine, agriculture, industry, and environmental monitoring. The traditional use of phenotypic characteristics for this purpose has now largely given way to a variety of molecular genetic techniques. Currently, the most widely accepted of these techniques involves restriction endonuclease cleavage of the whole genome with rare-cutting enzymes, followed by separation of the resulting large DNA fragments by pulsed field gel electrophoresis.[43] The number and size distribution of these fragments is commonly referred to as a "fingerprint" and is normally characteristic for different strains.

Another widely used fingerprinting technique is ribotyping, which is based on the number and size distribution of restriction endonuclease-digested ribosomal DNA fragments that are detected on an electrophoretic gel blot by hybridization with a labeled total ribosomal DNA–specific probe.[44] As with other microbial identification methods, many newer approaches to microbial strain typing utilize nucleic acid amplification techniques such as PCR. Included among these PCR-based genomic fingerprinting methods are several variations of the random primed PCR technique[45–47] and amplification of a number of different repetitive DNA elements present in microbial genes such as the characteristic REP, ERIC, and BOX elements in many bacterial species[48,49] and mini- and microsatellite sequences in both bacteria and fungi.[50,51] Another PCR-based genomic fingerprinting approach that is being increasingly used because of its repeatability, high resolving power, and applicability with both bacteria and fungi is the amplified fragment length polymorphism (AFLP) technique.[52,53] A number of additional microbial typing methods based on PCR amplification, nucleic acid fingerprinting, and various physical properties of the amplicons have been developed, including single-strand conformational polymorphism analysis, temperature and gradient gel electrophoresis, and terminal restriction fragment length polymorphism analysis.[54] Because of problems in comparing data between different laboratories and over time, there is a growing call for the augmentation or replacement of the image-based fingerprinting techniques described above. Newer approaches such as multilocus sequencing avoid these problems and are increasingly being used for microbial typing as well for basic research studies of microbial evolution.[55–57]

7.3. APPLICATIONS OF GENETICS-BASED METHODS FOR INDOOR MICROBIOLOGICAL ANALYSES

7.3.1. Detection of Pathogenic, Allergenic, and Toxigenic Bacteria and Fungi Using Nucleic Amplification and Hybridization Probe Techniques

Genetics-based methods for the detection of pathogenic microorganisms have been most widely developed and applied in clinical settings for the diagnosis and management of infections.[58,59] Normal indoor environments are not considered as reservoirs for microbial pathogens and therefore are not ordinarily monitored for these organisms. A notable exception is the specialized indoor environments of hospitals and hospital intensive care units. In these settings, monitoring for the occurrences of common and/or readily disseminated opportunistic pathogens is both warranted and often practiced, due to the increased infection risks posed by these organisms to immunocompromised and immunosuppressed patients.[60,61] PCR methods for detecting a number of microbial pathogens with the abovementioned characteristics in different environmental samples have been reported (Table 7.2). Molecular tests of this nature may often be more desirable in hospital monitoring situations than traditional culture-based detection methods because of their relatively short turnaround times of 1–2 days or less. Some relevant pathogens that cannot be presently cultivated, such as *Pneumocystis juroveci*, can be detected in environmental samples only by tests such as these. More specific applications of direct PCR based detection methods have included the identification of environmental reservoirs for pathogens in hospital settings[62–66] as well the monitoring of airborne pathogens originating from infected patients.[67–72]

Additional special situations requiring the detection of pathogenic microorganisms in indoor environments involve instances of real or threatened acts of bioterrorism. The need for rapid and sensitive methods for detecting bioterrorism agents in buildings was amply illustrated by the series of anthrax attacks that occurred in 2001. PCR-based tests for a wide range of highly infectious bacteria and viruses that are considered to be the most likely agents in potential future attacks have been developed.[73] Application of the PCR testing technique for direct air and surface monitoring following the 2001 attacks revealed some limitations in sensitivity compared with culture analysis; however, it proved to be highly useful for confirmatory analyses of suspected *B. anthracis* colonies isolated by culture.[74]

TABLE 7.2. PCR Methods for Detection of Different Pathogenic Microorganisms in Environmental Samples

Organism(s) Detected	Analysis Method	Sample Type	Reference
Asperillus fumigatus	Real-time PCR	Air	135
Pathogenic *Candida*	Real-time PCR	Water	90
Bordetella pertussis	PCR	Air	67
Pathogenic Fusaria	PCR	Not reported	136
Legionella pneumophila	PCR	Air	137
Legionella pneumophila	PCR	Water	138
Legionella pneumophila	Nested PCR	Water	62
Legionella pneumophila	Real-time PCR	Water	66, 139
Methicillin-resistant bacteria	PCR	Surfaces	140
Mycobacterium tuberculosis	PCR	Air	69, 71, 72, 141
Pneumocystis juroveci (formerly *P. carinii*)	PCR	Air	70, 142, 143
Pneumocystis juroveci	PCR	Dust	144
Pathogenic Zygomycetes	PCR	Not reported	145, 146

The possible roles of indoor fungi and bacteria as causative agents of a variety of residential and occupational building-related health complaints have also received a heightened amount of attention.[75,76] Deleterious health effects other than infection that are known or suspected to be associated with excessive exposures to certain fungal or bacterial groups include allergic responses such as inflammation, rhinitis, hypersensitivity pneumonitis, and asthma as well as a wide range of toxicoses.[77–80] Despite the widespread concern and an abundance of anecdotal case reports, cause–effect associations between specific exposures to indoor microorganisms and many of these illnesses are still considered to be unproved at this time.[81] PCR-based methods can provide several advantages over the normal culture/microscopy-based methods currently in use for the detection and quantification of general or specific groups of indoor microorganisms.[13,82–86] These methods may aid in the further demonstration of associations between specific microbes in the indoor environment and illnesses.[87]

Other practical applications of indoor dust sample analyses employing the real-time PCR technique have been reported for the characterization of normal and water-damaged buildings, based on the mold and bacterial populations present.[83,87,88] A number of environmental sampling and sample preparation methods have been developed and evaluated to support PCR and other genetics-based analyses of diverse indoor microorganisms[65,86,89–95] (see also Table 7.2). In addition, specific PCR primer sets and probes for a large number of species or phylogenetic groups of common indoor microorganisms are now available [13,82,83,96–100] (see also Table 7.2). While thus far employed primarily for agricultural and food testing applications, PCR primer sets and probes for genes involved in fungal toxin production and their RNA transcripts are also becoming available and should see increased use in the arena of indoor environmental analysis.[101–104]

An issue that often arises in PCR analyses is related to viability of the detected organism. Although any detection of a pathogen's DNA may warrant concern, often it is more useful to know whether the organism is viable. Various approaches have been used to detect viability by real-time PCR methods, including the use of reverse transcriptase PCR to detect short-lived messenger RNA targets. One of the newer approaches has been the use of the intercalating dye ethidium monoazide (EMA).[105,106] Exposure of a mixture of viable and nonviable *Campylobacter jejuni* to EMA in combination with photoactivation resulted in a procedure by which real-time PCR could be used to quantify only the viable cells. Testing of this approach with a wide variety of organisms in different environmental matrices may establish whether it is generally useful.

7.3.2. Microbial Strain Typing and Identification by Nucleic Acid Sequencing

As with genetics-based detection methods, the most widely employed uses of genetic strain typing and nucleic acid sequencing have been for the identification of microbial isolates from hospital environments. Analyses of this nature most commonly address the question of whether nosocomial infections have originated from the physical hospital environment as opposed to their spread from other patients or caregivers, or from latent community-acquired organisms already carried by the patients. This information is used in turn to determine the best measures for controlling exposures to other patients by the infectious organisms. A number of studies have successfully used molecular strain typing techniques to link nosocomial bacterial and fungal infections with air, water, and other specific reservoirs within hospitals.[107–118] In other studies, however, difficulties with this approach have been reported. Some of these difficulties have been attributed to a high degree of genetic diversity in certain commonly encountered species such as *Aspergillus fumigatus*.[119,120] This diversity can necessitate the testing of extensive numbers of environmental isolates with still no assurance of ultimately establishing the source of infection. Selection of a typing method with an appropriate level of resolution for the species under investigation, specifically, low resolution for species with high levels of genetic diversity and high resolution for species with low levels of genetic diversity, can increase the likelihood of successful source identification.

While applications of molecular typing techniques and DNA sequencing for the identification of microorganisms in general indoor environments are fairly limited to date, several studies have demonstrated the usefulness of these approaches for characterizing indoor populations of selected opportunist and nonpathogenic fungal and bacterial groups with potential health implications.[121–123] More studies of this nature would undoubtedly facilitate the development of new and potentially more sensitive primer sets and probes for the direct detection of additional microbial groups that may be of interest in indoor environments.

7.4. QUALITY CONTROL/QUALITY ASSURANCE AND OTHER CHALLENGES

It is well recognized that genetics-based microbial detection, identification, and characterization methods can pose unique challenges that must be addressed to ensure acceptable data quality. PCR and other techniques employing nucleic acid amplification are particularly susceptible to both false-positive results, namely, detection of target sequences when none are present in the original sample, and false-negative results, specifically, nondetection of target sequences when they are present in the original sample. These factors represent the largest technical limitations to the use of nucleic acid amplification techniques in clinical and environmental testing laboratories.

False-positive results are generally associated with laboratory contamination by target nucleic acids either from the natural environment or from amplicons generated by previous analyses. Consistent avoidance of these problems generally requires strict adherence to a number of control measures including personnel training, protective clothing, laboratory design, and workflow procedures that have been reviewed previously.[124–126] Several methods have been developed for the inactivation of PCR amplicons from previous analyses, but these do not eliminate the need for the more general control practices referred to above. In association with these practices, negative control samples should be prepared and analyzed in parallel with each set of test samples to confirm the efficacy of these practices.

False-negative results can be associated with poor assay (i.e., primer and/or probe) design, poor reagents, poor recovery of target nucleic acids from the sample matrix, or, most commonly, the presence of inhibitory compounds in the samples analyzed. Inhibitory compounds most often originate from the original sample matrix but may also be introduced during sample preparation. The first step in addressing false-negative problems is to develop validated sample preparation and analysis procedures that define the expected performance of the overall method. Positive control samples should be prepared and analyzed in parallel with each set of test samples to confirm the efficacy of these sample preparation and analysis procedures. Since these external positive controls may not detect the influences of specific sample matrices on the outcome of the analyses, however, a common practice is to also include analyses of positive control sequences that are added both prior to and after preparation of the samples. General guidelines for the development and effective use of positive control sequences that can be distinguished from the target sequences in the analyses have been reported.[127,128]

Another challenge faced by nucleic acid amplification–based microbial detection techniques relates to the potential differences that can exist between the theoretical and *de facto* sensitivities of analytical methods that incorporate them. While it is known, for example, that the PCR technique has the capability to detect extremely low numbers of nucleic acid target molecules in an analyzed sample, the need to recover and concentrate these nucleic acids from microorganisms in large and/or complex environmental samples down to the small volumes typically used for analysis can present formidable problems.

Another major factor influencing the *de facto* sensitivity of the PCR technique is the commonly encountered requirement of separating the recovered nucleic acids from a wide range of potential PCR-inhibitory substances that may be coextracted from environmental as well as other sample matrices.[4] While considerable effort has gone into the development of nucleic acid extraction and purification methods, and an assortment of products are commercially available for these purposes, the portions of extracted nucleic acid samples that are analyzed at one time still often represent frequently only a small fraction of the total sample volume. In addition, purification procedures that are intended to increase the fractions of the extracted samples that can be analyzed without inhibition often tend to lower the recovery efficiencies of the nucleic acids from the original samples.

Quantitative techniques such as real-time PCR present additional challenges in terms of maintaining quality control and assurance.[3] Paramount among these is the absence of accurate and universally accepted quantitative standards and controls. Added to this problem is the plethora of different reagent systems, platforms, and often primer and/or probe assays that are used by different laboratories to analyze for the same microorganisms. While strides toward the standardization and multilab validation of test methods, reagents, and quantitative standards have been made for the analyses of certain high-profile microbial and viral pathogens in the clinical and bioterrorism arenas, further work toward this goal is still needed for most environmental analysis applications.

7.5. OUTLOOK FOR THE FUTURE

Despite technical problems that still persist, the use of genetics-based analytical methods is already well established and is widely expected to play an increasingly important role in clinical diagnostic microbiology and environmental microbiology as well as other areas in the future. Further improvements and progress toward the elimination of many of these technical problems will undoubtedly continue to occur. Future progress in the development and use of molecular methods for indoor environmental analyses, however, will likely occur only to the extent that they demonstrate socioeconomic value. PCR analyses, for example, remain relatively expensive and require major investments for instrumentation compared with currently used culture-based and microscopic analysis methods. Therefore, it is likely that this and other nucleic acid–based testing methods will have to demonstrate a value-added factor over the traditional methods. The groundwork for this has already been established by the high speed, and potential for greater specificity, sensitivity, and standardization that these methods possess. Studies using real-time PCR analyses are currently being directed toward the development of rapid, standardized approaches for predicting the water damage status of residential buildings as well as for determining the associations between specific groups of microorganisms in buildings and important illnesses such as asthma. Success in these efforts will clearly provide an impetus for the continued development and more widespread use of these methods.

(*Note*: The United States Environmental Protection Agency, through its Office of Research and Development, partially funded and collaborated in the research described here. It has been subjected to Agency review and approved for publication.)

REFERENCES

1. Alcorn, T. M. and S. M. Anderson, Automated DNA sequencing, in *Molecular Microbiology, Diagnostic Principles and Practice*, D. L. Persing, F. C. Tenover, J. Versalovic, Y.-W. Tang, E. R. Unger, D. A. Relman, and T. J. White, eds., ASM Press, Washington, DC, 2004, pp. 153–159.
2. Heid, C. A., J. Stevens, K. J. Livak, and P. M. Williams, Real time quantitative PCR, *Genome Res.* **6**:986–994 (1996).
3. Shepley, D. P. and D. M. Wolk, Quantitative molecular methods: Results standardization, interpretation and laboratory quality control, in *Molecular Microbiology, Diagnostic Principles and Practice*, D. L. Persing, F. C. Tenover, J. Versalovic, Y.-W. Tang, E. R. Unger, D. A. Relman, and T. J. White, eds., ASM Press, Washington, DC, 2004, pp. 95–129.
4. Hayden, R. T., In vitro nucleic acid amplification techniques, in *Molecular Microbiology, Diagnostic Principles and Practice*, D. L. Persing, F. C. Tenover, J. Versalovic,

Y.-W. Tang, E. R. Unger, D. A. Relman, and T. J. White, eds., ASM Press, Washington, DC, 2004, pp. 43–69.

5. Manoni, M., R. Pergolizzi, M. Luzzana, and G. De Bellis, Dideoxy linear PCR on a commercial fluorescent automated DNA sequencer, *Biotechniques* **12**:48–53 (1992).
6. Mitchel, L. G. and C. R. Merril, 1989. Affinity generation of single stranded DNA for dideoxy sequencing following the polymerase chain reaction, *Anal. Biochem.* **178**: 239–242 (1989).
7. Mullis, K. B. and F. A. Faloona, Specific synthesis of DNA in vitro via a polymerase-catalyzed chain reaction, *Meth. Enzymol.* **155**:335–350 (1987).
8. Saiki, R. K., S. Scharf, F. Faloona, K. M. Mullis, G. T. Horn, H. A. Erlich, and N. Arnheim, Enzymatic amplification of beta-globin genomic sequences and restriction site analysis for diagnosis of sickle cell anemia, *Science* **230**:1350–1354 (1985).
9. Fredericks, D. N., and D. A. Relman, Applications of polymerase chain reaction to the diagnosis of infectious diseases, *Clin. Infect. Dis.* **29**:475–486 (1999).
10. White, T. J., R. Madej, and D. H. Persing, The polymerase chain reaction: Clinical applications, *Adv. Clin. Chem.* **29**:161–196 (1992).
11. Birch, D. E., Simplified hot start PCR, *Nature* **381**:445–446 (1996).
12. Sharkey, D. J., E. R. Scalice, K. G. Christy, Jr., S. M. Atwood, and J. L. Diass, Antibodies as thermal labile switches: High temperature triggering for the polymerase chain reaction, *Bio/Technology* **12**:506–509 (1994).
13. Haugland, R. A., M. Varma, L. J. Wymer, and S. Vesper. Quantitative PCR analysis of selected *Aspergillus, Penicillium* and *Paecilomyces* species, *System. Appl. Microbiol.* **27**:192–210 (2004).
14. Uhl, J. R., and F. R. Cockerill III, The fluorescence resonance transfer system, in *Molecular Microbiology, Diagnostic Principles and Practice*, D. L. Persing, F. C. Tenover, J. Versalovic, Y.-W. Tang, E. R. Unger, D. A. Relman, and T. J. White, eds., ASM Press, Washington, DC, 2004, pp. 295–306.
15. Livak, K. J., S. J. A. Flood, J. Marmaro, W. Giusti, and K. Deetz, Oligonucleotides with fluorescent dyes at opposite ends provide a quenched probe system useful for detecting PCR product and nucleic acid hybridization, *PCR Meth. Appl.* **5**:57–362 (1995).
16. Tyagi, S. and F. R. Kramer, Molecular beacons: Probes that fluoresce upon hybridization, *Nat. Biotechnol.* **14**:303–308 (1996).
17. Thelwell, N., S. Millington, A. Solinas, J. Booth, and T. Brown, Mode of action and application of Scorpion primers to mutation detection, *Nucleic Acids Res.* **28**:3752–3761 (2000).
18. Wittwer, C. T., M. G. Herrmann, A. A. Moss, and R. P. Rassmussen, Continuous fluorescence monitoring of rapid cycle DNA amplification, *Biotechniques* **22**:130–138 (1997).
19. Wittwer, C. T. and N. Kusukawa, Real-time PCR, in *Molecular Microbiology, Diagnostic Pricinples and Practice*, D. L. Persing, F. C. Tenover, J. Versalovic, Y.-W. Tang, E. R. Unger, D. A. Relman, and T. J. White, eds., ASM Press, Washington, DC, 2004, pp. 71–84.
20. Wu, D. Y. and R. B. Wallace, The ligation amplification reaction (LAR)—amplification of specific DNA sequences using sequential rounds of template-dependent ligation, *Genomics* **4**:560–569 (1989).
21. Walker, G. T., M. S. Frasier, J. L. Schram, M. C. Little, J. G. Nadeau, and D. P. Malinowski, Strand-displacement amplification—an isothermal in vitro DNA amplification technique, *Nucleic Acids Res.* **20**:1691–1696 (1992).

22. Walker, G. T., M. C. Little, J. G. Nadeau, and D. D. Shank, Isothermal in vitro amplification of DNA by a restriction enzyme/DNA polymerase system, *Proc. Natl. Acad. Sci. USA* **89**:392–396 (1992).
23. Lui, D., S. L. Daubendiek, M. A. Zillman, K. Ryan, and E. T. Kool, Rolling circle DNA synthesis: Small circular oligonucleotides as efficient templates for DNA polymerase, *J. Am. Chem. Soc.* **118**:1587–1594 (1996).
24. Nilsson, M., H. Malmgren, M. Samiotaki, M. Kwiatkowski, B. P. Choudhary, and U. Landegren, Padlock probes: Circularizing oligonucleotides for localized DNA detection, *Science* **265**:2085–2088 (1994).
25. Compton, J., Nucleic acid sequence-based amplification, *Nature* **350**:91–92 (1991).
26. Fahy, E., D. Y. Kwoh, and T. R. Gingeras, Self-sustained sequence replication (3SR): An isothermal transcription-based amplification system alternative to PCR, *PCR Meth. Appl.* **1**:25–33 (1991).
27. Kwoh, D. Y., G. R. Davis, K. M. Whitfield, H. L. Chapelle, L. J. DiMichelle, and T. R. Gingeras, Transcription-based amplification system and detection of amplified human immunodeficiency virus type 1 with a bead-based sandwich hybridization format, *Proc. Natl. Acad. Sci. USA* **86**:1173–1177 (1989).
28. Wolk, D., P. S. Mitchell, and R. Patel, Principles of molecular microbiology testing methods, *Infect. Dis. Clin. N. Am.* **15**:1157–1204 (2001).
29. Hawkins, T. L., J. C. Detter, and P. M. Richardson, Whole genome amplification—applications and advances, *Curr. Opin. Biotechnol.* **13**:65–67 (2002).
30. Sanchez-Pescador, R., M. S. Stempien, and M. S. Urdea, Rapid chemiluminescent nucleic acid assays for detection of TEM-1 beta-lactamase-mediated penicillin resistance in *Neisseria gonorrhoeae* and other bacteria, *J. Clin. Microbiol.* **26**: 1934–1938 (1988).
31. Kramer, F. R. and P. N. Lizardi, Replicatable RNA reporters, *Nature* **339**:401–402 (1989).
32. Lomeli, H., S. Tyagi, C. G. Pritchard, P. M. Lizardi, and F. R. Kramer, Quantitative assays based on the use of replicatable hybridization probes, *Clin. Chem.* **35**: 1826–1831 (1989).
33. Eis, P. S., M. C. Olson, T. Takova, M. L. Curtis, S. M. Olson, T. L. Vener, H. S. Ip, K. L. Vednik, C. Tr. Bartholomay, H. T. Allawi, W. P. Ma, J. G. Hall, M. D. Morin, T. H. Rushmore, V. I. Lyamichev, and R. W. Kwiatkowski, An invasive cleavage assay for direct quantification of specific RNAs, *Nat. Biotechnol.* **19**:673–676 (2001).
34. Duck, P., G. Alvarado-Urbina, B. Burdick, and B. Collier, Probe amplifier system based on chimeric cycling oligonucleotides, *BioTechniques* **9**:142–148 (1990).
35. Li, J. and B. A. Hanna, DNA probes for culture confirmation and direct detection of bacterial infections: a review of technology, in *Molecular Microbiology, Diagnostic Principles and Practice*, D. L. Persing, F. C. Tenover, J. Versalovic, Y.-W. Tang, E. R. Unger, D. A. Relman, and T. J. White, eds., ASM Press, Washington, DC, 2004, pp. 19–26.
36. Juretschko, S., A. M. Buccat, and T. R. Fritsche, Applications of fluorescence in situ hybridization in diagnostic microbiology, in *Molecular Microbiology, Diagnostic Principles and Practice*, D. L. Persing, F. C. Tenover, J. Versalovic, Y.-W. Tang, E. R. Unger, D. A. Relman, and T. J. White, eds., ASM Press, Washington, DC, 2004, pp. 3–18.
37. Dunbar, S. A., Applications of Luminex XMAP technology for rapid, high-throughput multiplexed nucleic acid detection, *Clin. Chim. Acta* **363**:71–82 (2006).

38. Stöver, A. G., E. Jeffrey, J. Xu, and D. Persing, Hybridization array technology, in *Molecular Microbiology, Diagnostic Principles and Practice*, D. L. Persing, F. C. Tenover, J. Versalovic, Y.-W. Tang, E. R. Unger, D. A. Relman, and T. J. White, eds., ASM Press, Washington, DC, 2004, pp. 619–639.

39. Anthony, R. M., T. M. Brown, and G. L. French, Rapid diagnosis of bacteremia by universal amplification of 23S ribosomal DNA followed by hybridization to an oligonucleotide array, *J. Clin Microbiol.* **38**:781–788 (2000).

40. Kelly, J. J., S. Siripong, J. McCormack, L. R. Janus, H. Urakawa, S. El Fantroussi, P. A. Noble, L. Sappelsa, B. E. Rittman, and D. A. Stahl, DNA microarray detection of nitrifying bacterial 16S rRNA in wastewater treatment plant samples, *Water Res.* **39**:3229–3238 (2005).

41. Kolbert, C. P., P. N. Rys, M. Hopkins, D. T. Lynch, J. J. Germer, C. E. O'Sullivan, A. Trampuz, and R. Patel, 16S ribosomal DNA sequence analysis for identification of bacteria in a clinical microbiology laboratory, in *Molecular Microbiology, Diagnostic Principles and Practice*, D. L. Persing, F. C. Tenover, J. Versalovic, Y.-W. Tang, E. R. Unger, D. A. Relman, and T. J. White, eds., ASM Press, Washington DC, 2004, pp. 361–377.

42. Hall, L., K. A. Doerr, S. L. Wohlfiel, and G. D. Roberts, Evaluation of the MicroSeq system for identification of mycobacteria by 16S ribosomal DNA sequencing and its integration into a routine clinical mycobacteriology laboratory, *J. Clin. Microbiol.* **41**:1447–1453 (2003).

43. Goering, R. V., Pulsed-field gel electrophoresis, in *Molecular Microbiology, Diagnostic Principles and Practice*, D. L. Persing, F. C. Tenover, J. Versalovic, Y.-W. Tang, E. R. Unger, D. A. Relman, and T. J. White, eds., ASM Press, Washington, DC, 2004, pp. 185–196.

44. Pfaller, M. A. and R. J. Hollis, Automated ribotyping, in *Molecular Microbiology, Diagnostic Principles and Practice*, D. L. Persing, F. C. Tenover, J. Versalovic, Y.-W. Tang, E. R. Unger, D. A. Relman, and T. J. White, eds., ASM Press, Washington, DC, 2004, pp. 245–258.

45. Caetano-Annolés, G., B. J. Bassam, and P. M. Gresshoff, DNA amplification using very short arbitrary oligonucleotide primers, *Bio/Technology* **9**:553–556 (1991).

46. Welsh J. and M. McClelland, Fingerprinting genomes using PCR with arbitrary primers, *Nucleic Acids Res.* **18**:7213–7218 (1990).

47. Williams, J. G. K., A. R. Kubilek, K. J. Livak, J. A. Rafalski, and S. V. Tingey, DNA polymorphisms amplified by arbitrary primers are useful as genetic markers, *Nucleic Acids Res.* **18**:6831–6535 (1990).

48. Versalovic, J., M. Schneider, F. J. de Bruijn, and J. R. Lupski, Genomic fingerprinting of bacteria using repetitive sequence based PCR (rep-PCR), *Meth. Cell. Mol. Biol.* **5**:25–40 (1994).

49. Versalovic, J., T. Koeuth, and J. R. Lupski, Distribution of repetitive DNA sequences in eubacteria and applications to fingerprinting of bacterial genomes, *Nucleic Acids Res.* **19**:6823–6831 (1991).

50. Meyer, W., E. Lichtfeldt, K. Kuhls, E. Z. Freedman, T. Borner, and T. G. Mitchell, DNA- and PCR-fingerprinting in fungi, *EXS* **67**:311–320 (1993).

51. Van Belkum, A., S. Scherer, L. van Alphen, and H. Verbrugh, Short-sequence DNA repeats in prokaryotic genomes, *Microbiol. Mol. Biol. Rev.* **62**:275–293 (1998).

52. Savelkoul, P. H. M., H. Aarts, B. Duim, L. Dijkshoorne, J. De Haas, M. Otsen, J. Rademaker, L. Schouls, and J. A. Lenstra, Amplified fragment length polymorphism (AFLP): The state of the art, *J. Clin. Microbiol.* **24**:227–231 (1999).
53. Vos, P., R. Hogers, M. Bleeker, M. Reijans, T. van de Lee, M. Hornes, A. Frijters, J. Pot, J. Peleman, J. Kuiper, and M. Zabeau, AFLP: A new technique for DNA fingerprinting, *Nucleic Acids Res.* **23**:4407–4414 (1995).
54. Rademaker, J. L. W. and P. Savelkoul, PCR amplification-based microbial typing, in *Diagnostic Molecular Microbiology: Principles and Applications*, D. H. Persing, T. F. Smith, F. C. Tenover, and T. J. White, eds., ASM Press, Washington, DC, 2004, pp. 197–221.
55. Gaia, V., N. K. Fry, T. G. Harrison, and R. Peduzzini, Sequence-based typing of *Legionella pneumophila* serogroup 1 offers the potential for true portability in Legionellosis outbreak investigations, *J. Clin. Microbiol.* **41**:2932–2939 (2003).
56. Maiden, M. C. J., J. A. Bygraves, E. Feil, G. Morelli, J. E. Russel, R. Urwin, Q. Zhang, J. Zhou, K. Zurth, D. A. Caugant, I. M. Feavers, M. Achtmen, and B. G. Spratt, Multilocus sequence typing: A portable approach to the identification of clones within populations of pathogenic microorganisms, *Proc. Natl. Acad. Sci. USA* **95**:3140–3145 (1998).
57. O'Donnell, K., D. A. Sutton, M. G. Rinaldi, K. C. Magnon, P. A. Cox, S. G. Revankar, S. Sanche, D. M. Geiser, J. H. Juba, J. Van Burik, A. Padhye, E. J. Anaissie, A. Francesconi, T. J. Walsh, and J. S. Robinson, Genetic diversity of human pathogenic members of the *Fusarium oxysporum* complex inferred from multilocus DNA sequence data and amplified fragment length polymorphism analyses: Evidence for the recent dispersion of a geographically widespread clonal lineage and nosocomial origin, *J. Clin. Microbiol.* **42**:5109–5120 (2004).
58. Chen, S. C. A., C. L. Halliday, and W. Meyer, A review of nucleic acid-based diagnostic tests for systemic mycoses with an emphasis on polymerase chain reaction-based assays, *Med. Mycol.* **40**:333–357 (2002).
59. Post, J. C. and G. D. Ehrlich, The impact of the polymerase chain reaction in clinical medicine, *JAMA (J. Am. Med. Assoc.)* **283**:1544–1546 (2000).
60. Fiore, A. E., J. C. Butler, T. G. Emori, and R. P. Gaynes, A survey of methods used to detect nosocomial legionellosis among participants in the National Nosocomial Infections Surveillance System, *Infect. Control Hosp. Epidemiol.* **20**:412–416 (1999).
61. Gerberding, J. L., Nosocomial transmission of opportunistic infections, *Infect. Control Hosp. Epidemiol.* **19**:574–577 (1998).
62. Miyamota, H., H. Yamamoto, K. Arima, J. Fujii, K. Maratu, K. Izu, T. Shiomori, and S. Yoshida, Development of a new seminested PCR method for detection of *Legionella* species and its application to surveillance of legionellae in hospital cooling tower water, *Appl. Environ. Microbiol.* **63**:2489–2494 (1997).
63. Morrison, J., C. Yang, K.-T. Lin, R. A. Haugland, A. N. Neely, and S. J. Vesper, Monitoring *Aspergillus* species by quantitative PCR during construction of a multi-story hospital building, *J. Hosp. Infect.* **57**:85–87 (2004).
64. Neely, A., V. Gallardo, E. Barth, R. A. Haugland, G. D. Warden, and S. J. Vesper, Rapid monitoring by quantitative polymerase chain reaction for pathogenic *Aspergillus* during carpet removal from a hospital, *Infect. Control Hosp. Epidemiol.* **25**:350–352 (2004).
65. Schafer, M. P., K. F. Martinez, and E. S. Mathews, Rapid detection and determination of the aerodynamic size range of airborne mycobacteria associated with whirlpools, *Appl. Occup Environ. Hyg.* **18**:41–50 (2003).

66. Wellinghausen, N., C. Frost, and R. Marre, Detection of legionellae in hospital water samples by quantitative real-time LightCycler PCR, *Appl. Environ Microbiol.* **67**:3985–3993 (2001).

67. Aintablian, N., P. Walpita, and M. H. Sawyer, Detection of *Bordetella pertussis* and respiratory synctial virus in air samples from hospital rooms, *Infect. Control Hosp. Epidemiol.* **19**:918–923 (1998).

68. Bartlett, M. S., S. H. Vermund, R. Jacobs, P. J. Durant, M. H. Shaw, J. W. Smith, X. Tang, J. Lu, B. Li, S. Jin, and C. Lee, Detection of *Pneumocystis carinii* DNA in air samples: Likely environmental risk to susceptible persons, *J. Clin. Microbiol.* **35**:2511–2513 (1997).

69. Mastorides, S. M., R. L. Oehler, J. N. Greene, J. T. Sinnott, M. Kranik, and R. L. Sandin, The detection of airborne *Mycobacterium tuberculosis* using micropore membrane air sampling and polymerase chain reaction, *Chest* **115**:19–25 (1999).

70. Olsson, M., C. Lidman, S. Latouche, A. Björkman, P. Roux, E. Linder, and M. Wahlgren, Identification of *Pneumocystis carinii* f. sp. *Hominis* gene sequences in filtered air in hospital environments, *J. Clin. Microbiol.* **36**:1737–1740 (1998).

71. Vadrot, C., V. Bex, A. Mouilleseaux, F. Squinazi, and J. C. Darbord, Detection of *Mycobacterium tuberculosis* complex by PCR in hospital air samples, *J. Hosp. Infect.* **58**:262–267 (2004).

72. Wan, G. H. and Y. H. Tsai, Polymerase chain reaction used for the detection of airborne *Mycobacterium tuberculosis* in health care settings, *Am. J. Infect. Control* **32**:17–22 (2004).

73. Henchal, E. A., J. D. Teska, G. V. Ludwig, D. R. Shoemaker, and J. W. Ezzell, Current laboratory methods for biological threat agent identification, *Clin. Lab. Med.* **21**:661–678 (2001).

74. Higgins, J. A., M. Cooper, L. Schroeder-Tucker, S. Black, D. Miller, J. S. Karns, E. Manthey, R. Breeze, and M. L. Perdue, A field investigation of *Bacillus anthracis* contamination of U.S. Department of Agriculture and other Washington D.C. buildings during the anthrax attack of 2001, *Appl. Environ. Microbiol.* **69**:593–599 (2003).

75. Husman, T., Health effects of indoor-air microorganisms, *Scand J. Work Environ. Health* **22**:5–13 (1996).

76. Nevalainen, A. and M. Seuri, Of microbes and men. *Indoor Air* **15**(Suppl. 9):58–64 (2005).

77. Ammann, H. H., Indoor mold contamination—a threat to health? *J. Environ Health* **64**:43–44 (2002).

78. Ammann, H. H., Is indoor mold contamination a threat to health? Part Two. *J. Environ Health* **66**:47–49 (2003).

79. Peltola, J., M. A. Andersson, T. Haahtela, H. Mussalo-Rauhamaa, F. A. Rainey, R. M. Kroppenstedt, R. A. Sampson, and M. S. Salkinoja-Salonen, Toxic metabolite producing bacteria and fungus in an indoor environment, *Appl. Environ. Microbiol.* **67**:3269–3274 (2001).

80. Sorenson, W. G., Mycotoxins: Toxic metabolites of fungi, in *Fungal Infections and Immune Response*, J. W. Murphy, ed., Plenum Press, New York, 1993, pp. 469–491.

81. American College of Occupational and Environmental Medicine, *Adverse Human Health Effects Associated with Molds in the Indoor Environment*, ACOEM Evidence-based statement. (http://www.acoem.org/guidelines/article.asp?ID=52), 2002.

82. Haugland, R. A., S. J. Vesper, and L. J. Wymer, Quantitative measurement of *Stachybotrys chartarum* conidia using real time detection of PCR products with the TaqMan™ fluorogenic probe system, *Mol. Cell. Probes* **13**:329–340 (1999).

83. Meklin, T., R. A. Haugland, T. Reponen, M. Varma, Z. Lummus, D. Bernstein, L. J. Wymer, and S. J. Vesper, Quantititive PCR analysis of house dust can reveal abnormal mold conditions, *J. Environ. Monit.* **6**:615–620 (2004).

84. Roe, J., R. A. Haugland, S. J. Vesper, and L. J. Wymer, Quantification of *Stachybotrys chartarum* conidia in indoor dust using real time, fluorescent probe-based detection of PCR products, *J. Expos. Anal. Environ. Epidemiol.* **11**:12–20 (2001).

85. Wu, Z., G. Blomquist, S. Westermark, and X. Wang, Application of PCR and probe hybridization techniques in detection of airborne fungal spores in environmental samples, *J. Environ. Monit.* **4**:673–678 (2002).

86. Zhou, G., W. Z. Whong, T. Ong, and B. Chen, Development of a fungus-specific assay for detecting low level fungi in an indoor environment, *Mol. Cell. Probes* **14**:339–348 (2000).

87. Vesper, S. J., M. Varma, L. J. Wymer, D. G. Dearborn, J. Sobolewski, and R. A. Haugland, Quantitative polymerase chain reaction analysis of fungi in dust from homes of infants who develop idiopathic pulmonary hemorrhaging, *J. Occup. Environ. Med.* **46**:596–601 (2004).

88. Rintala, H., A. Hyvarinen, L. Paulin, and A. Nevalainen, 2004. Detection of streptomycetes in house dust—comparison of culture and PCR methods, *Indoor Air* **14**:112–119 (2004).

89. Belgrader, P., C. J. Elkin, S. B. Brown, S. N. Nasarabadi, R. G. Langlois, F. P. Milanovich, B. W. Colston Jr., and G. D. Marshall, A reusable flow-through polymerase chain reaction instrument for the continuous monitoring of infectious biological agents, *Anal. Chem.* **75**:3446–3450 (2003).

90. Brinkman, N. E., R. A. Haugland, L. J. Wymer, M. Byappanahalli, R. L. Whitman, and S. J. Vesper, Evaluation of a rapid, quantitative real-time PCR method for cellular enumeration of pathogenic *Candida* species in water, *Appl. Environ. Microbiol.* **69**:1775–1782 (2003).

91. Buttner, M. P., P. Cruz, and L. D. Stetzenbach, Enhanced detection of surface-associated bacteria in indoor environments by quantitative PCR, *Appl. Environ. Microbiol.* **67**:2567–2570 (2001).

92. Dean, T. R., D. Betancourt, and M. Y. Menetrez, A rapid extraction method for PCR identification of fungal air contaminants, *J. Microbiol. Meth.* **56**:431–434 (2004).

93. Haugland, R. A., N. Brinkman, and S. J. Vesper, Evaluation of rapid DNA extraction methods for the quantitative detection of fungi using real time PCR analysis, *J. Microbiol. Meth.* **50**:319–323 (2002).

94. Ishimatsu, S., H. Miyamoto, H. Hori, I. Tanaka, and S. Yoshida, Sampling and detection of *Legionella pneumophila* aerosols generated from an industrial cooling tower, *Ann. Occup. Hyg.* **45**:421–427 (2001).

95. Keswani, J., M. L. Kashon, and B. T. Chen, Evaluation of interference to conventional and real-time PCR for detection and quantification of fungi in dust, *J. Environ. Monit.* **7**:311–318 (2005).

96. Dean, T. R., B. Roop, D. Betancourt, and M. Y. Menetrez, 2005. A simple multiplex polymerase chain reaction assay for the identification of four environmentally relevant fungal contaminants, *J. Microbiol. Meth.* **61**:9–16 (2005).

97. Rintalla, H., A. Nevalainen, E. Ronka, and M. Suutari, PCR primers targeting the 16S rRNA gene for the specific detection of streptomycetes, *Mol. Cell. Probes* **15**:337–347 (2001).

98. Wu, Z., X. Wang, and G. Blomquist, Evaluation of PCR primers and PCR conditions for specific detection of common airborne fungi, *J. Environ. Monit.* **4**:377–382 (2002).

99. Wu, Z., Y. Tsumura, G. Blomquist, and X. Wang, 18S rRNA gene variation among common airborne fungi, and development of specific oligonucleotide probes for the detection of fungal isolates, *Appl. Environ. Microbiol.* **69**:5389–5397 (2003).

100. Zeng, Q., Å. Rasmuson-Lestander, and X. Wang, Extensive set of mitochondrial LSU rDNA-based oligonucleotide probes for the detection of common airborne fungi, *FEMS Microbiol. Lett.* **237**:79–87 (2004).

101. Geisen, R., Z. Mayer, A. Karolewiez, and P. Farber, Development of real time PCR system for detection of *Penicillium nordicum* and for monitoring ochratoxin A production in foods by targeting the ochratoxin polyketide synthase gene, *Syst. Appl. Microbiol.* **27**:501–507 (2004).

102. Mule, G., M. T. Gonzalez-Jaen, L. Hornok, P. Nicholson, and C. Waalwijk, Advances in molecular diagnosis of toxigenic *Fusarium* species: A review, *Food Addit. Contam.* **22**:316–323 (2005).

103. Niessen, L., H. Schmidt, and R. F. Vogel, The use of tri5 gene sequences for PCR detection and taxonomy of trichothecene-producing species of *Fusarium* section *Sporotrichiella, Internatl. J. Food Microbiol.* **95**:305–319 (2004).

104. Scherm, B., M. Palomba, D. Serra, A. Marcello, and Q. Migheli, Detection of transcripts of the aflatoxin genes aflD, aflO and aflP by reverse transcription-polymerase chain reaction allows differentiation of aflatoxin-producing and non-producing isolates of *Aspergillus flavus* and *Aspergillus parasiticus, Internatl. J. Food Microbiol.* **98**:201–210 (2005).

105. Nogva, H. K., S. M. Dromtorp, H. Nissen, and K. Rudi, Ethidium monoazide for DNA-based differentiation of viable and dead bacteria by 5′-nuclease PCR, *Biotechniques* **34**:804–813 (2003).

106. Rudi, K., B. Moen, S. M. Dromtorp, and A. L. Holck, Use of ethidium monoazide and PCR in combination for quantification of viable and dead cells in complex samples, *Appl. Environ. Microbiol.* **71**:1018–1024 (2005).

107. Anaissie, E. J., R. T. Kucher, J. H. Rex, A. Francesconi, M. Kasai, F. C. Müller, M. Loranzo-Chiu, R. C. Sumerbell, M. C. Dignani, S. J. Chanock, and T. J. Walsh, Fusariosis associated with *Fusarium* species colonization of a hospital water system: a new paradigm for the epidemiology of opportunistic mold infections, *Clin. Infect. Dis.* **33**:1871–1878 (2001).

108. Anaissie, E. J., S. L. Stratton, M. C. Dignani, R. C. Sumerbell, J. H. Rex, T. P. Monson, T. Spencer, M. Kasai, A. Francesconi, and T. J. Walsh, Pathogenic *Aspergillus* species recovered from a hospital water system: A 3-year prospective study, *Clin. Infect. Dis.* **34**:780–789 (2002).

109. Buffington, J., R. Reporter, B. A. Lasker, M. M. McNeil, J. M. Lanson, L. A. Ross, L. Mascola, and W. R. Jarvis, Investigation of an epidemic of invasive aspergillosis: utility of molecular typing with the use of random amplified polymorphic DNA probes, *Pediatr. Infect. Dis. J.* **13**:386–393 (1994).

110. El Sahly, H. M., E. Septimus, H. Soini, J. Septimus, R. J. Wallace, X. Pan, N. Williams-Bouyer, J. M. Musser, and E. A. Graviss, *Mycobacterium simiae*

pseudo-outbreak resulting from a contaminated hospital water supply in Houston, Texas, *Clin. Infect Dis.* **35**:802–807 (2002).

111. Girardin, H., J. Sarfati, F. Traoré, J. Dupouy Camet, F. Derouin, and J. P. Latgé, Molecular epidemiology of nosocomial invasive aspergillosis, *J. Clin. Microbiol.* **32**:684–690 (1994).

112. Lass-Florl, C., P. Rath, D. Niederwieser, G. Kofler, R. Wurzner, A. Krezy, and M. P. Dierich, *Aspergillus terreus* infections in haematological malignancies: Molecular epidemiology suggests association with in-hospital plants, *J. Hosp. Infect.* **46**:31–35 (2000).

113. Loudon, K. W., A. P. Coke, J. P. Burnie, A. J. Shaw, B. A. Oppenheim, and C. Q. Morris, Kitchens as a source of *Aspergillus niger* infection, *J. Hosp. Infect.* **32**:191–198 (1996).

114. Menotti, J., J. Waller, O. Meunier, V. Letscher-Bru, R. Herbrecht, and E. Candolfi, Epidemiological study of invasive pulmonary aspergillosis in a haematology unit by molecular typing of environmental and patient isolates of *Aspergillus fumigatus*, *J. Hosp. Infect.* **60**:61–68 (2005).

115. Moore, J. E., B. McIlhatton, J. Buchanon, D. Gilpin, A. Shaw, V. Hall, P. G. Murphy, and J. S. Elborn, Occurrence of *Burholderia cepacia* in the hospital environment, *Ir. J. Med. Sci.* **171**:131–133 (2002).

116. Morris, T., S. M. Brechner, D. Fitzsimmons, A. Durbin, R. D. Arbet, and J. N. Maslow, A pseudoepidemic due to laboratory contamination deciphered by molecular analyses, *Infect. Control Hosp. Epidemiol.* **16**:82–87 (1995).

117. Squier, C., V. L. Yu, and J. E. Stoudt, Waterborne nosocomial infections, *Curr. Infect. Dis. Rep.* **2**:490–496 (2000).

118. Warris, A., C. H. W. Klaassen, J. F. G. M. Meis, M. T. de Ruiter, H. A. de Valk, T. G. Abrahamsen, P. Gaustad, and P. E. Verweij, Molecular epidemiology of *Aspergillus fumigatus* isolates recovered from water, air, and patients shows two clusters of genetically distinct strains, *J. Clin. Microbiol.* **41**:4101–4106 (2003).

119. Debeaupuis, J.-P., J. Sarfati, V. Chazalet, and J.-P. Latgé, Genetic diversity among clinical and environmental isolates of *Aspergillus fumigatus*, *Infect. Immun.* **65**:3080–3085 (1997).

120. Symoens, F., J. Burnod, B. Lebeau, M. A. Viviani, M. A. Piens, A. M. Tortorano, N. Nolard, F. Chapuis, and R. Grillot, Hospital-acquired *Aspergillus fumigatus* infection: Can molecular typing methods identify an environmental source? *J. Hosp. Infect.* **52**:60–67 (2002).

121. Angenent, L. T., S. T. Kelley, A. St. Amand, N. R. Pace, and M. T. Hernandez, Molecular identification of potential pathogens in water and air of a hospital therapy pool, *Proc. Natl. Acad. Sci. USA* **102**:4860–4865 (2005).

122. Rintalla, H., A. Nevalainen, and M. Suutari, Diversity of streptomyces in water-damaged building materials based on 16S rDNA sequences, *Lett. Appl. Microbiol.* **34**:439–443 (2002).

123. Scott, J., W. A. Untereiner, B. Wong, N. Strauss, and D. Malloch, Genotypic variation in *Penicillium chyrsogenum* from indoor environments, *Mycologia* **96**:1095–1105 (2004).

124. Borst, A., A. T. Box, and A. C. Fluit, False-positive results and contamination in nucleic acid amplification assays: suggestions for a prevent and destroy strategy, *Eur. J. Clin. Microbiol. Infect. Dis.* **23**:289–299 (2004).

125. Burkhardt, H. J., Standardization and quality control of PCR analyses, *Clin. Chem. Lab Med.* **38**:87–91 (2000).

126. Mitchell, P. S., J. J. Germer, and R. Patel, Nucleic acid amplification methods: laboratory design and operations, in *Molecular Microbiology, Diagnostic Principles and Practice*, D. L. Persing, F. C. Tenover, J. Versalovic, Y.-W. Tang, E. R. Unger, D. A. Relman, and T. J. White, eds., ASM Press, Washington, DC, 2004, pp. 85–93.

127. Dubois, D. B. and J. T. Brown, Laboratory controls and standards, in *Molecular Microbiology, Diagnostic Principles and Practice* D. L. Persing, F. C. Tenover, J. Versalovic, Y.-W. Tang, E. R. Unger, D. A. Relman, and T. J. White, eds., ASM Press, Washington, DC, 2004, pp. 697–703.

128. Hoorfar, J., B. Malorny, A. Abdulmawjood, N. Cook, M. Wagner, and P. Fach, Practical considerations in design of internal amplification controls for diagnostic PCR assays, *J. Clin. Microbiol.* **42**:1863–1868 (2004).

129. Myers, T. W. and D. H. Gelfand, Reverse transcription and DNA amplification by a *Thermus thermophilus* DNA polymerase, *Biochemistry* **30**:7661–7666 (1991).

130. Haqqi, T. M., G. Sarkar, C. S. David, and S. S. Sommer, Specific amplification with PCR of a refractory segment of genomic DNA, *Nucleic Acids Res.* **16**:11844 (1988).

131. Chamberlain, J. S., R. A. Gibbs, J. E. Rainier, P. N. Ngyen, and C. T. Caskey, Detection screening of the Duchenne muscular dystrophy locus via a multiplex DNA amplification, *Nucleic Acids Res.* **16**:11141–11156 (1988).

132. Relman, D. A., Universal 16S rDNA amplification and sequencing, in *Diagnostic Molecular Microbiology: Principles and Applications*, D. H. Persing, T. F. Smith, F. C. Tenover, and T. J. White, eds., ASM Press, Washington, DC, 1993, pp. 489–495.

133. White, T. J., T. Bruns, S. Lee, and J. W. Taylor, Amplification and direct sequencing of fungal ribosomal RNA genes for phylogenetics, in *PCR Protocols: A Guide to Methods and Applications*, M. A. Innis, D. H. Gelfand, J. J. Sninsky, and T. J. White, eds., Academic Press, New York, 1990, pp. 315–322.

134. Orlando, C., P. Pinzani, and M. Pazzagli, Developments in quantitative PCR, *Clin. Chem. Lab. Med.* **36**:255–269 (1998).

135. McDeavitt, J. J., P. S. Lees, W. G. Merz, and K. J. Schwab, Development of a method to detect and quantify *Aspergillus fumigatus* conidia by quantitative PCR for environmental air samples, *Mycopathologia* **158**:325–335 (2004).

136. Mishra, P. K., R. T. Fox, and A. Culham, Development of a PCR-based assay for rapid and reliable identification of pathogenic Fusaria, *FEMS Microbiol. Lett.* **218**:329–332 (2003).

137. Pascual, L., S. Perez-Luz, A. Amo, C. Moreno, D. Apraiz, and V. Catalan, Detection of *Legionella pneumophila* in bioaerosols by polymerase chain reaction, *Can. J. Microbiol.* **47**:341–347 (2001).

138. Maiwald, M., K. Kissel, S. Srimuang, M. von Knebel Doeberitz, and H. G. Sonntag, Comparison of polymerase chain reaction and conventional culture for the detection of legionellas in hospital water samples, *J. Appl. Bacteriol.* **76**:216–225 (1994).

139. Yanez, M. A., C. Carrasco-Serrano, V. M. Barbera, and V. Catalan, Quantitative detection of *Legionella pnuemophila* in water samples by immunomagnetic purification and real-time PCR amplification of the dotA gene, *Appl. Environ. Microbiol.* **71**:3433–3441 (2005).

140. Klingenberg, C., G. T. Glad, R. Olsvik, and T. Flaegstad, Rapid PCR detection of the methicillin resistance gene, mecA, on the hands of medical and non-medical personnel

and healthy children and on surfaces in a neonatal intensive care unit, *Scand J. Infect. Dis.* **33**:494–497 (2001).

141. Schafer, M. P., J. E. Fernback, and P. A. Jensen, Sampling and analytical method development for qualitative assessment of airborne mycobacterial species of the *Mycobacterium tuberculosis* complex, *Am. Indust. Hyg. Assoc. J.* **59**:540–546 (1998).
142. Maher, N., H. K. Dillon, S. H. Vermund, and T. R. Unnasch, Magnetic bead capture eliminates PCR inhibitors in samples collected from the airborne environment, permitting detection of *Pneumocystis carinii* DNA, *Appl. Environ. Microbiol.* **67**:449–452 (2001).
143. Phillipe, L., C. Rene, J. Guillot, M. Berthalemy, B. Polack, V. Laine, P. Lacube, R. Chermette, and P. Roux, Impaction versus filtration for the detection of *Pneumocystis carinii* DNA in air, *J. Eukaryot. Microbiol.* **46**:94S (1999).
144. Maher, N., H. K. Dillon, A. Awooda, J. H. Lee, S. H. Vermund, and T. R. Unnasch, A comparison of two surface sample collection devices for use in polymerase chain reaction based detection of *Pneumocystis carinii* in house dust, *Appl. Occup. Environ. Hyg.* **17**:416–423 (2002).
145. Nagao, K., T. Ota, A. Tanikawa, Y. Takae, T. Mori, S. Udagawa, and T. Nishikawa, Genetic identification and detection of human pathogenic *Rhizopus* species, a major mucormycosis agent, by multiplex PCR based on internal transcribed spacer region of rRNA gene, *J. Dermatol. Sci.* **39**:23–31 (2005).
146. Voigt, K., E. Cigelnik, and K. O'Donnell, Phylogeny and PCR identification of clinically important Zygomycetes based on nuclear ribosomal-DNA sequence data, *J. Clin. Microbiol.* **37**:3957–3964 (1999).

CHAPTER 8

WOOD IN THE BUILT ENVIRONMENT—CONDITIONS FOR MOLD AND DECAY

SUSAN E. ANAGNOST

8.1. INTRODUCTION

This chapter was written with the objective to answer the question "Why do fungi grow on wood and wood products?" More specifically, when are conditions within buildings conducive to the growth of wood-inhabiting fungi? Of most concern are two types of fungi found in the built environment: molds and wood decay fungi. Not only is the presence of mold in buildings that we inhabit unsightly; mold can emit unpleasant odor (volatile organic compounds) and can cause allergic reaction, infection, or other illness if inhaled, contacted, or ingested. Wood decay fungi can also exist in buildings under conditions of high wood moisture content. If left unchecked, decay fungi can cause significant destruction. The presence of mold or wood decay fungi in buildings is often a sign of a serious moisture intrusion problem.

Fungal spores are ubiquitous in air and settled on materials. However, certain basic requirements must be met in order for these spores to germinate and develop. Water, oxygen, and specific nutrients are essential and temperature can limit growth if out of the optimal range. Nutrients include simple sugars, or more complex compounds such as cellulose and hemicelluloses, and for white-rot fungi, in addition, phenolic compounds including wood lignin. Food requirements are easily met in wood or wood-based building materials.

8.2. MOLDS AND WOOD DECAY FUNGI

Many wood decay fungi (brown-rot and white-rot fungi) are in the taxonomic group Basidiomycetes, while molds include Hyphomycetes, Coelomycetes, Zygomycetes,

Sampling and Analysis of Indoor Microorganisms, Edited by Chin S. Yang and Patricia A. Heinsohn

TABLE 8.1. Fungal Taxonomic Groups that Contain Mold Fungi and Wood Decay Fungi

Molds or Fungi	Common Name	Taxonomic Name
Molds	—	Hyphomycetes
	—	Coelomycetes
	—	Zygomycetes
	—	Ascomycetes
Wood decay fungi	Brown-rot fungi	Basidiomycetes
	White-rot fungi	Basidiomycetes
	Soft-rot fungi	Hyphomycetes, Coelomycetes, Ascomycetes
Sapstain fungi	—	Ascomycetes
	—	Hyphomycetes
	—	Coelomycetes

or Ascomycetes (Table 8.1). The group *microfungi* refers to fungi having small (microscopic) sporocarps.[1] Microfungi include molds, sapstain fungi, and wood-decaying soft-rot fungi. When we talk about mold, we generally think of surface contamination, in the form of discoloration, either on cellulose-based materials like wood or wallpaper, or on dirt and organic debris that has accumulated on non-organic surfaces such as paint. Fungi that cause mold growth may grow not only on surfaces of building materials (Fig. 8.1) but also into porous materials such as wood if there is sufficient moisture (Fig. 8.2). Certain mold fungi, such as *Aureobasidium*

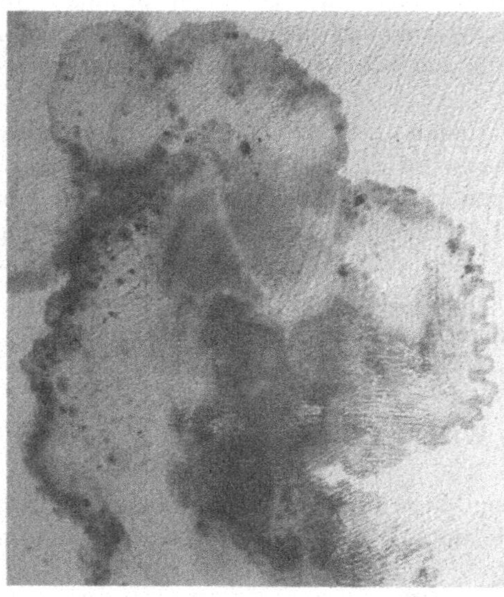

Fig. 8.1. Surface mold on a ceiling beneath a water leak in the roof.

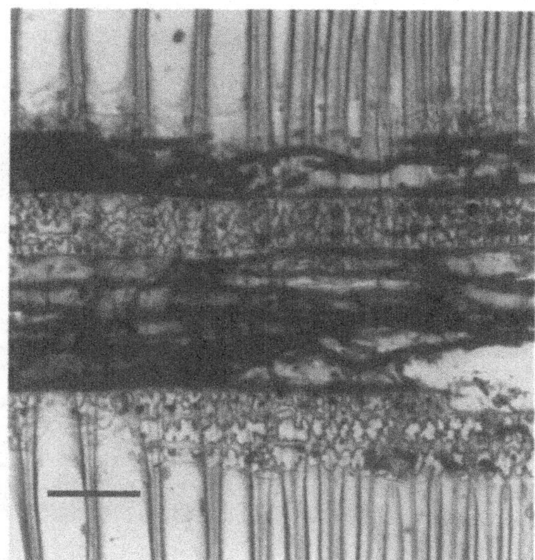

Fig. 8.2. Growth of *Aureobasidium pullulans* into the parenchyma cells of wood; this "mold" fungus discolors wood, but does not cause wood decay (bar = 50 μm).

pullulans, have the ability to grow into the parenchyma cells of wood, especially sapwood, where the contents of the wood cells, such as simple sugars and starch, support the growth of these fungi. It is generally assumed that molds will not attack cellulose, hemicellulose, and lignin found in the wood cell wall; thus the integrity of the wood remains unaffected. This is not always true, as many molds have exhibited the capability to cause soft rot in laboratory tests[2,3] (Fig. 8.3*a*,*b*) and have been recovered from utility poles with soft-rot damage.[4,5] Soft rot is a type of wood decay characterized by the formation of diamond-shaped cavities in the wood cell wall (Fig. 8.3*a*). Soft-rot fungi appear to tolerate high moisture levels, and soft rot is typically found in wood exposed outdoors, often in soil contact on structures such as utility poles, cooling towers, decks, and window sills. In relative terms it is a slow decay, characterized by a gradual erosion of wood surfaces that are exposed to moisture, such as utility poles in contact with soil. In a typical indoor environment significant deterioration by soft-rot fungi will be minimal, although the occurrence of soft-rot fungi in indoor air is very common. Soft-rot fungi include such ubiquitous species as *Alternaria alternata* and some species of *Aspergillus*, *Penicillium*, and *Cladosporium*.[2,5,6]

The wood decay fungi of greater concern in buildings, the brown-rot fungi and white-rot fungi, are in the taxonomic group Basidiomycetes. A large number of isolates of Basidiomycetes (870) were recovered from air samples taken within 74 of 103 homes sampled in Syracuse, New York,[7] and may be an important, yet often overlooked, factor in indoor air quality. Recovery of high indoor levels of basidiospores relative to outdoors may be an indication of a moisture or decay problem

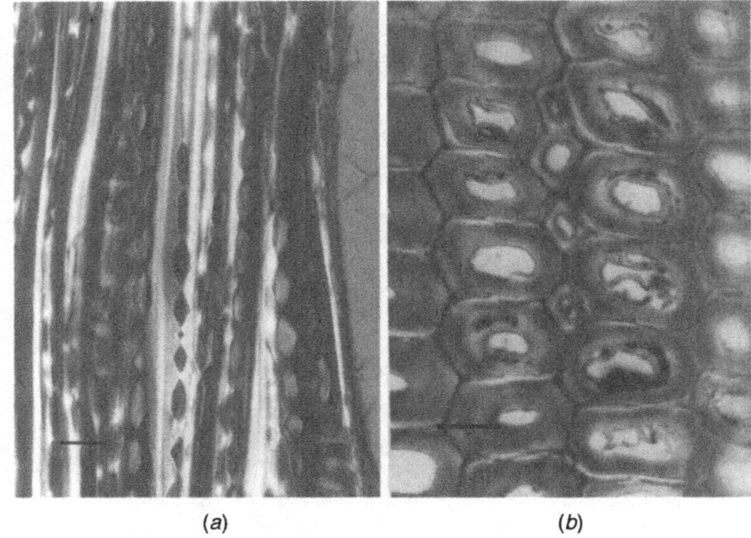

Fig. 8.3. (*a*) Soft-rot diamond-shaped cavities in the S2 layer of the wood secondary cell wall of yellow birch caused by *Phialophora melinii* (bar = 20 μm); (*b*) soft-rot cavities in a cross section of southern pine caused by the fungus *Alternaria alternata* (bar = 20 μm). (Figure 3*b* reprinted from *Wood Microbiology*, R. A. Zabel and J. J. Morrell, page 187, copyright 1992, with permission from Elsevier.)

within a building. At 5–10% weight loss from decay fungi, strength properties can be reduced from 20% to 80%, depending on the type of decay and the strength property measured.[8]

8.3. MOISTURE REQUIREMENTS FOR GROWTH OF MOLDS AND WOOD DECAY FUNGI

Excess moisture is the single most important factor that will ensure the growth of fungi in buildings. Although food sources for fungi are abundant in most buildings and include materials such as wood and wood-based composite products (Table 8.2), without sufficient moisture, fungi will fail to grow. In any climate the possibility for molds or wood decay fungi to grow and even thrive exists when a problem occurs in a building that allows excess moisture to accumulate. Typical indoor ambient conditions of relative humidity and temperature limit or prohibit the growth of many species of fungi, especially wood decay fungi. However, xerophilic molds prefer dryer conditions and can grow at ambient, although high, relative humidity conditions. Xerotolerant fungi, typically molds, have the ability to tolerate such drier conditions. Thus, in some locations, such as the southeastern United States or other locations where high relative humidity and temperature conditions exist, the

TABLE 8.2. Common Building Materials

Wood-based building materials
Wood lumber
Surface-treated wood
Preservative-treated wood
Laminated wood
Parquet
Cork
Plywood boards
Fiberboard
Sawdust insulation
Cellulose insulation
Paper
Wallpaper
Non-wood-based building materials
Gypsum board [$CaSO_4-2(H_2O)$, hydrated calcium sulfate]
Ceramics
Paints and glues
Plastics
Wood adhesives
Concrete

growth of mold will occur under ambient conditions. Also, in cases where temperature gradients exist, condensation can occur when moist, warm air contacts a cooler surface, thus providing an environment conducive to mold growth.

8.4. WATER AND WOOD—RELATIVE HUMIDITY VERSUS MOISTURE CONTENT

With regard to wood and decay or mold conditions, there are two ways to express moisture as it exists in buildings: relative humidity (RH) and moisture content (MC). Relative humidity is an expression of the amount of water vapor in air, while moisture content is the weight percent of water within a material, in this case wood. Relative humidity is the percent of moisture in air, expressed as $RH = 100(p/p_0)$, where p equals the actual amount of water present in air and p_0 equals the maximum water vapor that the air could hold at the same temperature.

Wood moisture content is the percent of water in wood expressed as a percent of the oven-dry weight of wood or [(weight of wood in the wet condition − weight of oven dry wood)/weight of oven-dry wood] in percentage.[9] In normal indoor ambient conditions, wood moisture content is low, generally below 20% (Table 8.3) and is constantly striving toward equilibrium with the surrounding air. At wood moisture content below the fiber saturation point (FSP), the relative humidity of the surrounding air controls wood moisture. Below FSP (below 20–30% MC)

TABLE 8.3. Calculated Moisture Content[a] of Wood at Various Temperature and Relative Humidity Levels

Relative Humidity (%)	Ambient Air Temperature (°F) 30	40	50	60	70	80	90	100	110	120	130
5	1.4	1.4	1.4	1.3	1.3	1.3	1.2	1.2	1.1	1.1	1.0
10	2.6	2.6	2.6	2.5	2.5	2.4	2.3	2.3	2.2	2.1	2.0
15	3.7	3.7	3.6	3.6	3.5	3.5	3.4	3.3	3.2	3.0	2.9
20	4.6	4.6	4.6	4.6	4.5	4.4	4.3	4.2	3.0	3.9	3.7
25	5.5	5.5	5.5	5.4	5.4	5.3	5.1	5.0	4.9	4.7	4.5
30	6.3	6.3	6.3	6.2	6.2	6.1	5.9	5.8	5.6	5.4	5.2
35	7.1	7.1	7.1	7.0	6.9	6.8	6.7	6.5	6.3	6.1	5.9
40	7.9	7.9	7.9	7.8	7.7	7.6	7.4	7.2	7.0	6.8	6.6
45	8.7	8.7	8.7	8.6	8.5	8.3	8.1	7.9	7.7	7.5	7.2
50	9.5	9.5	9.5	9.4	9.2	9.1	8.9	8.7	8.4	8.2	7.9
55	10.4	10.4	10.3	10.2	10.1	9.9	9.7	9.5	9.2	8.9	8.7
60	11.3	11.3	11.2	11.1	11.0	10.8	10.5	10.3	10.0	9.7	9.4
65	12.4	12.3	12.3	12.1	12.0	11.7	11.5	11.2	11.0	10.6	10.3
70	13.5	13.5	13.4	13.3	13.1	12.9	12.6	12.3	12.0	11.7	11.3
75	14.9	14.9	14.8	14.6	14.4	14.2	13.9	13.6	13.2	12.9	12.5
80	16.5	16.5	16.4	16.2	16.0	15.7	15.4	15.1	14.7	14.4	14.0
85	18.5	18.5	18.4	18.2	17.9	17.7	17.3	17.0	16.6	16.2	15.8
90	21.0	21.0	20.9	20.7	20.5	20.2	19.8	19.5	19.1	18.6	18.2
95	24.3	24.3	24.3	24.1	23.9	23.6	23.3	22.9	22.4	22.0	21.5
98	26.9	26.9	26.9	26.8	26.6	26.3	26.0	25.6	25.2	24.7	24.2

[a]Actual moisture content (mc) varies according to drying history of the wood and the wood species.

Source: After Table 3.4 in Ref. 10.

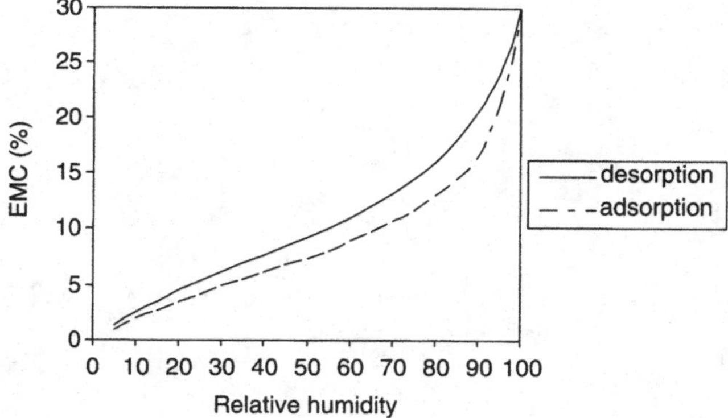

Fig. 8.4. Theoretical desorption and adsorption curves for the moisture content of wood. Equilibrium moisture content (EMC) is based on percent moisture based on the oven-dry weight of wood. Initial desorption from the green condition yields higher moisture content than does subsequent absorption. (After *The Wood Handbook.*[10])

water exists in wood in a chemically bound state; water molecules are hydrogen-bonded to the cellulose and hemicelluloses in the wood. As the relative humidity of the indoor air increases, as is typical in summer months, the number of bound water molecules in wood increases, and wood swells, while the opposite occurs in winter months, as heating systems generally lower the relative humidity of indoor air. At the same relative humidity and temperature conditions, wood will attain different equilibrium moisture content (EMC) levels depending on its drying history (Fig. 8.4).[10] For example, wood equilibrating from the green condition attains an EMC of 16% at conditions of 70°F and 80% RH, while wood that was previously dried to 6% EMC, when exposed to the same conditions, (70°F and 80% RH) will be unable to attain EMC of 16%, but instead may reach EMC of only 12–14%. This hysteresis effect occurs when, on drying, bound water molecules leave the wood cell wall and hydrogen bonds form between adjacent cellulose molecules, rendering those sites inaccessible to water.

8.4.1. Wood Moisture Content and the Fiber Saturation Point

The fiber saturation point (FSP) is the moisture content of wood where bound water is at a maximum and any additional water must exist as free water. The wood cell wall is highly porous; this porosity allows absorption of water vapor into the cell wall. The amount of water molecules that can chemically react with the cell wall is limited by the cell wall structure. Somewhere between 20% and 30% MC, the cell wall becomes saturated with bound water. At this point, as wood is exposed to water, the cell lumens begin to fill with "free water" (Fig. 8.5). Below the fiber saturation point wood is striving for equilibrium with the surrounding air and only

Fig. 8.5. Scanning electron micrographs of Douglas fir showing the porous nature of wood: (*a*) the largest pores are the cell lumens and are readily visible in the cross section and longitudinal sections through the tracheids; (*b*) cell lumens are devoid of free water; (*c*) the white color in the cell lumens indicates that the lumens are completely saturated with free water, or wood at its maximum moisture content (50× magnification). (Courtesy of the N. C. Brown Center for Ultrastructure Studies, SUNY College of Environmental Science and Forestry, Syracuse, NY.)

bound water exists. The FSP occurs somewhere in the range of 20–30% MC and corresponds to a relative humidity that is approaching 100%. Below the FSP the relative humidity and temperature conditions determine the wood moisture content. For example, at 70°F, as RH increases from 30% to 80%, the wood moisture content will increase from 6% to 16% (Table 8.3).

Within a building, for wood to exist in a state above fiber saturation, there must be a source of liquid water other than water vapor in the surrounding air, or changing conditions may cause condensation of water vapor on surfaces that are cooler than the surrounding air.

The porosity of wood is defined at three levels, and these levels help define bound versus free water.[11] The largest openings are the cell lumens with size ranging from 5 to 200 μm (Fig. 8.5). The second level includes the pit apertures and openings in pit membranes that range in size from 1×10^{-2} to 5 μm, and the smallest are the voids in the cell wall, which are less than 1×10^{-2} μm.[11] Bound water occurs in the smallest cell wall voids. According to Griffin,[11] corresponding relative vapor pressure (RH/100) for voids to begin to fill with water are 0.9998 or greater for the smallest voids, 0.9997–0.90 for openings in pits and pit membranes, and <0.90 for cell lumens.

For lumber or furniture, or any wood destined for indoor use, the wood is typically dried to a moisture content that will match the EMC range that the wood will attain in service. For this reason, recommended moisture content values for framing lumber are 15% to ≤19%.[10] For furniture and millwork, the target moisture content is close to that found under typical indoor conditions (relative humidity 50%, temperature 70°F) or a moisture content of 6–9%.

The fiber saturation point is also the defining point for the wood properties of strength and dimensional stability. Above FSP, strength and wood dimension do not change. Below FSP, as bound water leaves the wood cell wall, wood properties, especially strength, generally improve (Fig. 8.6). As RH decreases, wood MC decreases, and most strength properties increase gradually below FSP. At FSP and higher, shown in Figure 8.6 as ~22% MC, strength properties are at the minimum and stay the same up to full saturation. Also, wood reaches its maximum dimension at this point.

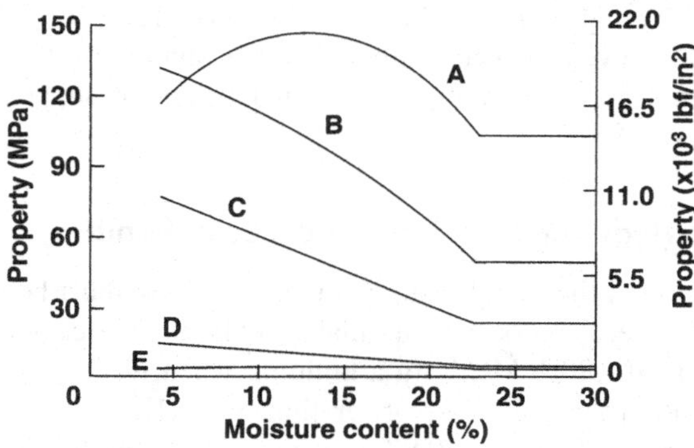

Fig. 8.6. The effect of wood moisture content on wood strength properties at and below the fiber saturation point: *A*—tension parallel to grain; *B*—bending; *C*— compression parallel to grain; *D*—compression perpendicular to grain; *E*—tension perpendicular grain. (Reprinted from the *Wood Handbook*.[10])

8.5. MOISTURE REQUIREMENTS FOR FUNGAL GROWTH

Moisture requirements for fungal growth can be defined in terms of water activity a_w, which is the relative vapor pressure of water in solution compared to that of pure water (p/p_0), which is equal to RH/100.[11] Another definition of water activity is the available water in a substrate as a decimal fraction of the amount present when that substrate is in equilibrium with a saturated atmosphere.[12] The availability of water to an organism varies with temperature and the amount of free water present. For most fungi the optimal water activity for growth is high, approaching $a_w = 1.00$.

Fungi are categorized into four groups—hydrophilic fungi ($a_w > 0.90$), mesophilic fungi ($0.80 \geq a_w \leq 0.90$, optimum >0.90), xerotolerant (minimum a_w <0.80, optimum >0.80), and xerophilic fungi (minimum a_w <0.80)—according to minimum and optimum moisture required for growth.[13] According to Griffin,[14] xerophilic fungi require a water activity (a_w) of <0.97 and have a minimum water activity that is <0.90. Xerotolerant fungi can tolerate low a_w, but have a maximum a_w that approaches 1.00; their minimum tolerance is also <0.90.[14] Most xerophilic or xerotolerant fungi are microfungi. Examples are *Wallemia sebi*, Xeromyces, and certain species of *Aspergillus* and *Penicillium*. Most wood decay fungi fall into the hydrophilic group and require a water activity of at least 0.97.[11]

The minimal moisture requirement for growth changes with temperature.[15,16] Clarke et al.[16] defined six groups of mold fungi ranging from highly hydrophilic (growth limited to >98% RH) to highly xerophilic (growth limited to >75% RH) (Fig. 8.7). Grant et al.[15] defined three groups of fungi as primary, secondary, or tertiary colonizers according to their minimal a_w for growth. Minimal a_w for primary colonizers is <0.80, for secondary colonizers 0.80–0.90, and for tertiary colonizers >0.90. According to Shaughnessy and Morey,[17] the lower limit of moisture content of a material to support microbial growth is theoretically $a_w = 0.65$. A water activity of 0.75 [equilibrium relative humidity (ERH) = 75%] is suggested as the useful limit to prevent growth, although other studies by Grant et al.[15] and Nielsen et al.[18] indicate slightly higher minimal values to support growth (0.80 and 0.78, respectively).

8.5.1. Moisture Requirements for Wood Decay Fungi

It is generally accepted that the moisture content of wood must be above the fiber saturation point for decay to occur. Optimal conditions for decay are when wood moisture content is 40–80% MC.[19] In a building when wood moisture content is below the fiber saturation point, the lower limit of RH for wood decay fungi to grow has been shown to be at $a_w > 0.97$, or when RH is >97%.[11] This corresponds to a wood moisture content of about 25–27% (Table 8.3). When free water is present and the wood moisture content reaches FSP (20–30% MC), the RH of the surrounding air will no longer be the limiting factor to growth. When wood becomes totally saturated with water, the lack of oxygen limits fungal growth.[19] Studies of the growth of sapstain fungi have shown that sapstain development can occur at or

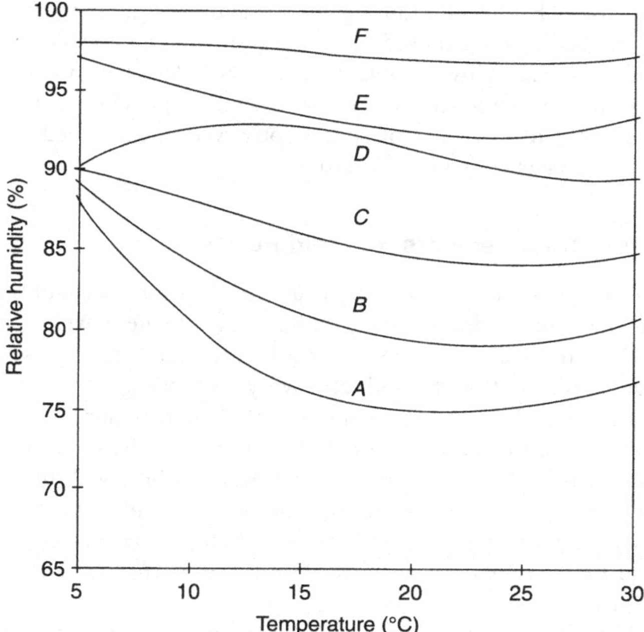

Fig. 8.7. Lower limits of relative humidity to support the growth of fungi grouped in six categories: [16] *A*—highly xerophilic; *B*—xerophilic; *C*—moderately xerophilic; *D*—moderately hydrophilic; *E*—hydrophilic; *F*—highly hydrophilic. (Reprinted from *Building and Environment*, Vol. 34, Clarke, J. A., C. M. Johnstone, N. J. Kelly, R. C. McLean, J. A. Anderson, N. J. Rowan, and J. E. Smith, A technique for the prediction of the conditions leading to mould growth in buildings, pages 515–521, copyright 1999, with permission from Elsevier.)

very near the fiber saturation point.[20] The minimum relative humidity tolerated by the sapstain fungus *Ophiostoma piceae* was 93–94% RH, which corresponded to a wood moisture content of 21–22% at 15°C.[20] Because the fiber saturation point can fall between 20% and 30% wood moisture content, there is a window of uncertainty of when decay can occur. Attempts to control wood moisture content within this range have shown that decay occurred at 27% MC in oriented strand board.[21]

The moisture content of green lumber ranges from as low as 30% to greater than 200%, and varies with wood specific gravity, amounts of sapwood and heartwood, and whether the wood is a softwood or hardwood.[10] Prior to drying, decay can develop if lumber is exposed outdoors to moisture contents >20%.[10] Use of green (unseasoned or wood that has not been air-dried or kiln-dried but allowed to dry in situ) lumber in construction can lead to mold and decay problems, as well as structural problems related to warping as the wood dries. Extended storage of lumber at moisture contents greater than 20% without drying can allow decay to develop.[10] After lumber is kiln-dried or air-seasoned to the moisture required for its use, what happens during storage and construction is critical. Wood lying outside on the ground exposed to rain is providing an excellent scenario for fungal colonization and growth. The potential for decay or mold to develop increases with time of exposure, as was evidenced by greater amounts of fungal colonization and greater diversity of species observed in kiln-dried southern pine after as little as 4 weeks of outdoor exposure.[22]

8.5.2. Moisture Requirements for Mold Fungi

According to studies in the literature, moisture requirements for molds differ from that of most wood decay fungi; many molds can tolerate lower a_w than wood decay fungi. The optimal and minimal relative humidity conditions for growth vary for each species of mold and also vary depending on temperature and growth substrate.[15,18,23] In results complied by Viitanen and Ritschkoff,[23] the optimal growth for most molds occurred at high relative humidity (98–99%), at temperatures of 20–30°C, while the minimum RH at which growth occurred was 65% at 30°C for *Aspergillus fumigatus*. For other common indoor molds, the minimum RH needed for growth was reported as 85% for *Alternaria* sp., *Cladosporium herbarum*, and *Penicillium chrysogenum*; 93% for *Stachybotrys atra*, and 80% for *Aspergillus flavus*.[23]

In controlled studies, the water activity required for minimal growth was observed to increase significantly with a decrease in temperature.[15] For *Penicillium chrysogenum* at 25°C the minimum a_w to support growth was 0.79 while at 12°C, a_w was 0.87.

The minima a_w for growth also varies depending on the test substrate. Growth on MEA (malt extract agar) occurred at lower a_w (0.79) compared to growth on woodchip wallpaper where the minima was 0.84 for both *Aspergillus versicolor* and *Penicillium* spp.[15] For *Stachybotrys atra*, the minimum a_w was 0.93 on MEA but was much greater, 0.98, on woodchip wallpaper containing added nutrients. *Stachybotrys atra* was observed to grow on gypsum wallboard at 95% RH ($a_w = 0.95$).[18]

By adding a carbon source to the surface of a substrate, Grant et al.[15] showed that greater nutrient availability decreased the minimal water activity required for growth. This provided evidence that organic "dirt" on a substrate will facilitate growth at lower relative humidity conditions.

Experiments to test growth conditions require highly controlled environments; it is difficult to control relative humidity conditions that are greater than 95% during experiments on mold growth or wood decay.[11] If surfaces become cooler than the controlled air, condensation on these surfaces could lead to false-positive results for mold growth under nominal conditions that would not allow condensation. Despite the potential difficulties, careful control of relative humidity and temperature conditions can be achieved at a lower relative humidity range such as 70–95%.

More recently there has been considerable interest in developing mathematical models to predict mold growth based on relative humidity and temperature.[16,24,25] These models are an attempt to predict the lowest relative humidity at which mold will grow for each temperature. Although knowledge of the lowest moisture tolerance has importance in predicting mold growth, the incidence of mold is much greater, and has much greater impact, when relative humidity is much higher, greater than 90%, or in cases where surfaces are wet because of condensation or where there is excess moisture due to leakage.

A "mold index" was developed by Hukka and Viitanen[24] that derived two formulas for critical RH, one when T is $>20°C$ and the other when T is $\leq 20°C$. According to their mold index, at temperatures greater than 20°C, the critical RH is always 80%, while at temperatures less than or equal to 20°C critical relative humidity for mold growth is described by the formula

$$RH_{crit} = -0.00267T^3 + 0.160T^2 - 3.13T + 100.0$$

According to their model, below 20°C the minimum relative humidity required for growth increases to 82% at 10°C and 88% at 5°C[24].

The biohygrothermal model incorporates non-steady-state relative humidity and changing surface temperatures to predictions of mold growth, particularly spore germination.[26,27] This model considers the moisture content of a model spore, as it exists on different substrates in equilibrium with the air, and determines the boundary conditions at which germination will occur. The model appears to be accurate at predicting the latent periods of spore germination and subsequent growth rates when compared with actual spores.[26]

8.6. THE EFFECT OF CHANGING MOISTURE AND TEMPERATURE CONDITIONS

Below the fiber saturation point of wood, relative humidity of the indoor air directly influences the moisture content of the wood. Only when free water is present can wood reach fiber saturation. Obvious sources of free water are plumbing, foundation, roof, and window leaks. Another source of free water is condensed water vapor, which can form when warm air contacts a cooler surface. A drop in temperature, such as when warm moist air contacts a cold window, can cause air to reach its dewpoint. Using a psychometric chart, for each relative humidity–temperature combination, there is a dewpoint, or the temperature at which condensation will occur.[28] For example, when air is at 80°F and 50% relative humidity, and it comes into contact with a surface that is 60°F or below, water vapor will become liquid. Temperature gradients often exist in wall systems, and the type and extent of temperature gradients within walls will vary with climate, time of year, and a building's heating and air-conditioning systems and insulation. Condensation within wall systems can lead to mold growth and wood decay. This type of hidden growth can be further problematic if air moves through the walls and transports fungal spores and microbial volatile compounds into the indoor environment.

8.7. SUSCEPTIBILITY OF BUILDING MATERIALS TO MOLD

The types of materials used in building construction will influence the types and concentrations of fungi found in indoor air. The more porous and hygroscopic the building material, the more moisture it can absorb and retain. Wood and wood-based materials, which are hygroscopic, are susceptible to mold growth and decay. Paint or other sealants applied to wood can be effective at preventing moisture adsorption if properly applied and maintained. In gypsum wallboard it is the paper backing and additives (starches) that will absorb moisture and serve as a nutrient source for fungi. Gypsum itself will not decay, or support mold growth, although it provides minerals, such as calcium, that may aid fungal growth. There is a trend to develop and market new mold-resistant building materials. New gypsum wallboard products that contain biocides are available on the market. However, the effectiveness and usefulness of such products in preventing mold growth require a thorough evaluation.

In a study of molds growing on 1140 samples of seven building materials, the type and frequency of fungi varied between materials (Table 8.4).[20] *Penicillium* spp. were frequent on all materials as were nonsporulating isolates (isolates that lacked fruiting structures and thus were not identifiable). Most common on wood were *Penicillium* and yeasts, and also *Aureobasidium*, a genus that contains the common wood stain and paint disfiguring fungus, *Aureobasidium pullulans*. *Acremonium*, *Aspergillus*, and *Cladosporium* were more common on paper than on wood. On gypsum board, the genus with the greatest frequency was *Stachybotrys* (19.7%). On other materials, including wood, *Stachybotrys* was much less frequent. On ceramic products the frequencies of *Acremonium* and *Aspergillus* were very high. Mineral insulation was high in *Penicillium*, while paints and glues had greater frequencies of *Acremonium* and *Aspergillus* than on either paper or wood, and were also high in the genus *Tritirachium*. Plastics were high in yeast, *Acremonium*, *Aspergillus*, and *Fusarium*.

Nielsen et al.[18] studied the growth of molds on 21 building materials subjected to several levels of equilibrium relative humidity at each of four temperatures. After inoculation by 16 strains of fungi, the most common fungi recovered after 4 or 7 weeks incubation were *Penicillium chrysogenum* and four species of *Aspergillus*, which grew on all materials. On wood, the minimum RH observed to support growth was 78% RH at 20°C and 25°C and 90% RH at 5°C. Gypsum board was the only material to support the growth of *Stachybotrys* and only at RH of at least 86%.

8.7.1. Visible versus Concealed Mold

Visible mold can be a good indicator of a moisture problem (Fig. 8.1); however, the presence of visible mold does not always correlate with high concentrations of bioaerosols. For example, in a study of 103 inner-city homes, no correlation was observed between visible mold growth and the amount of airborne fungi

TABLE 8.4. Frequency (%) of 24 Taxa of Fungi Isolated from Seven Types of Building Materials

	Wood ($n = 438$)	Paper ($n = 49$)	Gypsum Boards ($n = 43$)	Ceramic Products ($n = 161$)	Mineral Insulation ($n = 217$)	Paints and Glues ($n = 86$)	Plastics ($n = 125$)
Acremonium	5.1	7.5	2.7	17.5	7.5	15.0	8.1
Aspergillus spp.	5.1	6.9	5.0	5.0	7.5	12.1	8.2
A. versicolor	2.5	2.1	4.8	11.9	3.8	13.5	4.8
Aureobasidium	6.7	6.7	1.9	1.2	4.5	1.4	3.3
Chaetomium	1.4	1.0	0	0.4	0.7	3.1	0.2
Cladosporium	1.7	10.8	0.1	4.6	4.2	2.2	1.3
Eurotium	0.3	0.3	0	0.1	0.1	0	0
Exophiala	2.2	0.4	0.3	0.1	0.1	0.1	0.5
Fusarium	0.8	0.2	0	1.3	0	0.1	2.7
Geomyces	0.3	0	0	0.5	0.1	0	0.9
Mucor	0	0	0	0	0.3	0	0
Nonsporulating isolates[a]	13.2	10.6	12.6	14.0	11.3	7.6	11.4
Oidiodendron	2	1.0	0	0	1.4	0	0
Paecilomyces	2.3	0.1	5.7	0.7	0.7	1.3	0.2
Penicillium	22.9	18.9	23.4	17.4	29.8	16.1	17.3
Phialophora	3	1.1	3.5	2.0	0.2	1.0	3.4
Rhizopus	0	0.8	0	0.1	0.5	0	0
Scopulariopsis	0.9	1.9	0.6	3.1	0.6	0.1	2.1
Sphaeropsidales	6.3	1.6	5.4	2.1	4.8	2.6	8.1
Stachybotrys	0.7	4.0	19.7	1.9	1.0	0.9	1.0
Trichoderma	0.6	0	0	0.9	1.3	0.2	0.2
Tritirachium	0	0	0	3.1	0.3	8.4	3.1
Ulocladium	0.5	0.5	0	0	0.7	0	0.9
Wallemia	0	0	0	0.1	0	0	0
Yeasts	16.7	14.9	14.3	5.9	13.6	13.2	17.7

[a]Nonsporulating isolates could not be identified.

Source: Reprinted from *International Biodeterioration & Biodegradation*, Vol. 49, Hyvärinen, A., Meklin, T., Vepsalainen, A., and A. Nevalainen, "Fungi and actinobacteria in moisture-damaged building materials—concentrations and diversity," pages 27–37. Copyright 2002, with permission from Elsevier.

(CFU/m^3).[30] Visible colonies may be producing few or no spores or the spores that are produced are slimy or "sticky" and do not readily become airborne.

On the other hand, the absence of visible mold growth is not always an indication of good air quality as growth may be occurring within walls or other spaces concealed from view. In a study of homes damaged during El Niño, Morey et al.[31] found greater levels of concealed mold in rooms in which significant moisture intrusion had occurred. Greater amounts of concealed mold corresponded to differences in the frequency of fungal species recovered from indoor air samples. High concentrations of certain species of indoor bioaerosols may indicate a hidden moisture problem. Because the spores from hidden mold growth significantly altered the fungal composition of indoor air[31] and visible mold does not always correlate with bioaerosol concentration, the presence or absence of visible mold is not always a good indication of indoor air quality.

8.8. THE EFFECTS OF BUILDING DESIGN AND CONSTRUCTION

The easiest way to prevent mold or wood decay is to limit availability to water, either water vapor or free water. Evidence suggests that relative humidity as low as 75–80% can support mold growth. In cooler climates ambient relative humidity and temperature conditions would rarely support the growth of molds, except near water sources that provide localized high RH conditions, such as laundryrooms, sinks, and bathrooms, particularly showers, or when temperature gradients exist within buildings, causing condensation of water vapor. In certain climates, such as the southeastern United States, ambient high RH (>80%) are possible and can lead to mold growth. The use of vinyl wall coverings or wall barriers can hold moisture within walls, prevent drying, and promote mold growth. To prevent this type of moisture accumulation, it has been suggested by some that vinyl wallpaper or other moisture "barriers" should not be used. Proper design is a necessary first step in eliminating excessive moisture in buildings. During building construction proper installation of foundations, wall systems, and windows should prevent water intrusion. However, nail holes or openings in wall barriers or improper installation of window flashings can occur, leading to a leakage of the moisture barrier. Although effective in most climates, a vapor retarder was shown to be ineffective when colder air-conditioned air meets a warmer surface, such as a vapor barrier within a wall.[32] Condensation occurs within the wall, moisture cannot escape, and fungal growth is likely. Climate should be considered in designing wall systems,[32] and a moisture index based on moisture (actual rainfall in five North American cities) and evaporation potential has been proposed to assist in the assessment of wall systems.[33] Even if a building is designed to minimize or prevent moisture intrusion, errors during construction can lead to breaks in the protective building envelope and allow water to penetrate (Figs. 8.8 and 8.9).

Fig. 8.8. Decay fungi growing inside the wall of a gymnasium. During construction, the wall's interior moisture barrier was intentionally torn, allowing moisture to intrude. Decay was apparent after removal of the wall panels. (Photo courtesy of Dr. Edson Setliff, Adjunct Professor, State University of New York College of Environmental Science and Forestry, Syracuse, NY.)

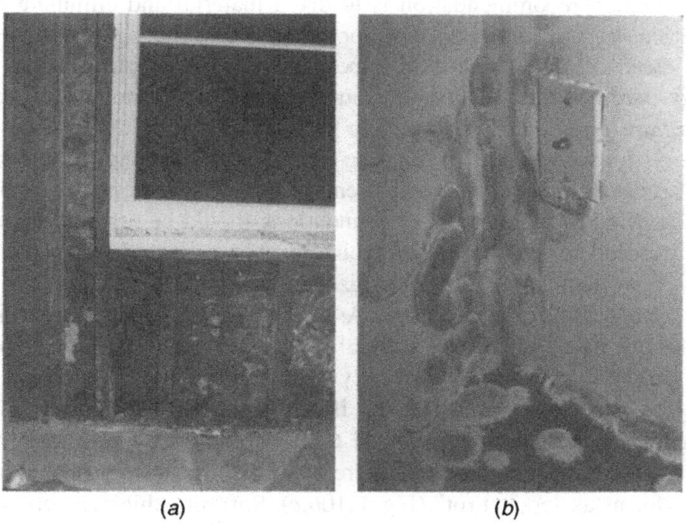

(a) (b)

Fig. 8.9. Problems in buildings caused by poor construction practices. (*a*) improper installation of a window that allowed water to seep into the wall beneath the window and decay to occur; (*b*) fungal fruiting bodies indicating extensive, prolonged growth of the fungus that started within the wall cavity, eventually growing through cracks in the wall and through openings by the electric outlet. (Photos courtesy of Dr. Edson Setliff, Adjunct Professor, State University of New York College of Environmental Science and Forestry, Syracuse, NY.)

8.9. TECHNIQUES FOR ASSESSMENT OF WOOD DECAY AND MOLD IN BUILDINGS

A first step to diagnosing a mold or decay problem can be to look for any visible signs of fungal growth such as fungal structures (rhizomorphs, even fruiting bodies in some cases) or discoloration, and signs of associated moisture, such as condensation, water stains, peeling paint, and warping. For molds, fungal colonies can be light or dark; either white, gray, black or brownish black; or brightly colored, such as green, red, yellow, or pink. Colonies of wood decay fungi are typically white or tan, sometimes brown or orange-brown. Rhizomorphs are large strands of aggregate hyphae that are indicative of the building rot fungi, *Meruliporia incrassata, Serpula lacrymans*, and *S. himantioides*. Rhizomorphs function to transport water, and give this fungus the unique ability to utilize a water source that is a considerable distance from the wood that it is attacking. This fungus rapidly degrades wood; thus it is critical to remove not only the decayed wood but also any other wood in close proximity as it may also be infected.

Early in the decay process, when it is most beneficial to recognize their presence, the signs of decay are the most difficult to detect. If left these signs are undetected, the probability of decay is increased over time if the conditions for decay continue to be available. Hidden decay has greater potential to proliferate if the source of moisture is left uncorrected. It is generally thought that if drying has not begun within 48 h of initial wetting, a material capable of supporting fungal growth will become colonized. A common recommendation is to dry a material and eliminate a moisture source within 48 h to minimize mold occurrence and damage. Surface mold can be easily cleaned, but a long-term exposure to moisture can allow a fungus to grow into the material, making it more difficult to remove. An example is *Aureobasidium pullulans*, growing into parenchyma cells discoloring the wood (Fig. 8.2). As for wood decay fungi, long-term exposures to high levels of moisture can have significant effects on wood strength. For these reasons, any leaks through roofs, walls, windows, plumbing, or foundations should be repaired promptly.

Wood exposed to high levels of moisture over time will most likely develop extensive decay that is easy to recognize. There may be a heavy mat of fungal mycelium and a strong, musty odor. As brown rot develops, it causes wood to shrink and split into a distinctive cubical pattern (Fig. 8.10*a*). Brown-rotted wood crumbles easily and is usually slightly reddish-brown in color. An extremely harmful type of brown rot is caused by the building rot fungus, *Meruliporia incrassata* which is commonly referred to as "dry rot." White-rotted wood has a white or streaked appearance; the texture is sometimes stringy or flaky; white rot sometimes forms as "pocket rot" (Fig. 8.10*b*,*c*). Soft rot exhibits a very fine shrinkage pattern, and is gray (Fig. 8.10*d*).

As it is a desirable goal to eliminate decay or mold before considerable damage or extensive colonization has occurred, it is preferable in this case to use methods that are sensitive to detecting incipient decay or hidden fungal colonies. These methods can also be applied to determining the type of fungi present in visible fungal colonies. These include nondestructive sampling regimes, such as air sampling, tape

Fig. 8.10. Appearance of wood at advanced stages of brown rot, white rot, and soft rot: (*a*) brown rot containing large cubical fractures due to decay and subsequent shrinkage; (*b*) white-rotted wood that is light in color and streaked; (*c*) white pocket rot in which decay is localized to the lighter areas; (*d*) soft rot as evidenced by surface checking due to decay. The checking pattern is smaller and shallower than that of brown rot. (All micrographs 1× magnification.)

lifts, and culturing; and destructive sampling methods such as a pick test, core boring, or drilling. Borescopes, which are useful for viewing fungal colonies inside walls or under flooring, can be mildly destructive, requiring drilling a small hole to insert the probe.

Moisture detection, as opposed to detection of fungi, serves to define the source and extent of moisture intrusion into a building. Air temperature and relative humidity indicators can measure indoor air conditions, while moisture meters can measure the moisture content of wood and other materials. Care must be taken for proper use and interpretation of both types of indicators.

When measuring the conditions of indoor air, most commercial temperature and relative humidity meters can record relative humidity in the range of 0–95%. Accuracy is typically ±2%, but accuracy declines at higher humidities to ±3 or 4%. As a guideline, a persistent relative humidity of 75% or greater can be an indication that mold growth is occurring. Brief episodes of high RH should not lead to the development of fungi, as long as ventilation adequately eliminates the excessive moisture.

To measure the moisture content of wood or other materials, electrical resistance moisture meters are typically used. Many commercial moisture meters can provide readings of 6–60% moisture and also provide correction factors for temperature and species of wood. Some moisture meters include scales for other building materials. Moisture contents should be taken at a number of locations within a structure to compare normal moisture content with any problematic areas. As a guide, <20% moisture content would indicate that decay fungi would not be growing, although mold growth may be possible. Any moisture content >20% indicates that decay is possible, although at 20–25% it will probably be minimal. It must also be considered that moisture content is not always uniform within a piece of wood and that conditions that lead to condensation can rapidly alter the moisture content.

8.9.1. Nondestructive Testing

Cultures can be made from visible mold colonies by taking tape lift samples, taking scrapings of surfaces, or removing small pieces of wood or other material. The presence of fungi can also be evaluated from air samples. Fungal spores and fragments are collected in active samplers in which a discrete, reproducible amount of air is forced through a sample collector. The type of air sampler suitable for culturing must contain growth media for the collection of viable samples. An example is the Andersen N-6 single-stage sampler for the collection of bioaerosols onto plastic Petri dishes containing a growth medium. The type of growth media used can affect the population of viable cultures.[13,34] To avoid contamination it is critical that sterile sampling materials be used.

Air sampling can be a tool to find hidden sources of mold or decay, or to determine whether spore counts are excessively high. Results of air sampling can be expressed in two ways: (1) as the total concentration of spores per volume of air or, (2) as the presence of identifiable species or genera of fungi. Airborne data must be interpreted carefully to determine whether there could be hidden sources of indoor molds and decay. High total concentration of spores indoors relative to outdoors can indicate an indoor source.

To detect advanced wood decay, such as decay hollows within building timbers, a number of nondestructive techniques are available, such as stress-wave timers. These and other wood inspection devices have been evaluated for accuracy and ease of use.[35–37]

8.9.2. Destructive Testing

A probe or pick test is useful if there is no concern with destruction or appearance of the wood. The probe, similar to an ice pick, is inserted into the wood across the grain, and tilted to lift a small piece of the wood. A splintering, uneven failure indicates sound wood, while a brash, smooth failure indicates weakened wood. The relative amount of pressure applied to fracture the wood can be an indication of decay.

Cores can be bored out of wood for culturing and/or microscopic analysis. Culturing of cores can not only reveal the presence of fungi but can also determine whether the fungus is destructive (a decay fungus) or nondestructive to the wood (a mold or stain fungus). Identification techniques can classify the fungi to its type, such as basidiomycete or ascomycete, or sometimes to genus or species. The methods for sample collection and culturing require that sterile materials be used in handling, such as sterile foil or bags for sample collection and handling. Alcohol (95% ethanol) and flame are effective for sterilizing borer bits and other tools used during sample collection.

Microscopic examination of wood samples is another technique that can determine the types of decay that are present. Samples can be examined for fungal hyphae, spores, and other structures with a stereomicroscope, light microscope, or scanning electron microscope. The stereomicroscope readily reveals fungal colonies and reproductive structures (Fig. 8.11) and is useful for distinguishing mold colonies from other types of surface debris. Fine white crystals have been mistaken for mold on a surface, if wood contains salts or other treatments that can migrate to the surface and crystallize. Stereomicroscopic examination can reveal whether this fine white coating is composed of sharp, angular crystals or threadlike fungal hyphae.

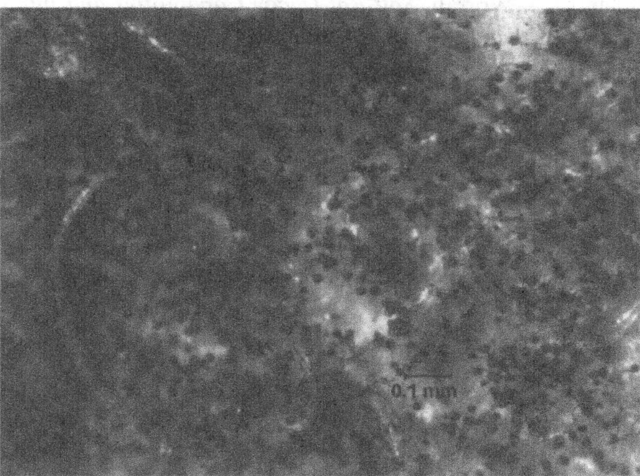

Fig. 8.11. Fungal colony on the surface of a wall behind the rubber baseboard molding, as observed with a stereomicroscope. The fungus was later identified as *Stachybotrys sp.* (bar = 0.10 mm).

Both the extent and type of decay can be assessed with light microscopy or scanning electron microscopy. Thin sections of wood can reveal erosion in the wood cell walls, such as fungal boreholes, soft-rot cavities, and other distinguishing features.[38] Scanning electron microscopy is also useful for observing most decay features, such as hyphae, boreholes, and erosion with the exception of embedded soft-rot cavities that are readily visible under the light microscope.

8.10. SUMMARY

- Wood decay and mold have different moisture requirements.
- Many building materials support mold growth.
- Prolonged moisture exposure can cause structural damage.
- Certain moisture conditions favor specific fungi.
- Different building materials support different species of molds.
- Proper building construction and maintenance are essential; identification of a moisture problem is the critical first step.
- Techniques for detecting mold or decay fungi can involve either nondestructive or destructive sampling of wood or indoor air.

REFERENCES

1. Kirk, P. M., P. F. Cannon, J. C. David, and J. A. Stalpers, eds., *Dictionary of the Fungi*, 9th ed., CAB International, Wallingford, UK, 1999.
2. Wang, C. J. K. and J. J. Worrall, *Soft Rot Decay Capabilities and Interactions of Fungi and Bacteria from Fumigated Utility Poles*, Electric Power Research Institute Publication EPRI TR-101244, Electric Power Research Institute, Palo Alto, CA, 1992.
3. Worrall, J. J., S. E. Anagnost, and R. A. Zabel, Comparison of wood decay among diverse lignicolous fungi, *Mycologia* **89**(2):199–219 (1997).
4. Wang, C. J. K. and R. A. Zabel, eds., *Identification Manual for Fungi From Utility Poles in the Eastern United States*, Allen Press, Lawrence, KS, 1990.
5. Zabel, R. A., C. J. K. Wang, and S. E. Anagnost, Soft-rot capabilities of the major microfungi isolated from Douglas-fir utility poles in the Northeast, *Wood Fiber Sci.* **23**(2):220–237 (1991).
6. Anagnost, S. E., J. J. Worrall, and C. J. K. Wang, Diffuse cavity formation in soft rot of pine, *Wood Sci. Technol.* **28**:199–208 (1994).
7. Anagnost, S. E., C. M. Catranis, C. J. K. Wang, L. Zhang, A. A. Fernando, S. R. Morey, P. DeStefano, C. Garback, M. LaMoy, G. Hall, J. A. Crawford, D. Naishadham, A. Hunt, and J. L. Abraham, *Survey of Fungi Found in Syracuse AUDIT Study Homes*, Center of Excellence in Environmental Systems, Syracuse Univ., Syracuse, NY, 2004.
8. Wilcox, W. W., Review of the literature on the effects of early stages of decay on wood strength, *Wood Fiber* **9**(4):252–257 (1978).

9. Panshin, A. J. and C. de Zeeuw, *Textbook of Wood Technology*, 4th ed., McGraw-Hill, New York, 1981.
10. Forest Products Laboratory, *Wood Handbook—Wood as an Engineering Material*, General Technical Report FPL-GTR-113, U.S. Department of Agriculture, Forest Service, Forest Products Laboratory, Madison, WI, 1999.
11. Griffin, D. M. Water potential and wood decay fungi, *Ann. Rev. Phytopathol.* **15**:319–329 (1977).
12. Kendrick, B., *The Fifth Kingdom*, Mycologue Publications, Newburyport, MA, 2000.
13. Burge, H. A. and J. Otten, Fungi, in *Bioaerosols: Assessment and Control*, J. Macher, ed., American Conf. Governmental Industrial Hygienists, Cincinnati, OH, 1999, pp.19-1–19-13.
14. Griffin, D. H., *Fungal Physiology*, Wiley, New York, 1981.
15. Grant, C., C. A. Hunter, B. Flannigan, and A. F. Bravery, The moisture requirements of moulds isolated from domestic dwellings, *Internat. Biodeter. Biodegrad.* **25**:259–284 (1989).
16. Clarke, J. A., C. M. Johnstone, N. J. Kelly, R. C. McLean, J. A. Anderson, N. J. Rowan, and J. E. Smith, A technique for the prediction of the conditions leading to mould growth in buildings, *Build. Environ.* **34**:515–521 (1999).
17. Shaughnessy, R. J. and P. R. Morey. Prevention and control of microbial contamination, in *Bioaerosols: Assessment and Control*, J. Macher, ed., American Conf. Governmental Industrial Hygienists, Cincinnati, OH, 1999, pp.10-1–10-13.
18. Nielsen, K. F., G. Holm, L. P. Uttrup, and P. A. Nielsen, Mould growth on building materials under low water activities: Influence of humidity and temperature on fungal growth and secondary metabolism, *Internat. Biodeter. Biodegrad.* **54**:325–336 (2004).
19. Zabel, R. A. and J. J. Morrell, *Wood Microbiology: Decay and Its Prevention*, Academic Press, New York, 1992.
20. Payne, C., J. A. Petty, and S. Woodward, The softwood staining fungus *Ophiostoma piceae*: Influence of relative humidity, temperature and timber drying method on mycelial growth and coremiophore production in vitro and on wood, *J. Inst. Wood Sci.* **15**:165–172 (2000).
21. Morris, P. I. and J. E. Winandy, *Limiting Conditions for Decay in Wood Systems*, International Research Group for Wood Preservation Document IRG/WP 02-10421, International Research Group on Wood Preservation, Stockholm, Sweden, 2002.
22. Anagnost, S. E., S. Zhou, H. Yeo, C. J. K. Wang, W. B. Smith, and D. M. Roberts, Fungi inhabiting southern pine utility poles during manufacture, *Forest Products J.* **56**(1):53–59 (2006).
23. Viitanen, H. and A. C. Ritschkoff, *Mold Growth in Pine and Spruce Sapwood in Relation to Air Humidity and Temperature*, Swedish Univ. Agricultural Sciences Department of Forest Products Report 221, Uppsala, Sweden, 1991.
24. Hukka, A. and H. A. Viitanen, A mathematical model of mould growth on wooden material, *Wood Sci. Technol.* **33**:475–485 (1999).
25. Sedlbauer, K, Prediction of mould growth by hygrothermal calculation, *J. Thermal Envelope Build. Sci.* **25**(4):321–336 (2002).
26. Sedlbauer, K., *Prediction of Mould Fungus Formation on the Surface of and inside Building Components*, Ph.D. dissertation, Univ. Stuttgart, Germany, 2001.

27. Sedlbauer, K., M. Krus, W. Zillig, and H. M. Kunzel, *Mold Growth Prediction by Computational Simulation*, ASHRAE IAQ 2001, San Francisco, 2001.
28. Shelton, D. P. and G. R. Bodman, *Air Properties: Temperature and Relative Humidity*, Publication G82-626-A, Cooperative Extension, Institute of Agriculture and Natural Resources, Univ. Nebraska, Lincoln, NE, 1983.
29. Hyvärinen, A., T. Meklin, A. Vepsalainen, and A. Nevalainen, Fungi and actinobacteria in moisture-damaged building materials—concentrations and diversity, *Internat. Biodeter. Biodegrad.* **49**:27–37 (2002).
30. Rosenbaum P. F., T. M. Hargrave, J. L. Abraham, J. A. Crawford, A. Hunt, C. Liu, G. Hall, S. E. Anagnost, C. M. Catranis, A. A. Fernando, S. R. Morey, S. Zhou, and C. J. K. Wang, Indoor mold and the risk of wheeze in the first year of life for infants at risk of asthma, *Am. J. Epidemiol.* **161**(11):134 (2005).
31. Morey, P. R., M. C. Hull, and M. Andrew, El Niño water leaks identify rooms with concealed mould growth and degraded indoor air quality, *Internat. Biodeter. Biodegrad. Bull.* **52**:197–202 (2003).
32. Burch, D. M. and A. TenWolde, *A Computer Analysis of Moisture Accumulation in the Walls of Manufactured Housing*, ASHRAE Transaction 99, 1993.
33. Cornick, S. and W. A. Dalgliesh, A moisture index to characterize climates for building envelope design, *J. Thermal Envelope Build. Sci.* **27**(2):151–128 (2003).
34. Duchaine, C., A. Meriaux, and P. Comtois, Usefulness of using three different culture media for mold recovery in exposure assessment studies, *Aerobiologia* **18**:245–251 (2002).
35. Brashaw, B. K., R. J. Vataralo, J. P. Wacker, and R. J. Ross, *Condition Assessment of Timber Bridges 1. Evaluation of a Micro-drilling Resistance Tool*, General Technical Report FPL-GTR-159, U.S. Department of Agriculture, Forest Service, Forest Products Laboratory, Madison, WI, 2005.
36. Brashaw, B. K., R. J. Vataralo, J. P. Wacker, and R. J. Ross, *Condition Assessment of Timber Bridges 2. Evaluation of Several Stress-Wave Tools*, General Technical Report FPL-GTR-160, U.S. Department of Agriculture, Forest Service, Forest Products Laboratory, Madison, WI, 2005.
37. Ross, R. J., B. K. Brashaw, X. Wang, R. H. White, and R. F. Pellerin, *Wood and Timber Condition Assessment Manual*, Forest Products Society, Madison, WI, 2004.
38. Anagnost, S. E., Light microscopic diagnosis of wood decay, *Internat. Assoc. Wood Anatomists J.* **19**(2):141–167 (1998).

CHAPTER 9

USE OF STATISTICAL TOOLS FOR DATA PRESENTATION AND ANALYSIS OF INDOOR MICROORGANISMS

STELLA M. TSAI

9.1. INTRODUCTION

Many indoor microorganism studies generate a limited number of samples. The purpose of this chapter is to encourage field investigators to consider statistical principles in drawing conclusions even when the number of samples is small. In this chapter, we will start with descriptive analysis using commonly available software packages to acquaint readers with basic statistics, charts, and graphs that can be used to present datasets. Correlation analysis and hypothesis tests [*t*-test, analysis of variance (ANOVA), and nonparametric comparisons] are also included in this chapter. As appropriate, Microsoft Excel, S plus, and SAS (all registered trademarks) are included in the description of these data presentation and analysis methods.

Ideally, statisticians with experience in environmental databases should be consulted prior to conducting study design and field investigation, and on data interpretation afterward. One important point is that causality cannot be proved by statistical significance only; on the other hand, insignificant results of known hazards may still be important in identifying health hazards. For example, a single spore or colony of *Stachybotrys chartarum* (or other marker fungi) detected in air samples may be statistically insignificant, but may be important from a health perspective or to determine potential water damage and microbial infestation.

Texts on data analysis for bioaerosols are suggested for more details on calculations and inferences used in bioaerosol data analysis. Eudey et al.[1] have written a chapter (Chapter 13) in the book on bioaerosols describing basic summary statistics and some useful hypothesis tests. Macher[2] describes various statistical methods used in bioaerosol research. Several useful statistical packages are

Sampling and Analysis of Indoor Microorganisms, Edited by Chin S. Yang and Patricia A. Heinsohn

described in a book published by AIHA [e.g., Microsoft Excel and S plus (www.insightful.com)].[3]

The following sections provide a brief overview of some statistical procedures that can be applied to indoor microorganism investigations. Microsoft Excel is the primary software used as an application tool for data analysis and interpretation in this chapter. More advanced statistical packages, S plus and SAS, are described when functions in Microsoft Excel are not available for specific data presentation or analysis procedure.

9.2. DESCRIPTIVE ANALYSIS

When you have data from either field investigations or laboratory reports, you should be able to describe the data in a straightforward and easy-to-comprehend way to your clients or audience. One of the first steps in data analysis is to generate descriptive statistics for all variables. The results from the descriptive statistics can answer some of the very common and general questions, such as "How many samples did you collect?" and "What are the mean (arithmetic average), standard deviation, maximum and minimum concentrations of your data?" The standard deviation is a measure of how widely spread the data are. If the data points are sufficient and form a "bell-shaped" (normal) distribution, about 68% of the data will fall within 1 standard deviation of the mean, that is, mean ± 1 standard deviation. Roughly, about 95% and 99.7% of the data will be within 2 and 3 standard deviations, respectively.

A very commonly used data analysis and management package, Microsoft Excel, is described in this section to demonstrate how to generate a quick descriptive analysis report.

In Microsoft Excel, you can key in or transfer your data into the spreadsheet. Data in spreadsheet format can be presented by counts, concentrations, and/or percentage depending on the bioaerosol types and laboratory procedures. From the spreadsheet format, you can generate reports by using Auto Format under Format. A variety of choices are available to meet different reporting requirements.

If you want to conduct data analysis from the spreadsheet, open Tools. There, you will find the selection Data Analysis. By clicking Data Analysis, you will open up a window with all the analysis tools available in the Microsoft Excel package. If you do not see Data Analysis under Tools, click on Add-Ins and then click the Analysis ToolPak, then click OK to install the package. The Analysis ToolPak provides functions and interfaces for data analysis.

After opening the analysis tools, click Descriptive Statistics. This will open up a window in which you can input the range of data for analysis. The range of data can be put in by highlighting the area (Table 9.1a) in the spreadsheet, or directly put into the cell address with punctuations of colon (e.g., A3:B6) for cells labeled continuously (Table 9.1a). For cells not labeled continuously, highlight the first set of data first, then use the CTRL button to highlight the rest of data, or using both colon and comma (e.g., A3:B3, A5:B5) with the cell address to put in the data (Table 9.1b).

TABLE 9.1. Basic Spreadsheet Formats in Microsoft Excel

a. Highlighted Areas are Cells Labeled Continuously

	A	B
1	Total Fungal Concentration of Andersen Samples (CFU/m^3)	Total Fungal Concentration of Air-O-Cell Samples (structures/m^3)
2	1655	3675
3	1604	6189
4	1416	3627
5	1775	6189
6	1809	3240

b. Highlighted Areas Are Cells Not Labeled Continuously

	A	B
1	Total Fungal Concentration of Andersen Samples (CFU/m^3)	Total Fungal Concentration of Air-O-Cell Samples (structures/m^3)
2	1655	3675
3	1604	6189
4	1416	3627
5	1775	6189
6	1809	3240

Source: Data in both tables from Tsai et al.[4]

Then, click on the box for Summary Statistics to get the output table. In the output table, you will find many terms used to describe the data. Mean and median are approximately the same for normally distributed data. Kurtosis is the parameter that describes whether the data distribution is in a central peak or in the tails with a flat shape. The value for kurtosis can be as small as -3 for flat distribution with short tails, 0 for a normal distribution, and positive for tails heavier than a normal distribution.[5] Skewness is the parameter that describes the symmetry relative to the mean of the data distribution. Table 9.2 shows an example of Microsoft Excel summary statistics output table using data in Table 9.1.

A chart or graph allows you to visually display data, compare data, and view trends. Microsoft Excel is an excellent package to describe data in various types of charts and graphs [e.g. column, bar, line, pie, *XY* (scatter), and area]. Each chart or graph type has several subtype selections. Under Insert, click Chart Wizard. A list of chart/graph types will show on the window. Column, bar, and line charts can be useful to present data collected from different locations and times to observe trends. Figure 9.1 presents an example of a line chart from Microsoft Excel. The *XY* (scatter) plot can be used to view paired samples to check consistency of repeated measurements and check to see if there are any extreme high or low data points (outliers). Try your data to choose the best chart or graph to describe and present the dataset.

TABLE 9.2. An Example of Summary Statistics Output from Microsoft Excel

	A	B	C	D
1	Total Fungal Concentration of Andersen Samples (CFU/m^3)		Total Fungal Concentration of Air-O-Cell Samples (Structures/m^3)	
2	Mean	1,652	Mean	2,619
3	Standard error	70	Standard error	377
4	Median	1,655	Median	2,100
5	Mode	#N/A	Mode	3,537
6	Standard deviation	156	Standard deviation	843
7	Sample variance	24,447	Sample variance	710,246
8	Kurtosis	0	Kurtosis	−3
9	Skewness	−1	Skewness	1
10	Range	393	Range	1,685
11	Minimum	1,416	Minimum	1,851
12	Maximum	1,809	Maximum	3,537
13	Sum	8,259	Sum	13,097
14	Count	5	Count	5

Source: Data from Tsai et al.[4]

Panel graphs can display multiple sets of variables sampled with various X and/or Y. Figure 9.2 shows an example panel graph created by S plus for two sets of samples collected during a period of time to display the trend over time. S plus can be an added-in with Microsoft Excel spreadsheet. S plus has the same 2D and 3D graph functions as Microsoft Excel has.

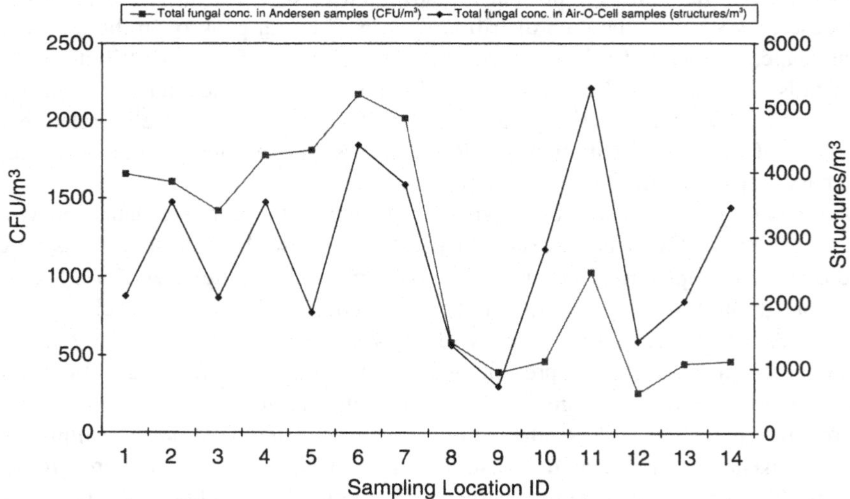

Fig. 9.1. An example of a line chart from Microsoft Excel (data from Tsai et al.[4]).

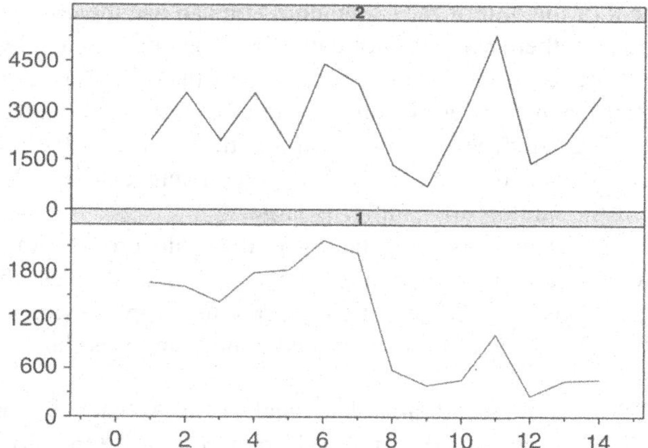

Fig. 9.2. An example of a panel graph using S plus for paired samples collected using both Andersen sampler and Air-O-Cell cassettes. Panel 1 shows Andersen samples in units of CFU/m^3; panel 2, Air-O-Cell samples in units of structures/m^3 (data from Tsai et al.[4]).

9.3. CORRELATION

From the *XY* scatterplot, you can roughly visualize the correlation between two sets of paired samples (Fig. 9.3). You can compute the Pearson's correlation coefficient (r) by using the Microsoft Excel Tools. Under Data Analysis, select Correlation. This analysis tool measures the relationship between two datasets that are scaled

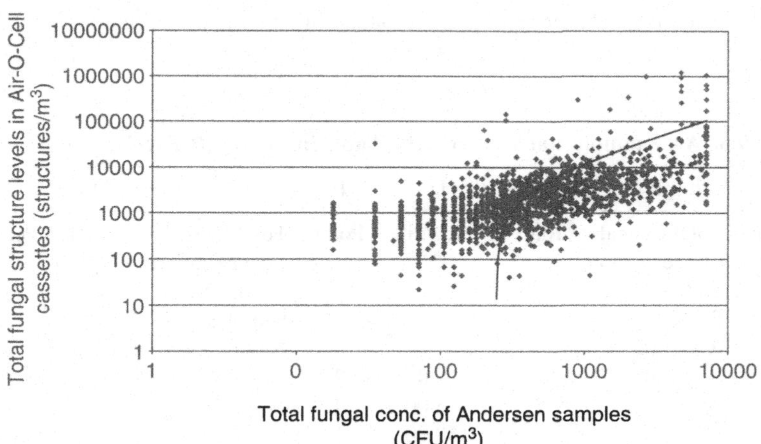

Fig. 9.3. Total fungal structure levels in Air-O-Cell samples versus total fungal concentrations in Andersen samples ($r = 0.32$); a trend line was added in the chart (data from Tsai et al.[4]).

to be independent of the unit of measurement. You can use the correlation analysis tool to determine whether two ranges of data move together—that is, whether large values of one set are associated with large values of the other (positive correlation, $0 < r < 1$), whether small values of one set are associated with large values of the other (negative correlation, $-1 < r < 0$), or whether values in both sets are unrelated (correlation near zero). The correlation coefficient can be relatively small, yet significant, if the number of samples is large.

Spearman's rank correlation, especially useful for bioaerosol samples, has been applied and evaluated in several publications.[1,2,6–8] for samples collected from two different locations or collected at two different sampling times. The sample size of each group, sampling conditions and laboratory analysis should be consistent between groups.

Microsoft Excel spreadsheet and data analysis tools can be applied to this analysis. First, input groups of data into different columns. Open Tools in Excel; you will find the selection Data Analysis. Clicking Data Analysis, then click Rank and Percentile. This will open up a window in which you can input the range of data for analysis. In the output table (e.g., see Table 9.3), you will find rank and percentile of each data point in each group of data. By using the rank information, a table for Spearman rank correlation can be created in the Microsoft Excel spreadsheet (Table 9.4). The rank difference (d_i) of each interested bioaerosol can be calculated in the spreadsheet by creating a formula (e.g., = C2 − E2). Another column can be created for d_i^2. In any blank cell of Excel spreadsheet, you can calculate the Spearman rank correlation coefficient r_s, by using the following formula[2]:

$$r_s = 1 - \frac{6^* \sum (d_i)^2}{n(n^2 - 1)}$$

where d_i is the rank difference for each bioaerosol and n is the number of bioaerosol groups.

TABLE 9.3. An Example Rank–Percentile Table in Microsoft Excel

	A	B	C	D	E	F	G	H
1	Point	Total Fungal Concentration of Andersen Samples (CFU/m^3)	Rank	Percentile (%)	Point	Total Fungal Concentration of Air-O-Cell Samples (structures/m^3)	Rank	Percentile (%)
2	5	1809	1	100.00	2	3537	1	75.00
3	4	1775	2	75.00	4	3537	1	75.00
4	1	1655	3	50.00	1	2100	3	50.00
5	2	1604	4	25.00	3	2073	4	25.00
6	3	1416	5	0.00	5	1851	5	0.00

Soure: Data from Tsai et al.[4]

TABLE 9.4. An Example Spearman Rank Correlation Test

	A	B	C	D	E	F	G	H
1	Allergen	Day 1 Concentration	Rank	Day 2 Concentration	Rank	d_i = Rank (Day 1) − Rank (Day 2)	d_i^2	r_s
2	a	12	2	10	4	−2	4	−0.6
3	b	8	5	17	1	4	16	—
4	c	10	4	14	2	2	4	—
5	d	11	3	9	5	−2	4	—
6	e	15	1	12	3	−2	4	—

In Table 9.4, r_s is calculated as −0.6. From the Spearman correlation analysis, if the two groups of samples are similar in composition, r_s is close to 1. As r_s decreases below 1, the chance of less similarity in two groups increases. If the number of samples is large, the correlation coefficient can be relatively small to be significant.

It is common that during the IAQ investigations, investigators need to collect more than one environmental factor, and various biological agents. Factor analysis, principal-component analysis (PCA), canonical correlation analysis, and path analysis have been used to assess correlations between fungal taxa/compositions (or concentrations) of biological agents and various environmental parameters. A study in 2002 used PCA to evaluate airborne fungi in large office buildings and its correlations with other environmental parameters to identify the fungal taxa that covaried in air.[9] Factor analysis and PCA may allow researchers to predict what agents will be found in particular types of environments.[2] S plus has functions under Statistics–Multivariate–Factor Analysis and Principal Components for these types of data analysis.

9.4. *t*-TEST AND NONPARAMETRIC METHODS

Each indoor microbial contamination investigation is conducted to answer several questions. One of the most common questions is "Is the indoor microbial concentration detected from the area with complaints different from the no complaint areas?" In order to answer this question, sufficient numbers of samples should be collected to represent microbial levels in both areas. The standard *t*-test can be conducted to test the hypothesis if the following three assumptions are met. The first assumption is that the two groups of data must be independent. Second, the distribution of each group of data should be normally distributed, and third, the variances of two groups should be approximately equal. If the assumptions are not met, nonparametric methods are suggested for the hypothesis test. Normality tests that check a given set of data for similarity to the normal distribution can be performed using several statistical analysis software packages, such as SAS, SYSTAT, and NCSS.

9.4.1. Using *t*-Test to Test Differences between Two Means (Two Groups of Samples)

The *t*-test is a procedure to determine whether the probability of observed difference in the means is due to chance. In Microsoft Excel Data Analysis, you can find three different ways of doing *t*-tests: (1) paired two-sample for means, (2) two-sample assuming equal variances, and (3) two-sample assuming unequal variances. Types 2 and 3 can be used to determine whether the sample means from two independent groups are equal under aforementioned assumptions. If two groups of samples are not independent, type 1 is used. For example, if you want to compare the pre- and post-remediation microbial levels from the same sampling points before and after remediation, the paired *t*-test can be applied when the other two assumptions are met. The paired *t*-test computes the mean and standard error of the pre-versus-post differences, and determines the probability that the absolute value of the mean differences greater than zero is by chance alone.

t-Test results from Microsoft Excel present p $(T <= t)$ in the table. The commonly used significance level is 0.05: that is, if the p value is below 0.05, then the means between two groups of data are considered statistically significantly different.

9.4.2. Nonparametric Test for Two Groups of Samples

One common problem in indoor microbial contamination sampling is that the sample size is small and data distributions are very difficult to determine. When the assumptions for using the *t*-test are not met, several distribution-free tests (nonparametric tests) can be used to test hypotheses. For comparing multiple samples for a single biological agent collected at two sampling locations or times (e.g., day 1 and day 2), the Wilcoxon nonparametric test is useful for paired data not normally distributed. The parallel parametric test is a *t*-test of two sample means. The equation is discussed in the following section with the *H*-test.

Other nonparametric tests, including sign test, sign test for paired data, rank statistics, Wilcoxon signed rank test, and Wilcoxon–Mann–Whitney *U*-test, are described by Eudey et al.[1]

9.5. ANALYSIS OF VARIANCE (ANOVA) AND CHI-SQUARE STATISTICS

The *t*-test (or its nonparametric equivalent) is used to test the mean differences of two groups of samples. If more than two groups of samples are in the study design, analysis of variance (ANOVA) is used to test whether all groups have the same mean. This method has the advantage of testing whether there are any differences between the groups with a single probability associated with the test. The same assumptions for a *t*-test should be met when conducting an ANOVA—that is, all groups of data must be independent (unless a repeated measures design is used); all groups of data must be normally distributed and come from populations with

TABLE 9.5. An Example Dataset for One-Way ANOVA

	A	B	C	D
1	Livingroom	Bedroom	Children's Bedroom	Basement
2	24	12	10	46
3	20	8	17	45
4	21	10	14	45
5	15	11	9	35
6	22	15	12	49

equal variances. One question that may be asked during the investigations is, "Do all rooms, including livingroom, master bedroom, children's bedroom, and basement have the same levels of airborne microbial levels?" Microsoft Excel can be used to run ANOVA. In Data Analysis, you will find three ways to run ANOVA: single factor (one-way ANOVA), two-factor with replication, and two-factor without replication (two-way ANOVA). To run one-way ANOVA, first, enter data into data table, then go to Tools, Data Analysis, ANOVA: Single Factor. Enter input range, check how the data are grouped, check the box that says "Labels in first row" if applicable, set alpha (preset to 0.05), and then set output range.

Table 9.5 shows an example dataset for one-way ANOVA with five samples collected at each location. Table 9.6 shows the ANOVA results from Microsoft Excel. The value for F is calculated by dividing the mean square for between groups (locations) by the mean square for within groups, that is, $MS_{(\text{between groups})}/MS_{(\text{within groups})}$. The ANOVA output shows the F value and the critical value of F. In Table 9.6, the F value is greater than the F critical value; therefore the null hypothesis is rejected. The probability (p) of accepting the null hypothesis is well below 0.05, the means between groups of data (livingroom, master room, children's bedroom, and basement) are considered statistically significantly different.

After finding that there are differences between groups of data, we now want to know what the differences are. There are many methods available in other statistical packages (e.g., SAS) to investigate the differences including Duncan's multiple-range test (DUNCAN), the Student–Newman–Keuls multiple-range test

TABLE 9.6. An Example One-Way ANOVA Table Computed Using Microsoft Excel

	A	B	C	D	E	F	G
1	Source of Variation	Sum of Squares (SS)	Degree of Freedom (df)	Mean Squares (MS)	F	P Value	F Critical Value
2	Between groups	3476.8	3	1158.933	82.34	6.09×10^{-10}	3.24
3	Within groups	225.2	16	14.075	—	—	—
4	Total	3702	19	—	—	—	—

TABLE 9.7. An Example of Multiple Comparison Results Using SAS

	A	B	C	D	E	F
1	Groups	Count	Sum	Average	Variance	Difference from Groups
2	Livingroom	5	102	20.4	11.3	B
3	Master bedroom	5	56	11.2	6.7	C
4	Chilidren's bedroom	5	62	12.4	10.3	C
5	Basement	5	220	44	28	A

(SNK), least–significant-difference test (LSD), Tukey's studentized range test (TUKEY), Scheffe's multiple-comparison procedure (SHEFFE), and others. In order to do these tests in SAS, place the test name following a slash (/) on the SAS PROC ANOVA MEANS statement. Applying these methods using data in Table 9.5, all methods show grouping as follows: A (basement) > B (livingroom) > C (master room and children's bedroom) (Table 9.7). Locations with the same letters are not significantly different.

When the assumptions for using ANOVA are not met, several distribution-free tests (nonparametric tests) can be used to test hypotheses. The Friedman procedure, for more than two groups of samples from different locations or collected at different times, is equivalent to ANOVA. To test whether there is no difference existing among groups, a chi-square statistic is used, χ_r^2:

$$\chi_r^2 = \frac{12}{nk(k+1)} \sum_{m=1}^{k} (R_m)^2 - 3n(k+1)$$

where R_m is sum of ranks of each group (e.g., days or locations), n is the number of bioaerosol groups, and k is the number of groups.

As mentioned in the Spearman rank correlation, the rank of bioaerosol is calculated *within* each group. In the Friedman procedure for more than two groups of samples, the rank is calculated *among* groups; that is, if you have three groups, there will be only three ranks for each bioaerosol. A Microsoft Excel spreadsheet can be used to organize data for this test. Apply the Rank and Percentile function in Data Analysis, and remember to change of the defaulted "Grouped by" from Columns to Rows. R_m^2 and χ_r^2 can both be calculated in spreadsheet using the formula above (Table 9.8). The critical values of the χ^2 distribution have been listed by Macher[2] to compare with χ_r^2. In Table 9.8, a χ_r^2 was calculated as 4.67, which is less than a critical value of 5.99 for a degree of freedom f, as 2 (i.e., number of days minus 1) at the significant level ($p = 0.01$). The test shows that the differences among groups (day 1, day 2, and day 3) are not significant at the selected level.

TABLE 9.8. An Example Nonparametric Test Using Friedman Procedure

	A	B	C	D	E	F	G	H
1	Allergen	Day 1	Rank	Day 2	Rank	Day 3	Rank	χ_r^2
2	A	12	2	10	3	20	1	4.67
3	B	8	3	17	2	20	1	—
4	C	10	3	14	2	24	1	—
5	R_m	—	8	—	7	—	3	—

In some situations, we need to compare only one bioaerosol group, but sampled at different locations or times. The *H*-test or two-sample Wilcoxon test can be applied. The *H*-test (Kruskal–Wallis procedure) is a generalization of the two-sample Wilcoxon test, and is used to evaluate the differences among samples from more than two groups (e.g., different locations or times). *H* is calculated as the following. When $k = 2$, this equation is used for Wilcoxon test:

$$H = \frac{12}{N(N+1)} \sum_{i=1}^{k} \left(\frac{R_i}{n_i}\right)^2 - 3(N+1)$$

where R_i is the sum of ranks of observations, group; *N* is the total number of observations, sum of n_i; n is the number of observations in each *i* group; and *k* is the number of groups.

9.6. DISCUSSION

Several basic statistical methods are described in the sections above using Microsoft Excel, S plus, and SAS. The reader should consult standard statistical texts and refer to the user manual of each statistical package for more detailed and advanced information. Numerous studies have used these methods, and even more advanced statistical methods have been seen in the peer-reviewed literature. Methods used in those publications are not necessarily the best or the only approach to analyze data from your investigations. The presentations of the methods in this chapter provide the reader an opportunity to evaluate these methods. By using statistical tools to understand the database, this provides a great opportunity to better characterize exposure to indoor microorganisms.

Data missing because of unexpected events may occur in indoor microbial investigations. Imputation is one of the approaches to substitute some value for a missing data point. Once all missing values have been imputed, the dataset can be analyzed using standard statistical techniques. However, standard analysis techniques do not reflect the additional uncertainty due to imputing for missing data, and therefore further adjustments are necessary to account for this. Another approach to treating missing data is to omit them from data analysis. This may reduce sample size of data analysis, but may avoid errors or uncertainties from the data imputation procedure. However, if missing data were due to improper operation or malfunction of the equipment, the entire dataset collected would be questionable. Under such circumstances it is necessary to recollect the dataset, whenever time and cost are permitted.

The power of a *statistical test* is the probability that the test will reject a false *null hypothesis*. The higher the power means that there is a greater chance to obtain a *statistically significant* result when the null hypothesis is false. Increasing the sample size is the most commonly used method for increasing statistical power. Prior to investigation, a calculation of sample size is needed to enable the statistical test to generate accurate and reliable results that would be helpful. If the sample size is too low, the results will lack the precision to provide reliable answers to the questions that it is investigating. If the sample size is too large, time and resources will be wasted. However, because of the nature of wide variations of indoor microbiological sampling results, power analysis may not be practical to investigators in many situations.

REFERENCES

1. Eudey, L., H. J. Su, and H. A. Burge, Biostatistics and Bioaerosols, in *Bioaerosols*, H. A. Burge, ed., Lewis Publishers, Boca Raton, FL, 1995, pp. 269–307.
2. Macher, J. M., Data analysis, in *Bioaerosols: Assessment and Control*, J. Macher, ed., American Conf. Governmental Industrial Hygiene (ACGIH), Cincinnati, OH, 1999, pp. 13-1–13-16.
3. AIHA, Planning and conducting a survey, in *Field Guide for the Determination of Biological Contaminants in Environmental Samples*, L. L. Hung, J. D. Miller, and H. K. Dillon, eds., American Industrial Hygiene Association, Fairfax, VA, 2005, pp. 47–91.
4. Tsai, S. M., C. S. Yang, P. Moffett, and A. Puccetti, Comparative studies of collection efficiency of airborne fungal matter using andersen single-stage sampler and Air-O-Cell cassettes, in *Bioaerosol, Fungi and Mycotoxins: Health Effects, Assessment, Prevention and Control*, E. Johanning, ed., Boyd Printing, Albany, NY, 2001.
5. Kleinbaum, D. G., L. L. Kupper, and K. E. Muller, *Applied Regression Analysis and Other Multivariable Methods*, 2nd ed., Duxbury Press, 1987.
6. Spicer, R. C. and H. J. Gangloff, Limitations in application of Spearman's rank correlation to bioaerosol sampling data, *Am. Indust. Hyg. Assoc. J.* **61**:362–366 (2000).
7. Horner, W. E., A. G. Worthan, and P. R. Morey, Air- and dustborne mycoflora in houses free of water damage and fungal growth, *Appl. Environ. Microbiol.* **70**:6394–6400 (2004).
8. AIHA, Viable fungi and bacteria in air, bulk, and surface samples and *Legionella*, in *Field Guide for the Determination of Biological Contaminants in Environmental Samples*, L. L. Hung, J. D. Miller, and H. K. Dillon, eds., American Industrial Hygiene Association, Fairfax, VA, 2005, pp. 93–140.
9. Chao, H. J., J. Schwartz, D. K. Milton, and H. A. Burge, Populations and determinants of airborne fungi in large office buildings, *Environ. Health Perspect.* **110**:777–782 (2002).

CHAPTER 10

ECOLOGY OF FUNGI IN THE INDOOR ENVIRONMENT

CHIN S. YANG and DE-WEI LI

10.1. INTRODUCTION

Ecology is the scientific study of relations and interactions between organisms and their environment. Fungi are no different. Their growth indoors follows similar ecological principles. It is well understood that fungal spores are ubiquitous indoors and outdoors and that fungi require moisture, nutrients, reasonable temperature, oxygen, and a substrate to grow on. Fungal spores may exploit niches found in the indoor environments and grow when the appropriate requirements are available. Although observations have been made and described to a certain degree regarding fungi and the indoor environment, there has been very little or no systematic study or review of fungal ecology indoors by mycologists. An attempt was made to discuss "the fungal ecology of indoor environments."[1] The four-page section was so short and simplistic as to provide the reader a disservice. However, the intrinsic relationship between fungi and their growth requirements in the indoor setting have not been fully evaluated and discussed. A more recent review on indoor fungal ecology provides good and useful information.[2]

This chapter reviews existing mycology literature and discusses the ecology of indoor fungi on the basis of current knowledge as well as the authors experience and observations.

10.2. FACTORS AFFECTING FUNGAL GROWTH INDOORS

Many environmental parameters can influence fungal growth indoors. Some are biotic factors, and others are physical and chemical, or abiotic factors. Biotic factors include the presence of fungal propagules or spores, viability of spores, the nature of the fungal species, and competing fungi and other organisms.

Sampling and Analysis of Indoor Microorganisms, Edited by Chin S. Yang and Patricia A. Heinsohn

Abiotic factors include nutrients, temperature, moisture, pH, oxygen and carbon dioxide, and light. Fungi are not photosynthetic and require a wide range of nutrients, including carbon, nitrogen, sulfur, and various macro- and micronutrients for their survival and growth. Most fungi, including anamorphilic ones, and yeasts are little impacted in their growth by hydrogen ions, as measured in pH.[3] Fungi are generally considered mildly acidophilic, with a preferred pH of slightly acidic at pH ~5. Most fungi found in food and other substrates grew in a wide pH range of 3–8.[3,4] Some fungi, such as *Aspergillus niger* and *Penicillium italicum*, are able to grow at a pH as low as 2.[3] The pH levels in the substrates may affect competition between fungi and other microbes, such as bacteria. However, bacteria may outperform fungi at near-neutral pH and higher.[5] On the other hand, most building materials are within the range cited by Wheeler et al.[4] The impact of pH on indoor fungal growth is generally considered insignificant.

For practical purposes, moisture, nutrients, and temperature and their interactions with fungi indoors are discussed in detail below because they are the most important abiotic factors in determining whether fungal spores may germinate and grow. The impact of other environmental parameters is also discussed.

10.2.1. Biotic Factors

Four biological factors contribute to fungal growth indoors: presence of fungal propagules or spores, viability of spores, the nature of the fungal species, and competing fungi and other organisms.

It is generally agreed that fungal spores of a wide variety are ubiquitous in nature as well as in the indoor environment. There is no reason to doubt their presence indoors. However, only certain spores may find the right indoor niches and conditions for growth. Fungal spores in the environment may be dead or viable, either in dormancy or ready to germinate and grow. The viability of fungal spores is associated with a number of factors, such as age, UV light, and extreme conditions. Dead spores, although they may still be allergenic and contain secondary metabolites, cannot germinate and grow. Therefore, they do not cause infections. Some viable spores, however, may be in dormancy. Dormant fungal spores are usually either physically or chemically restricted for germination. A physical barrier, such as a thickened spore wall, restricts absorption of water for spore germination. Some fungal spores contain inhibitors, which prevent spores from germination. Some spores germinate only after stimuli are applied or available. There are also spores of fungi growing on burnted sites that require heat treatment for germination. Spores of obligate parasites (e.g., rust fungi) or of some symbionts (e.g., mycorrhizal fungi) will not germinate without the presence of appropriate hosts. Spores of xerophiles may not germinate if a_w (water activity) is not within its range. Although all these spores may be found indoors, the majority of fungi that grow indoors do not have such restrictions.

The majority of airborne fungal spores in a typical indoor environment are of outdoor origins, particularly from such sources as leaves and vegetation, or phylloplane.[6,7] Common phylloplane fungi include species of *Alternaria*,

Aureobasidium, Cladosporium, and *Epicoccum*.[7] In addition, soilborne fungi also contribute to airborne fungal spores. Spores of soilborne fungi are not uncommon indoors, particularly in carpet dust. Other sources of indoor fungal spora include wood products, foodstuff, vegetables, and fruits. It is important to point out that some fungi may come from more than one source. Species of *Alternaria, Aureobasidium, Cladosporium*, and *Epicoccum* have been described from multiple sources.[5,7–10] Scott correctly pointed out that many common indoor fungi have multiple sources.[7]

Fungi do not differ from other organisms. They compete with one another for space, nutrients, and moisture. Although little has been studied and reported about competition among fungi and other organisms, mycophagous mites and insects and their skeletons and feces have been observed among fungal colonies in samples collected from the indoor environment.[7,11–13] Samson and Lustgraaf observed the association of xerophilic fungi, *Aspergillus penicillioides* and *Eurotium halophilicum*, with house dust mites *Dermatophagoides pteronyssinus*.[14] Scott described fungal growth and association of mites along a vertical moisture gradient on a water-damaged drywall.[7] Hydrophilic fungi, such as *Stachybotrys chartarum* and *Acremonium* sp., have been observed near the lower extent of wallboard, in which the highest a_w is likely to exist. Mites and insects not only feed on fungal matter but also help dispersal of fungal spores when they crawl and move around.[12] Common storage mites *Acarus siro* and *Tyrophagus putrescentiae* are known to feed on common indoor mold, such as *P. chrysogenum*.[9] Mycoparasites, such as *Trichoderma harzianum*,[8] are commonly isolated from water-damaged, cellulose-containing building materials.[15] *Trichoderma harzianum* is also known as a cellulolytic fungus.[8] It is not clear whether *T. harzianum* grows independently, as a mycoparasite, or both.

10.2.2. Abiotic Factors

Nutrients. Fungi are achlorophyllous and nonphotosynthetic. Their survival and growth rely heavily on their ability to obtain nutrients, such as sugars, amino acids, vitamins, and macro- and micronutrients, from the substrates. Different species of fungi have different abilities to access and utilize simple or complex forms of carbohydrate, organics, and mineral nutrients.[16] Some fungi, such as species of *Aspergillus* and *Penicillium*, are called "sugar fungi" because they are fast-growing and prefer simple sugars. On the other hand, some fungi can break down complex carbohydrates (such as cellulose and lignin) or complex organics (such as wood) to obtain nutrients, and are usually late colonizers of the substrates.[17] Decomposition and degradation of a substrate is due to enzyme activities. The types of enzymes required depend on the substrates.[16]

The primary food source for indoor fungi is cellulosic matter. Cellulose is made of glucose units in a linear linkup with β-(1–4) bonds. Cellulose chains are further crosslinked by hydrogen bonds to form microfibrils and then cellulose fibers.[18] Cellulose fibers may come from several different plant sources: wood, herbaceous stem (i.e., flax and jute), leaf (i.e., manila), or seed (i.e., cotton). The majority of cellulose fibers used in building materials are processed, delignified wood fibers, which become much more susceptible to fungal attack, due to chemical or physical means used in the pulping process, than in their native form. The susceptibility of cellulose fibers to fungal attack is associated with their chemicophysical form. Cellulose in its native state consists mostly of crystalline form with a few amorphous sites. The number of amorphous sites increases along native cellulose during the pulping process, which increases the susceptibility to fungal attack.[18] This may also help explain the observation made by Chang et al. that used ceiling tiles were more susceptible to fungal growth than new ones in a moist-chamber study.[19] Aging of the ceiling tiles may also increase the number of amorphous sites. The widespread use of cellulosic materials in building materials as well as in building construction increases the availability of food sources for fungal growth. Karunasena et al. studied colonization and growth of *Cladosporium*, *Penicillium*, and *Stachybotrys* on ceiling tiles made of inorganics or containing organics (i.e., cellulose).[20] The results showed that inorganic ceiling tiles did not support the growth of fungi, while cellulose-containing ceiling tiles did. The authors suggested that inorganic ceiling tiles could serve as a replacement for cellulose-containing ceiling tiles to avoid fungal growth.

In wood, cellulose fibers are chemically bound to lignin. Lignin consists of large molecule units of complex polyphenolic polymers made of carbon, hydrogen, and oxygen elements.[18,21] Lignin helps protect cellulose from microbial attacks by reducing the accessibility of susceptible amorphous sites.[18] *Stachybotrys chartarum*, a known cellulolytic fungus[22] that grows mostly on paper products rather than wood products, is a good example.

In addition to wood and cellulose products, chemically modified cellulose derivatives are used in building finishing products. Emulsion paints may contain carboxymethyl cellulose. These derivatives are as susceptible to fungal attack as cellulose fibers.[18]

Although cellulose substrates are the primary carbon source for many cellulolytic fungi growing indoors, a study by Murtoneimi et al. demonstrated that other chemical components of building materials were found to influence growth and sporulation of *Stachybotrys chartarum* as well as cytotoxicity of the spores on 13 modified gypsum wallboards.[23] The spores produced on the wallboards were also tested for the inflammatory potential. It was reported that, in comparison to the reference board (nonmodified), growth and sporulation of the fungus decreased on wallboards in which (1) the paper liner was treated with a fungicide, (2) starch was removed from the plasterboard, or (3) desulfurization gypsum was used in the core. Spores collected from all wallboards exhibited cytotoxicity to macrophages. Biocide application did not reduce fungal growth; however, spores collected from biocide-treated board exhibited the highest degree of cytotoxicity. The conventional additives (a foam, an accelerator, and a water-reducing agent) used in the core were found to have some inhibitory effects on growth. Recycled wallboard and the board without the starch triggered spore-induced tumor necrosis factor alpha (TNF-α) production in macrophages. The study concluded that growth of a *Stachybotrys chartarum* strain on wallboard and the subsequent cytotoxicity of spores were affected by minor changes in the composition of the core or paper liners.

Most fungi identified as capable of growing indoors or on building materials[7,24–26] are also fungi known to cause food spoilage and biodeterioration.[3] With the existence of a wide range of organic matter in an indoor environment, one should not be surprised to find unusual fungi growing indoors.[27]

Temperature. Fungi, again like most organisms, usually grow in a wide range temperature. In this context, *range*, including minimum, optimum, and maximum temperatures, can be defined as a temperature profile. Each species has its own profile. Some are narrow and some are wide. Fungi that can grow in a wider temperature range may also have a competitive edge, for example, *A. fumigatus*. Most fungi, known as *mesophiles*, grow best at a temperature range near room temperature. Some fungi are classified as psychrophilic, literally cold-loving, or *thermophilic*, literally heat-loving. Psychrophilic fungi are defined as not growing above 20°C and having a minimum temperature at or below 0°C and an optimum temperature in the range of 0–17°C.[28] Thermophilic fungi, such as *Aspergillus fumigatus* and *Absidia corybifera*, have minima at or above 20°C, maxima at or above 50°C, and optima somewhere in the higher half of that range. Mesophiles are defined as those having minima above 0°C, maxima below 50°C, and optima between 10°C and 40°C. Unfortunately, fungi seldom follow this rigid artificial classification. *Aspergillus fumigatus* has a wide temperature range of 12–52°C and does not fit into a classification neatly. However, most fungi, particularly those commonly found growing indoors, are mesophilic.

Moisture. Fungi can utilize organics by secreting enzymes to break down organics into small molecules. Enzymes require water as a solvent for their activities as well as for the endproducts, such as simple sugars and amino acids. The most commonly utilized organics are sugars of various kinds, from simple sugars, such as glucose, to sugar polymers, such as cellulose. Some fungi, such as species of *Penicillium* and *Aspergillus*, grow quickly in the presence of simple sugars and are often called "sugar fungi."[17] *Stachybotrys chartarum, Trichoderma* species, *Chaetomium globosum, and Chrysosporium pannorum* (syn. *Geomyces pannorum*), to name only a few, are cellulolytic and capable of secreting cellulases to break down cellulose into simple sugars. Wood decay fungi are capable of producing cellulases, ligninases, or both to utilize complex organics, namely, wood.[21]

In the indoor environment food is usually abundant and temperature is usually moderate. Moisture is usually the critical factor limiting the germination of fungal spores and subsequent fungal growth.[26,29] It is biologically important for some fungi to regulate their spore release, such as some members of basidiomycetes and ascomycetes. Fungi, like all organisms, require water for various physiological as well as metabolic activities. Water serves as a solvent for carrying solutes in and out of hyphae and for enzymatic and other metabolic reactions.[28] It is also used to regulate and maintain the turgor pressure inside the cell so that it does not collapse as a result of high external pressure. Various terms, including *osmotic pressure*, *osmotic potential*, *water potential*, and *water activity*, have been used to define or describe the role that water plays in a biological system. Water activity (a_w), probably the most commonly used term, describes and defines the available free (not chemically or covalently bound) water in a substrate available for biological growth. It is measured when equilibrium is reached between atmospheric relative humidity and the water content in a substrate or a solution, or it expresses the available water in a substrate as a decimal fraction of the amount present when the substrate is in equilibrium with a saturated atmosphere (an equilibrium relative humidity of 70% around the substrate means that the substrate has a water activity of 0.70).[30] Water activity can be calculated as $a_w = p/p_0$, where p is the partial pressure of water pressure in the substrate and p_0 is the saturation vapor pressure of pure water under the same conditions.[5] Water activity is numerically equal to equilibrium relative humidity (ERH) expressed as a decimal. If a sample of substrate is held at constant temperature in a sealed enclosure until the water in the sample reaches equilibrium with the water vapor in the enclosure, then $a_w = \text{ERH}/100$. Another commonly used measurement of moisture in a substrate is moisture content in percentage of dry weight. It is often used in measuring moisture content of wood or wood products.

The use of ERH in calculation of a_w often gives the false impression that relative humidity (RH) is critically important and an overriding factor in indoor fungal growth. Relative humidity is the measurement of water molecules and water droplets in a given airspace. Since fungi do not and cannot grow in air, RH has only secondary effects on fungal growth, condensation, and hygroscopicity of materials. In fact, most indoor fungal growth occurs as a result of incoming water but not just high RH and condensation on indoor surfaces.[9,31,32] Scott suggested that *Stachybotrys chartarum* is a principal colonizer of paper products following water damage, but is rarely involved in materials subjected to condensation.[7] He also referenced that *Cladosporium cladosporioides, C. spheraospermum, Alternaria, Ulocladium, Aspergillus versicolor, Penicillium chrysogenum, P. griseofulvum*, and *P. spinulosum* were reported to cause disfiguration on interior surfaces of buildings due to condensation.[33] However, these observations may not be applied to every situation, since prolonged or large-scale condensation is no different from water damage. It is more likely that the difference in the observations was due to the duration of wetness and progession in fungal succession. Persistently high RH in a poorly ventilated condition does allow hygroscopic materials to increase a_w to a level that favors growth of xerophilic fungi.

The measurement of water activity indicates the minimal water requirement for a fungus to grow, which is seldom the optimal condition for fungi to compete and grow in the environment. In practice, it is also important to also look at the optimal a_w for each fungus. Clearly, fungi have competitive advantages when they are at their optimal growth conditions. *Aspergillus versicolor* has a reported a_w of 0.79 at 25°C[32] but an optimal a_w of 0.98 at 27°C.[34] It is, in fact, common to observe growth of *Aspergillus versicolor* on water-damaged materials, presumably at higher a_w. However, the range of a_w that a fungus can grow can be a competitive edge in finding an ecological niche. Fungi that can germinate and grow at a wider a_w range clearly have a competitive advantage. Water activities of specific fungi are available from many reference sources.[24,26,31,35,36]

Grant et al. found that minimal a_w for spore germination and growth of indoor fungi on building materials was much higher than on the fungal growth medium MEA.[32] Minimal a_w required was different for spore germination and growth of indoor fungi, and a_w for fungal growth is usually higher than for spore germination.[32] On building materials fungal growth and production of significant quantities of mycotoxins required significantly different water activity, 0.8 versus 0.95.[31] Moderately xerophilic fungi, such as *Penicillium* spp. and *Aspergillus* spp., will begin to grow at a_w between 0.78 and 0.90 depending on the composition of substrates of construction materials.[31]

Clarke et al., on the basis of an extensive review of mycological literature and laboratory validations with mold samples, defined six groups of fungi ranging from highly hydrophilic to highly xerophilic.[37] They used a growth limit curve, which is defined by the minimum combination of local surface temperature and humidity for which growth will occur on building surfaces, to define the six groups. The laboratory validation tests showed the minimum time for growth at the lowest RH to be 75 days. Highly xerophilic fungi have growth limited to >98% RH, while highly xerophilic fungi have growth limited to >75% RH. *Aspergillus repens* is considered highly xerophilic; *A. versicolor*, xerophilic; *Penicillium chrysogenum*, moderately xerophilic; *Cladosporium sphaerospermum*, moderately hydrophilic; *Ulocladium consortiale*, hydrophilic; and *Stachybotrys chartarum*, highly hydrophilic. The study did not determine the optimal RH for their growth. The authors suggested using such information to provide a design tool that can predict the likelihood and extent of mold growth.

In a similar but controlled laboratory study, Chang et al. found that new ceiling tiles supported the growth of *Penicillium chrysogenum* and *P. glabrum* at a_w 0.85 and a corresponding moisture content >2.2% and of *Aspergillus niger* at a_w 0.94 and a corresponding moisture content >4.3% on used ceiling tiles.[19] *Penicillium chrysogenum* is known to have a germination a_w at 0.78–0.85 and the minimum a_w for growth at 0.79,[26] while *Aspergillus niger* is reported to have a minimum germination a_w at 0.77.[5]

In a moist chamber study of new gypsum boards just off the production line, Doll and Burge showed that 11 fungal genera were present on new gypsum board without artificial inoculation. *Penicillium* spp. and *Aspergillus* spp. were found at 95% RH in moist chambers only on the paper sides, and the number of fungi found on the new gypsum board increased with increasing moisture content.[38] On one occasion *Stachybotrys* sp. was present on the gypsum boards. This study showed that new gypsum boards were not free from naturally occurring fungal spores, which will readily germinate and grow when a suitable moist condition was met.

Another common measurement of moisture content in a substrate or building material is percentage (%) of moisture on an oven-dry-weight basis.[21,39] Although moisture level on a percentage basis does not indicate the availability and amount of free water in a substrate and has no direct bearing on mold growth, the expression is still commonly used in the literature of certain industries. In the wood industry, the precent of moisture content is frequently used in practice as well as in the literature. Wood decay seldom occurs when the moisture of the wood products is below fiber saturation point, on average 30%.[21,40] However, the golden rule that is still in use today by many wood users to prevent the growth of microbes and the development of wood decay is not to allow moisture levels in wood to exceed 20%.[21]

To fungi, water is essential for growth whether it is clean, gray, or black.[1] Although contaminated water may have a higher content of organics to serve as nutrients, the requirement of water for fungal growth is otherwise the same.

Light. The effects of light on fungal development and ecology may be underestimated. According to Tan,[41] many fungi required light for the successful completion of morphogenesis or sporulation. He also opined that light may be one of the most crucial external factors controlling fungal development. Light elicits a number of fungal reactions from light-induced uptake of glucose in *Blastocladiella britannica* and the light-stimulated synthesis of protein and polychaccharides by *B. emersonii*,[42,43] through the influence on orientation and growth rate of hyphae and fruiting bodies, to the morphogenetic initiation of fruiting bodies.[44] Trinci and Banbury[45] noted that photoinduced changes in pigmentation and conidiophore growth occurred only in the part of a mycelium of *Aspergillus giganteus* exposed to light directly, but not the adjacent area kept in the dark. Light may affect the release of conidia of *Botrytis squamosa*.[46] A number of the perithecial ascomycetes need light to initiate ascospore discharge.[47,48] It was observed that light triggers spore release in several fungi.[49] A number of fungi have evolved phototropic mechanisms for their spore release so as to increase the efficiencies for spore dispersal.[30] In the Ascomycetes class, radiation, minimum humidity, changes in humidity, and minimum wind velocity were all directly correlated with levels of airborne ascospores.[48]

However, light in indoor environments is different from the light outdoors in spectra and intensity. It should be noted that there is literally no report on the responses of fungi to artificial light (fluorescent and incandescent light) and the light passing through window glass in indoor environments. Light normally is not considered a limiting factor for the development and reproduction of most indoor fungi. Further research on the effect of light on indoor fungi is necessary.

Oxygen and CO_2. Most fungi are obligate aerobes. Moore[50] noted that development of fungal fruit bodies requires oxidative metabolism [amplified glycolysis and the tricarboxylic acid (TCA) cycle activity] and, subsequently, oxygen supplies. Oxygen is not a limiting factor for indoor fungi in most situations. It was observed that fungi did not develop on the building materials continuously submerged in water. CO_2 is also very important to fungal development.[50] Buston et al.[51] found that elevated CO_2 concentration was necessary for perithecium formation of *Chaetomium globosum*, and the fungus developed the maximal number of perithecia in 10% of CO_2. It was found that high CO_2 concentration promoted mushroom stem elongation of agaricus whereas cap expanding was enhanced by lowered CO_2 concentration.[52,53] Poorly maintained or malfunctioning HVAC systems may lead to elevated CO_2 concentration and deteriorating indoor air quality. However, the effects of oxygen and CO_2 concentration on indoor fungi are not well understood because of lack of research.

10.2.3. Interactions Between Temperature, Moisture, and Fungi Indoors

Although little is known regarding interactions between temperature (T), moisture, and fungi in the indoor environment, studies on the interactions in the environment offer some useful information.

Although relative humidity (RH) and temperature were reported to be the most important environmental parameters regulating spore production,[54,55] current understanding of the relation between fungal growth and the availability of free water (a_w) in the growth substrate have significantly changed the importance of RH on spore production. Sufficient moisture is not only important in vegetative growth but also the most important factor in sporulation.[48] Spore release and dispersal of some hyphomycetes was correlated with increasing T and decreasing RH. Lower RH allows dry spores to become airborne. The moisture content of the indoor air was reported to have significant effects on all measurable airborne spore concentrations.[56] Spore release of *Botrytis squamosa*, which is most likely of outdoor origin, was promoted largely by declining RH, increasing T, and rain, but occasionally by increased RH.[46] On the other hand, Leach[49] found that spore release of *Drechslera turcica* and several other fungi was affected by decreasing RH but not by temperature changes. Drastic increases of conidia of *Cercospora asparagi* were trapped beginning at 7:00–8:00 a.m., when T increased and RH fell below 90%.[57] Beaumont et al.[58] reported a positive correlation between concentrations of total airborne spores of *Cladosporium* and *Botrytis* with higher temperatures.

Variations and fluctuations in indoor humidity and temperature were found to have significant effects on fungal growth,[33,59] such as in a bathroom situation with transient high humidities where dominant mycota included species of *Alternaria*, *Aureobasidium*, *Cladosporium*, *Phoma*, and *Ulocladium*.[25,60]

Basidiospores of *Paxillus panuoides* Fr., a mushroom likely found outdoors, were released when temperatures were above freezing, and daily peaks were usually correlated with increased T and decreased RH. Spore releases were found to increase from a temperature of 2°C, to a maximum at 37°C, but ceased at 45°C. RH treatments did not significantly affect spore release. Temperature was determined to be the stimulus for the natural spore release pattern.[61] Higher counts of basidiospores might be due to higher relative humidity and lower sunshine in 1978 in Galway, Ireland.[62] Li[63] found that RH and dew were positively correlated with release of basidiospores of *Amanita muscaria* var. *alba*, and T values showed no significant correlation with the release of the basidiospores. Less than 0.1% of the basidiospores released infiltrated a residence nearby. Because fluctuations and changes of RH are related to temperature variations as well as moisture sources, the effects of these two parameters could not be defined independently.

10.3. ECOLOGICAL INTERACTIONS BETWEEN FUNGI AND ENVIRONMENTAL FACTORS

A wide variety of fungal species have been documented to grow indoors.[24,25,27] The species are represented in all major fungal taxonomic groups: Ascomycetes, Basidiomycetes, the form class Deuteromycetes, Zygomycetes, and Myxomycetes. The taxonomic relationship between these groups is available from existing literature and is discussed only where appropriate.

10.3.1. Indoor Fungal Ecology

The observations and reports of fungi growing indoors are not new. Wood decay in buildings has been studied for generations.[21] In addition, fungi as biodeteriorating agents have been known. *Aspergillus versicolor* was reported as commonplace on the wallpaper surfaces in New Zealand public buildings in 1945.[64] Growth of *Scopulariopsis brevicaulis* in buildings leading to poison gas and illness in occupants was reported in the nineteenth century. Morton and Smith[65] reviewed the literature on *S. brevicaulis* and described how it was discovered to grow on building materials containing arsenic and produced a characteristic garliclike odor as far back as the early nineteenth century. The garliclike odor was identified as diethylarsine and later trimethylarsine. Cases of poisoning from exposure to volatile arsenic compounds derived from arsenic pigments in wallpaper or tapestry have been reported since 1820. The compounds were recognized by their typical garliclike odor, and the production of the odor was associated with the presence of "arsenic mold." The "mold" was later isolated and identified as *Scopulariopsis brevicaulis*. In 1932, a child died and several other people became seriously ill from living in damp rooms in which wallpapers contained Paris green, which had a small amount of arsenic. These suggest that damp indoor environments and mould growth are not a new problem.

In one of the early studies in the United States, Morgan-Jones and Jacobsen[66] identified and reported mold growth found on carpets, plasterboard, and wallpaper from three hotels in Florida and Georgia. Many fungi identified in these materials are known biodeterioration agents of paper, textiles, and plaster. The fungi identified from mold samples included species in the ascomycetous genus *Chaetomium*, the dematiaceous hyphomycete genera *Alternaria, Cladosporium, Stachybotrys*, and *Ulocladium*; the moniliaceous hyphomycete genera *Acremonium, Aspergillus*, and *Penicillium*; and the pycnidial genus *Phoma*. They reported 14 species, including two new species of *Cladosporium*, in 11 genera in the study.

Käpylä[67] of Finland found that *Aureobasidium pullulans* was the predominant fungus growing on wooden frames of insulated windows. In a study of toxicity of moldy building materials, Johanning et al.[68] not only detected cytotoxicity of the materials to cell cultures but also identified several groups of fungi and detected satratoxin H and spirolactone/lactams in the mouldy samples. The fungi identified included several of those described by Morgan-Jones and Jacobsen[66] and additional species of *Aspergillus*, *Paecilomyces*, and *Trichoderma* from gypsum wallboard and other building materials. In more recent studies, *C. sphaerospermum* was found to grow better than *P. chrysogenum* on various plaster materials, paints, and plasterboards under conditions with fluctuating a_w, but *P. chrysogenum* grew better than *C. sphaerospermum* did at constant a_w.[31,69]

Another source of fungal growth indoors is carpet flooring. Carpet is a sink for dust and fungal spores and a good reservoir for survival of fungal spores and their resuspension into air. Spores in the reservoir will germinate and grow when suitable conditions become available. The most common fungal genera identified in dust were *Aspergillus*, *Alternaria*, *Cladosporium*, *Epicoccum*, *Eurotium*, and *Penicillium* among 41 different genera and species.[70] In a comprehensive study of indoor fungi in dust samples collected from 369 houses in Wallaceburg, Ontario, Scott[7] found approximately 250 fungal taxa and identified the 10 most common species: *Alternaria alternata*, *Aureobasidium pullulans*, *Eurotium herbariorum*, *Aspergillus versicolor*, *Penicillium chrysogenum*, *Cladosporium cladosporioides*, *P. spinulosum*, *C. sphaerospermum*, *A. niger*, and *Trichoderma viride*.

Hodgson and Scott[71] studied 243 carpet dust samples collected from problem and control buildings in the United States. They found that average fungal concentrations, analyzed with the extraction–serial dilution method, in dust samples from problem buildings were higher than in control samples (1.8 million vs. 30,000 CFU/g). The most common dominant fungal group detected in problem buildings were species of *Penicillium*, followed by *Cladosporium*, and then *Aspergillus*. In the control buildings, *Cladosporium* was the most frequently detected dominant taxon, followed by *Penicillium*, *Phoma*, yeasts, and *Epicoccum nigrum* at equal percentages. *Aspergillus* was never a dominant taxon in the control buildings. Among individual *Aspergillus* species, *A. versicolor* was the most common species in samples from problem buildings. Other *Aspergillus* species of significance included *A. niger* and *A. sydowii*. They concluded that dominant taxa can also serve as an indicator of fungal contamination. *Penicillium* spp. and *Aspergillus* spp., especially *A. versicolor*, are generally the major or dominant taxa associated with problem building samples. The fact that *Penicillium* species are commonly isolated from both problem and control buildings makes interpretation of the data much more difficult, especially if the total concentration of fungi in the dust samples is in the range of 10^4 CFU/g. Knowing the species of *Penicillium* is very important in order to determine whether it is likely from an indoor source.

Scott[7] reported in the study on genotypic variation of two common dustborne *Penicillium* species, *P. brevicompactum* and *P. chrysogenum*. It was found that there were two genetically divergent groups of *P. brevicompactum* according to sequence analysis of the β-tubulin (benA) and rDNA loci. The predominant group, accounting for 86% of isolates, was found to cluster with the authentic strains of *P. brevicompactum* and *P. stoloniferum*.[7] Another report[72] revealed found that isolates of *P. brevicompactum* recovered from dust in houses in Ontario could be divided into two genetically divergent groups. They found that the two groups were distributed throughout the sample population and coexist at several sites. On the basis of their findings, they reasoned that the primary amplification sites of *P. brevicompactum* were outside the building environment, rather than the commonly held belief that it was of indoor origin.

A clonal pattern of inheritance in another common indoor fungus, *P. chrysogenum*, was observed and phylogenetic analyses of allele sequences found the population segregated into three divergent groups, accounting for 90%, 7%, and 3% of the house dust isolates, respectively.[7] Type isolates of *P. chrysogenum* and its synonymous species *P. notatum* clustering within the secondary lineage confirmed this synonym. No available species names could be applied to the predominant and minor lineages. This implied that *P. chrysogenum*, delineated previously, is a species complex including three taxa, two of which are currently unnamed. Further studies are necessary to determine the nomenclature status of the two unknown lineages of *P. chrysogenum*. Primary interest would be in the true identity of the dominant group currently identified as *P. chrysogenum*.

Horner et al.[73] reported that 50 single-family detached homes, built since 1945, with less than 2 ft^2 of known water damage, and located within a central city census tract in the metropolitan Atlanta city (DeKalb and Fulton Counties), Georgia were assessed to establish a baseline of "normal and typical" types and concentrations of airborne and dustborne fungi in urban homes. The homes were predetermined not to have noteworthy moisture problems or indoor fungal growth and were sampled twice (summer and winter) within a calendar year. Dust samples were sieved and inoculated by the "direct plating" method onto MEA and DG-18 media. *Cladosporium cladosporioides, C.* spp., and *C. sphaerospermum* were the top three fungal species in both indoor and outdoor air samples. Rankings by prevalence and abundance of the types of airborne and dustborne fungal spores were not found to differ from winter to summer, nor did the rankings differ when air samples taken indoors were compared with those taken outdoors. Water indicator fungi (such as *Chaetomium, Stachybotrys*, and *Ulocladium*) were essentially absent from both air and dust samples.

However, it should always be kept in mind that isolation and identification of fungi from carpet or carpet dust does not by itself indicate fungal growth. Spores, which are ubiquitous in indoor dust reservoirs, can germinate and grow on culture media. To confirm that a fungus is in fact growing in carpet flooring, a direct microscopic observation and examination of the carpet or dust for vegetative growth and reproductive structures is often necessary. In addition, if there is no history of wetting of the carpet or persistent excessive humidity and no water staining, then the carpet is unlikely to be supporting growth.

10.3.2. Specific Ecological Niche of Some Common Indoor Fungi

Most, if not all, common fungi found growing indoors are saprophytes, biodeterioration agents, or both. Species of *Aspergillus* and *Penicillium* are, in particular, well-known saprophytes and documented biodeteriorating agents of a wide variety of materials: fruits, foodstuff, produce, textiles, paints, and wood and paper products.[5,10,74–76] These two genera include more than 180 species for *Aspergillus* and more than 220 species for *Penicillium*.[77] With each genus, they include both xerophiles and hydrophiles and cover a wide range of a_w. On the other hand, some fungi have a unique and specific niche, while some can grow on a wide range of substrates.

Several *Aspergillus* species have commonly been reported from water-damaged buildings.[9,12,25,32] *Aspergillus versicolor*, one of the most common fungi in a wide range of water-damaged building materials, also occurs in many kinds of foodstuff, including spices, cereals, and nuts.[5,77] It was reported as commonplace on water-damaged wallpaper surfaces in New Zealand as early as 1945.[64] Additional reports of its occurrence in water-damaged environments are many.[9,12,25,32,68] Additional *Aspergillus* species reportedly recovered from water-damaged environments include *A. candidus, A. flavipes, A. fumigatus, A. flavus, A. niger, A. niveus, A. nidulans, A. ochraceus, A. terreus, A. sydowii*, and *A. ustus*.[78] *Aspergillus penicillioides* and *A. restrictus* were from a "dry environment, e.g., house dust."[78] *Aspergillus fumigatus, A. versicolor*, and *A. terreus* were isolated and identified from preservative-treated wood poles.[75] Two xerophilic species of *Aspergillus* growing on glass were isolated and identified.[79]

One interesting and unique species of *Aspergillus* is *A. oryzae*, which is traditionally used in oriental food fermentation processing, including the production of soy sauce, miso, and enzymes.[80] Its primary distribution is in soil of the tropics[77] and is seldom reported from the United States or elsewhere in North America, if ever. The fungus is morphologically and phylogenetically very close to *A. flavus* and *A. parasiticus*, two aflatoxin producers.[77,81] However, isolates of *A. oryzae* do not produce aflatoxin.[8] Therefore, one must be extremely careful when *A. oryzae* is reported in samples collected from North American or other temperate regions. Chances are that it is a misidentification by the laboratory and probably not *A. oryzae*.

Species of *Penicillium* are common contaminants of indoor environments. However, individual *Penicillium* species may thrive in specific niches because of their tolerance or their ability to utilize certain chemicals in the substrates. *Penicillium decumbens, P. fellutanum*, and several species of *Penicillium* are known to have significant tolerance to formaldehyde, and can commonly be found in wood and building products, which contain formaldehyde or phenolformaldehyde.[12,82] *Penicillium chrysogenum* and *P. waksmanii* can colonize particleboard and utilize phenolformaldehyde resins as a carbon source.[10,83] *Penicillium chrysogenum* is considered to be one of the most widespread *Penicillium* species and "among the most common eukaryotic life forms on earth,"[74] including indoors. *Penicillium spinulosum*, and *P. ochrochloron* can tolerate high concentrations of copper because they can produce oxalic acid, which precipitates copper ions into relatively insoluble copper oxalate as a mechanism of tolerance.[84] *Penicillium spinulosum* is also known to have a high degree of resistance to preservatives and other chemicals.[3] *Penicillium arenicola* was also reported to have a high copper tolerance.[74] *Penicillium glabrum* was identified as one of the dominant primary colonizers in a copper–chrome–arsenate (CCA)-treated sapwood of *Pinus radiata*.[10,85] Therefore, it should not be a surprise to recover these species from CCA-treated building lumber. *Penicillium diversum, P. funiculosum*, and *P. janthinellum* were isolated and identified from wood utility poles treated with wood preservatives.[75] A wide variety of *Penicillium* species were recorded from a study in Canada as well as from other studies of wood products.[10] Not surprisingly, many species on the list included those commonly found growing indoors, such as *P. brevicompactum, P. chrysogeum, P. decumbens, P. glabrum, P. commune, P. crustosum*, and *P. spinulosum.* Pitt[3] considered *P. purpurogenum, P. variabile, P. funiculosum, P. minioluteum, P. pinophilum*, and *P. verruculosum* to be common biodeteriogens that are capable of producing very diverse degradative enzymes. They can also grow in a wide range of moist environments. Textiles and canvas in tropical regions, mine equipment, cooling towers, corks, and rubber are all susceptible to invasion by these species.[3]

Many species of *Penicillium* are also common food contaminants and cause spoilage in foodstuff,[5] particularly in low-temperature storage at −2°C and 5°C.[3] Several species have specific affinity for certain foodstuff or fruits. *Penicillium italicum* and *P. digitatum* are well known to grow and cause rot on citrus fruits. *Penicillium italicum* is common on lemons and other citrus fruits, while *P. digitatum* has a preference for oranges. *Penicillium expansum* is also a well-known rotter of pomaceous fruits, such as apples, pears, and other fruits and plant tissues. *Penicillium solitum* is common in processed meat products and is a pathogen on apples.[74] *Penicillium camemberti* and *P. roqueforti* are known as the "cheese molds" and for their role in cheese manufacturing. However, *Penicillium roqueforti* is a soil fungus,[9] has a wide distribution, and is a common spoilage fungus, even at refrigeration temperature.[5] The authors have observed and identified this fungus at elevated concentrations from samples collected from a mine. *Penicillium camemberti*, on the other hand, is rare outside of cheese-associated environments. It is considered a domesticated strain of *P. commune.*[9]

Cladosporium is a large and diverse genus with over 60 species.[77] Common species include *C. cladosporioides, C. herbarum*, and *C. sphaerospermum.* They are considered phylloplane fungi but have been recovered from soil[8] and other substrates. *Cladosporium herbarum* can grow at a very low temperature (−5°C) and have caused problems in chilled beef.[3] The other two species are also common indoors and tend to grow on cold condensing surfaces[7] and on fibrous glass insulation in the HVAC system.[11]

Chaetomium is an ascomycetous genus and includes over 80 species.[77,86] They are cellulolytic and cause soft rot in wood. *Chaeomium globosum* is the most commonly encountered species in wood and paper products or cellulose-containing materials in moldy buildings;[31,78] however, one should not be surprised to find other species of *Chaetomium* in chronically water-damaged buildings. *Chaetomium globosum* and other species of *Chaetomium* are considered as tertiary colonizers because of their cellulolytic nature[17] and high water activity requirement.[31,32]

Scopulariopsis is a genus comprising about 20 species.[65] Several species, including *S. brevicaulis, S. candida*, and *S. fusca*, have been documented to grow indoors.[9,24,78] Two additional species, *S. chartarum* and *S. brumptii*, have been isolated and identified from moldy indoor materials by the authors. All the species have been reported from a wide variety of substrates, including paper, wallpaper, straw, soil, cheese, human tissues (such as toenail and skin), silkworm, stored meat, leather, awnings, and fabrics, or as a culture contaminant.[64,65] *Scopulariopsis brevicaulis* is the most common species detected in the genus and has been associated with the production and release of methylated arsenic gas and poisoning of building occupants.[65] Smith[87] reported that this species flourishes on high-protein substrates, such as meat and cheese.

Trichoderma harzianum, T. viride, T. citrinoviride, T. atroviride, T. longibrachiatum, and *T. koningii* have been recorded and reported from the indoor environment.[24,31,78] Many, if not all, species of *Trichoderma* are cellulolytic and grow well on wood and paper-based substrates. *Trichoderma viride* is known to produce powerful cellulases and is a common fungus growing on textiles, paper, and timber.[3] *Trichoderma reesei* is used in industrial-scale production of cellulase enzymes.[88] Bissett[89] identified species of *Trichoderma* from ureaformaldehyde–based foam insulation (UFFI). However, UFFI has not been a popular building insulation material in Canada or the United States for many years. The likely source of *Trichoderma* in indoor dust was soil.[7] *Trichoderma harzianum* is also a well-known mycoparasite, a fungus that parasitizes another fungus[8] and is commonly referred to as "green mold" in the mushroom cultivation industry. Occurrence of this fungus can lead to serious yield losses to production of cultivated mushroom *Agaricus bisporus*[90] and other species. Samuels considered many *Trichoderma* species to have been incorrectly identified and regional or limited to certain ecological or geographic locales.[88] *Trichoderma harzianum* and *T. asperellum* are considered two true cosmopolitan species.[88]

Stachybotrys and *Memnoniella* are currently recognized as two genera.[22] The primary characteristic for differentiation between these two genera is whether condia are in a slimy mass (e.g., *Stachybotrys*) or in dry chains (e.g., *Memnoniella*). However, many mycologists have long considered that the choice of using that characteristic to differentiate between these two genera is insufficient. In fact, studies have shown that some species can produce both dry spores in chains and slimy spores en masse.[27,91,92] Many *Stachybotrys* and *Memnoniella* species were shown to be cellulolytic.[22] *Stachybotrys chartarum, S. chlorohalonata*, and *M. echinata* are the primary species detected in moldy cellulose-containing building materials. *Stachybotrys chlorohalonata* is a new species segregated from *S. chartarum sensu lato* on the basis of morphological, three-gene fragment sequences, and mycotoxins profiles.[93] However, other unusual species, such as *S. yunanensis, S. nephrospora, S. microspora*, and *S. elegans*, have also been reported from moldy buildings.[27,94] *Memnoniella echinata*, although not as common as *Stachybotrys chartarum* and *S. chlorohalonata*, can be found on cellulose-containing materials suffering from long-term water damage. Li and Yang[94] found that the PCR primers and probe for detecting *S. chartarum* used by commercial laboratories were not able to differentiate *S. chartarum* from *S. chlorohalonata* and *S. yunnanensis*.

Although microfungi or hyphomycetes are the most common species to grow indoors, wood decay basidiomycetes, such as *Sistotrema brinkmannii, Meruliporia incrassata* (a.k.a. *Poria incrassata*), and *Serpula lacrymans* (a.k.a. *Merulius lacrymans*), are occasionally identified and reported in wood-constructed buildings.[24,78]

Sistotrema brinkmannii was reported growing on wet, decaying wood-framed windows.[24] *Meruliporia incrassata* is an important wood decay fungus, causing brown cubical rot, and is commonly found in the western United States, such as California. *Serpula lacrymans* is another wood decay fungus. It usually grows on soft wood in constant contact with wet masonary.[24] The latter two fungi are often called "dry-rot fungi" because they appear to grow on dry wood. They can send out rhizomorphs, a rootlike structure, and mycelial fans to search for and transport back water and nutrients.[21] The authors have occasionally isolated these fungi in decaying wood materials from buildings in the United States. A real-time polymerase chain reaction (real-time PCR) has been developed for the detection of *Meruliporia incrassata* and *Serpula lacrymans*.[95]

Fungi, depending on individual species, may grow at a wide range of moisture conditions, from xerophilic to hydrophilic. Xerophilic fungi, literally meaning dry-loving, were defined as those that can grow at a moisture content of the substrate below a_w 0.85.[96] Xerophilic fungi, by definition,[5] usually have an absolute requirement for reduced a_w. In addition, xerophilic fungi, such as *Aspergillus penicillioides* and *A. restrictus*,[3,5,76,81] are usually slow growers and cannot compete effectively with nonxerophilic fungi at high a_w conditions.[5] Xerophilic species of *Eurotium*, which are the sexual or the teleomorphic form of some *Aspergillus* species, are slow-growing on high-water-activity media but are relatively fast-growing on xerophilic or low-water-activity media, such as DG-18 or media supplemented with 20–40% sucrose or NaCl.[5,76,81] Therefore, xerophilic fungi are ecologically unique. Their survival and growth indoors rely on low or reduced a_w in which fast-growing non-xerophilic fungi are inhibited or limited in their competitiveness. The ideal conditions for xerophilic fungi are persistent high RH with stagnant ventilation, in which hygroscopic substrates can absorb sufficient water vapor to raise a_w high enough for xerophilic fungi, there is no direct contact with liquid water. *Aspergillus penicillioides*, *A. restrictus*, *Eurotium amstelodami*, *E. chevalieri*, and *Wallemia sebi* are common xerophilic species found growing indoors and were reported from "dry environment, e.g., house dust."[78] Scott[7] reported *Eurotium herbariorum* and *Wallemia sebi* in household dust samples from Ontario, Canada.

Some xerophilic fungi, which prefer growth in a high salt environment, are called halophiles.[5] They exhibit superior growth on media or substrates with high salt content. *Basipetospora halophia* (syn. *Scopulariopsis halophilica*) and *Polypaecilum pisce* are two known halophiles. Although there has been no report of their detection indoors, there is the possibility that they might grow on saltwater-damaged buildings by seashores.

10.4. RELEASE AND DISPERSAL OF FUNGAL SPORES

Fungi reproduce by means of spore production. There are two basic categories of spores, sexual and asexual. Sexual spores include ascospores, basidiospores, and zygospores. Asexual spores include mainly conidia and sporangiospores. Spores serve as propagules and are often released and dispersed through various means.

Fungal spores may be grouped into two types: dry or wet and slimy. Some spores are dry and powdery, while some are wet and slimy. Whether spores are dry or wet may affect their release and dispersal.

10.4.1. Fungal Spore Types and Their Release

In conidial fungi or molds, dry conidia are often released by passive mechanisms or by disturbance such as gravitation force, convection currents, deflation force, or mechanical disturbance. Some spores are actively released by osmotic force built up with water. Movement and feeding of arthropods, such as insects and mites, are also a common release–dispersal mechanism.[97–99] Drifting mist or tiny water droplets in air current can serve as a secondary process to carry and disperse spores of *Cladosporium*[97] and *C. herbarum.*[100] Dry-spored fungi include such common indoor fungi as species of *Alternatia, Aspergillus, Cladosporium*, and *Penicillium.* Wet-spored or slimy-spored fungi include species of *Acremonium, Aureobasidium, Fusarium, Phoma, Stachybotrys*, and *Trichoderma*, to name only a few. Active discharge mechanisms by osmotic force are not uncommon in fungi, but rare in molds or hyphomycetes. *Epicoccum nigrum* and *Arthrinium cuspidatum* are discharged when sudden rounding of the basal cell under the spore occurs.[97] Pressure buildup from water absorption is involved in spore discharge in many species of ascomycetes and basidiomycetes.[100] The well-known "cannonball" fungus, *Sphaerobolus stellatus*, absorbs moisture to build up pressure in the inner wall of the sporophore. The pressure and strain cause the inner wall to turn inside out and catapult the spore mass covered in a sporing tissue (the entire structure is called *gleba*) for a distance of several meters. The cannonball fungus is a basidiomycete and common on very rotten wood, dung of herbaceous animals, or mulch.[100] The black dots on exterior walls of houses and buildings are occasionally reported when rainfall occurs after new mulch is applied. The gleba is covered with a sticky liquid, which becomes tightly bound to the wall after drying.

Ascomycetes may actively discharge their ascospores by internal pressure of asci. Some ascomycetes, however, may dissolve their asci and release ascospores in wet masses for insect pickup and dispersal. *Chaetomium* species are wet-spore producers. Their ascospores are released in wet mass after their asci are dissolved.[86] Wind or air movement plays an important role in spore dispersal of most spore types and basidiospores. Air movement carries and disperses basidiospores after they are ejected from basidia and fall out from between the gills of the basidioma.[47] Basidiospores of *Coprinus* species can be carried away and dispersed by insects, which are attracted by odors produced by the mushroom and pick up slimy spore masses by contact. Although wet, slimy spore masses are likely to be picked up and dispersed by insects, they may become dry and airborne.

In hyphomycetes, dry-spore release is often influenced by air movement and wind speed.[48] The minimum wind speed needed to release dry conidia varies with each species, between 0.4 and 2.0 m/s.[97] Sutton et al.[46] reported that *Botrytis squamosa* spores were apparently released at very low wind speeds. Maximum wind speed was found to negatively correlate with spore concentrations of *Cladosporium* and

Alternaria and unidentified ascospores and basidiospores. It was the only factor among 10 weather parameters found to significantly correlate with all groups. Wind and air movement is known to play an important role in basidiospore dispersal. Minimum wind was directly correlated with spore counts, while maximum wind was inversely correlated.[48] High wind speed is also likely to disperse spore clouds and dilute spore concentrations. Li[63] found that wind speed was negatively correlated with basidiospore release of *Amanita muscaria* var. *alba*. A study of mushroom wood model and fresh *Agaricus bisporus* var. *bisporus* suggested that decreasing wind speed may reduce the number of the basidiospores being blown back into the cap.[101] Basidiospores are dispersed primarily by air movement after being released from their basidiomata.[47]

10.4.2. Fungal Spore Dispersal

Fungi have evolved to disperse their spores to increase their geographic area of distribution, chance of genetic development and survival,[100] and population diversity. Fungi can disperse their spores in short or long distances. However, the dispersal mechanism does have a significant implication on secondary contamination and human health when fungi are allowed to grow indoors.

Some fungi have their spores dispersed by other organisms, such as mycophagous animals, mites, and insects. The finding of mites and insects and their exoskeletons (molts) and feces on fungal colonies in indoor samples suggests a close association between animals and fungal growth.[11,12,22] Mites and insects not only feed on fungal matter but also help dispersal of fungal spores when they crawl and move around.[12] Common storage mites *Acarus siro* and *Tyrophagus putrescentiae* are known to feed on common indoor mold, such as *P. chrysogenum*.[9] The authors have frequently observed crawling mites and insects in association with mold growth on building materials subjected to long-term water damage. This has also been reported by other researchers.[7] New fungal colonies can be observed developing along the track of mites and insects, indicating dispersal of spores by them.

Spore dispersal by air movement is considered the primary route of spore dispersal for dry-spored fungi. Many fungal spores are adapted for aerial dispersal. In the outdoor environment, air movement is, however, often unpredictable in the transport and dispersal of fungal spores. How well spores are dispersed and survive over horizontal distance is determined largely by their ability to survive in the atmospheric environment.[102] However, little is understood on the airborne dispersal of fungal spores in the indoor environment. Human activities and the operation of the HVAC system are considered important factors in spore dispersal indoors.[26]

Although wind and air movement is known to assist in the release and dispersal of spores in nature, its importance in the indoor environment has not been studied and is poorly understood. The primary air mover in a mechanically ventilated environment is the HVAC system.[26] Since the HVAC system is powerful enough to move significant amounts of air through the system for distribution to occupied space, its impact on spore release and dispersal in the indoor environment cannot be overlooked.

10.5. CONCLUSIONS

Fungi require the availability of a number of important parameters, such as moisture, temperature, and nutrients, for their growth indoors. The primary limiting factor is the available free water, as measured in a_w or by moisture content in percent, in the substrate for spore germination and growth. Water and nutrient substrates, such as building materials, are the most important ecological factors for indoor fungi. The aerodynamics of airborne fungal spores is an area of future study for understanding the mechanism of fungal spore dispersal and subsequent distributions indoors. Fungal ecology in indoor environments is still poorly understood and only at the infancy stage of study because of the complexity of the indoor environment. Fungal succession on building materials and interactions among fungal contaminants in the indoor environment are largely unknown at this time except for observations and descriptions from field studies. There is no doubt that fungal ecology is an important issue for us to better understand fungal infestations, their spore dispersal, and relevant environmental factors. More research on fungal ecology in indoor environments is definitely needed. The current status of research in fungal ecology indoors is providing more opportunities and posing greater challenges for mycologists to face.

REFERENCES

1. IICRC, *IICRC Standard and Reference Guide for Professional Mold Remediation S520*, Institute of Inspection, Cleaning and Restoration, Vancouver, WA, 2003.
2. Hung, L.-L., J. D. Miller, and H. K. Dillon, *Field Guide for the Determination of Biological Contaminants in Environmental Samples*, 2nd ed., AIHA Publications, Fairfax, VA, 2005.
3. Pitt, J. I., Food spoilage and biodeterioration, in *Biology of Conidial Fungi*, Vol. 2, G. T. Cole and B. Kendrick, eds., Academic Press, New York, 1981, pp. 111–142.
4. Wheeler, K. A., B. F. Hurdman, and J. I. Pitt, Influence of pH on the growth of some toxigenic species of *Aspergillus, Penicillium* and *Fusarium*, *Internatl. J. Food Microbiol.* **12**:141–150 (1991).
5. Pitt, J. I. and A. D. Hocking, *Fungi and Food Spoilage*, Blackie Academic & Professional, London, 1997.
6. Dillon, H. K., P. A. Heinsohn, and J. D. Miller, *Field Guide for the Determination of Biological Contaminants in Environmental Samples*, AIHA Publications, Fairfax, VA, 1996.
7. Scott, J. A., *Studies on Indoor Fungi*, Ph.D. dissertation, Department of Botany, Univ. Toronto, 2001.
8. Domsch, K. H., W. Gams, and T.-H. Anderson, *Compendium of Soil Fungi*, Vol. 1, reprinted by IHW-Verlag with supplement by W. Gams, 1993.
9. Gravesen, S., J. C., Frisvad, and R. A., Samson, *Microfungi*, Munksgaard, Copenhagen, Denmark, 1994.
10. Seifert, K. A. and J. C. Frisvad, *Penicillium* on soild wood products, in *Integration of Moderan Taxonomic Methods for Penicillium and Aspergillus Classification*,

R. A. Samson and J. I. Pitt, eds., Harwood Academic Publishers, Amsterdam, Netherlands, 2000, pp. 285–298.

11. Yang, C. S., Fungal colonization of HVAC fiber-glass air-duct liner in the USA, Proc. *Indoor Air Conf.*, 1996, Vol. 3, pp. 173–177.
12. Samson, R. A. and J. Houbraken, Laboratory isolation and identification of fungi, in *Microorganisms in Home and Indoor Work Environments*, B. Flannigan, R. A. Samson, and J. D. Miller, eds., Taylor & Francis, London and New York, 2001, pp. 247–266.
13. Morey, P., Microbiological investigations of indoor environments: interpreting sample data-selected case studies, in *Microorganisms in Home and Indoor Work Environments*, B. Flannigan, R. A. Samson, and J. D. Miller, eds., Taylor & Francis, London and New York, 2001, pp. 275–284.
14. Samson, R. A. and B. v.d. Lustgraaf, *Aspergillus penicillioides* and *Eurotium halophilicum* in association with house-dust mites, *Mycopathologia* **64**:13–16 (1978).
15. Samson, R. A., J. Houbraken, R. C. Summerbell, B. Flannigan, and J. D. Miller, Common and important species of fungi and actinomycetes in indoor environments, in *Microorganisms in Home and Indoor Work Environments*, B. Flannigan, R. A. Samson, and J. D. Miller, eds., Taylor & Francis, London and New York, 2001, pp. 287–473.
16. Dighton, J., *Fungi in Ecosystem Processes*, Marcel Dekker, New York, 2003.
17. Singh, J., Nature and extent of deterioration in buildings due to fungi, in *Building Mycology*, J. Singh, ed., Chapman and Hall, London, 1994, pp. 34–53.
18. Allsopp, D., K. Seal, and C. Gaylarde, *Introduction to Biodeterioration*, 2nd ed., Cambridge Univ. Press, New York, 2003.
19. Chang, J. C. S., K. Foarde, and D. W. Vanosdell, Growth evaluation of fungi (*Penicillium* and *Aspergillus* spp.) on ceiling tiles, *Atmos. Environ.* **29**:2331–2337 (1995).
20. Karunasena, E., N. Markham, T. Brasel, J. D. Cooley, and D. C. Straus, Evaluation of fungal growth on cellulose-containing and inorganic ceiling tile, *Mycopathologia* **150**: 91–95 (2001).
21. Zabel, R. A. and J. J. Morrell, *Wood Microbiology*, Academic Press, San Diego, 1992.
22. Jong, S. C. and E. E. Davis, Contribution to the knowledge of *Stachybotrys* and *Memnoniella* in culture, *Mycotaxon* **3**:409–485 (1976).
23. Murtoneimi, T., A. Nevalainen, and M.-R. Hirvonen, Effect of plasterboard composition on *Stachybotrys chartarum* growth and biological activity of spores, *Appl. Environ. Microbiol.* **69**:3751–3757 (2003).
24. Flannigan, B., R. A. Samson, and J. D. Miller, *Microorganisms in Home and Indoor Work Environments*, Taylor & Francis, London and New York, 2001.
25. Samson, R. A., E. S. Hoekstra, J. C. Frisvad, and O. Filtenborg, *Introduction to Food- and Aiborne Fungi*, 6th ed., CBS, Utrecht, 2000.
26. Li, D.-W. and C. S. Yang, Fungal contamination as a major contributor of sick building syndrome, in *Advances in Applied Microbiology*, Vol. 55, *Sick Building Syndrome*, David Strauss, ed., Elsevier Academic Press, San Diego, 2004, pp. 31–112.
27. Li, D.-W. and C. S. Yang. Notes on indoor fungi I: New records and noteworthy fungi from indoor environments, *Mycotaxon* **89**:473–488 (2004).
28. Griffin, D. H., *Fungal Physiology*, 2nd ed., Wiley-Liss, New York, 1994.
29. Flannigan, B. and J. D. Miller, Microbial growth in indoor environments, in *Microorganisms in Home and Indoor Work Environments*, B. Flannigan, R. A. Samson, and J. D. Miller, eds., Taylor & Francis, London and New York, 2001, pp. 35–67.

30. Kendrick, B., *The Fifth Kingdom*, The Focus Publication, 2000.
31. Nielsen, K. F., *Mould Growth on Building Materials, Secondary Metabolites, Mycotoxins and Biomarkers*, Ph.D. thesis, BioCentrum-DTU Technical Univ. Denmark, 2002.
32. Grant, C., C. A. Hunter, B. Flannigan, and A. F. Bravery, The moisture requirements of moulds isolated from domestic dwellings, *Internat. Biodeter. Biodegrad.* **25**:259–284 (1989).
33. Adan, O. C. G. and R. A. Samson, Fungal disfigurement of interior finishes, J. Singh, ed., in *Building Mycology: Management of Decay and Health in Buildings*, Chapman & Hall, London, 1994, pp. 130–158.
34. Smith, S. L. and S. T. Hill, Influence of temperature and water activity on germination and growth of *Aspergillus restrictus* and *A. versicolor*, *Trans. Br. Mycol. Soc.* **79**:558–560 (1982).
35. Yang, C. S. and E. Johanning, Airborne fungi and mycotoxins, in *Manual of Environmental Microbiology*, 2nd ed., C. J. Hurst, G. R. Knudsen, M. J. McInerney, L. D. Stetzenbach, and M. V. Walter, eds., American Society for Microbiology, Washington, DC, 2002, pp. 839–852.
36. Yang, C. S. and E. Johanning, Airborne fungi and mycotoxins, in *Manual of Environmental Microbiology*, 3rd ed., C. J. Hurst, G. R. Knudsen, M. J. McInerney, L. D. Stetzenbach, and M. V. Walter, eds., American Society for Microbiology, Washington, DC. (in press).
37. Clarke, J. A., C. M, Johnstone, N .J. Kelly, R. C. McLean, J. A. Anderson, N. J. Rowan, and J.E. Smith, A technique for the prediction of the conditions leading to mould growth in buildings, *Build. Environ.* **34**:515–521 (1999).
38. Doll, S. C. and A. B. Harriet, Characterization of fungi occurring on "new" gypsum wallboard, *Proc. Indoor Air Quality Conf.*, 2001, pp. 1–8.
39. Simpson, W. T., Drying and control of moisture content and dimensional changes, in *Wood Handbook, Wood as an Engineering Material*, Forest Products Laboratory, USDA Forest Service, Madison, WI, 1999.
40. Highley, T. L., Biodeterioration of wood, in *Wood Handbook, Wood as an Engineering Material*, Forest Products Laboratory, USDA Forest Service, Madison, WI, 1999.
41. Tan, K. K., Light-induced fungal development, in *The Filanmenous Fungi*, Vol. 3, *Developmental Mycology*, J. E. Smith and D. R. Berry, eds., Wiley, New York, 1977, pp. 334–357.
42. Cantino, E. C., Light-stimulated development and phosphorous metabolism in the mold *Blastocladiella emersonii*, *Devel. Biol.* **1**:396–412 (1959).
43. Cantino, E. C. and E. A. Horenstein, The stimulating affect of light upon growth and carbon dioxide fixation in *Blastocladiella*. III. Further studies, in vivo and in vitro, *Physiol. Pantarum* **12**:251–263 (1959).
44. Page R. R., The physical environment for fungal growth. 3. Light, in *The Fungi*, Vol. 1, G. C. Ainsworth and A. S. Sussman, eds., Academic Press, New York, 1965, pp. 559–574.
45. Trinci, A. P. J. and G. H. Banbury, A study of the growth of the tall conidiophores of *Aspergillus giganteus*, *Trans. Br. Mycol. Soc.* **50**:525–538 (1967).
46. Sutton, J. C., C. J. Swanton, and T. J. Gillespie, Relation of weather variables and host factors to incidence of airborne spores of *Botrytis*, *Can. J. Bot.* **56**:2460–2469 (1978).
47. Moore-Landecker, E., *Fundamentals of the Fungi*, 4th ed., Prentice-Hall, Uppder Saddle River, NJ, 1996.

48. Lyon, F. L., C. L. Framer, and M. G. Eversmeyer, Variation of airspora in the atmosphere due to weather conditions, *Grana* **23**:177–181 (1984).

49. Leach, C. M., Influence of relative humidity and red-infrared radiation on violent spore release by *Drechslera turcica* and other fungi, *Phytopathology* **65**:1303–1312 (1975).

50. Moore, D., *Fungal Morphogenesis*, Cambridge Univ. Press, Cambridge, UK, 1998.

51. Buston, H. W., M. O. Moss, and D. Tyrrell, The influence of carbon dioxide on growth and sporulation of *Chaetomium globosum*, *Trans. Br. Mycol. Soc.* **49**:387–396 (1996).

52. Lambert, E. B., Effect of excess carbon dioxide on growing mushrooms, *J. Agric. Res.* **47**:599–608 (1933).

53. Turner, E. M., Development of sporocarps of *Agaricus bisporus* and its control by CO_2, *Trans. Br. Mycol. Soc.* **69**:183–186 (1977).

54. Smith, D. H. and F. L. Crosby, Aeromycology of two peanut leafspot fungi, *Phytopathology* **67**:1051–1056 (1973).

55. Mallaiah, K. V. and K. V. Rao, Aeromycology of two species of *Cercospora* pathogenic to groundnuts, *Proc. Ind. Nat. Sci. Acad. B* **46**:215–222 (1980).

56. Pessi, A. M., J. Suonketo, M. Pentti, M. Kurkilahti, K. Peltola, and A. Rantio-Lehtimaki. Microbial growth inside insulated external walls as an indoor air biocontamination source, *Appl. Environ. Microbiol.* **68**:963–967 (2002).

57. Cooperman, C. J., S. F. Jenkins, and C. W. Averre, Overwintering and aerobiology of *Cercospora asparagi* in North Carolina, *Plant Dis.* **70**:392–394 (1986).

58. Beaumont, F., H. F. Kauffman, T. H. vander Mark, H. J. Sluiter, and K. de Vries, Volumetric aerobiological survey of conidial fungi in the North-East Netherlands: I. Seasonal patterns and the influence of meteorological variables, *Allergy* **40**:173–180 (1985).

59. Vitanen, H. A. and J. Bjurman, Mould growth on wood at fluctuating humidity conditions, *Mater. Organismen* **29**:27–46 (1995).

60. Moriyama, Y., N. Nawata, T. Tsuda, and M. Nitta, Occurrence of moulds in Japanese bathrooms, *Internatl. Biodeter. Biodegrad.* **30**:47–55 (1992).

61. McCracken, F. I., Observations on the spore release of *Paxillus panuoides, Grana* **26**:174–176 (1987).

62. McDonald, M. S. and B. J. O'Driscoll, Aerobiological studies based in Galway. A comparison of pollen and spore counts over two seasons of widely differing conditions, *Clin. Allergy* **10**:211–215 (1980).

63. Li, D.-W., Release and dispersal of basidiospores from *Amanita muscaria* var. *alba* and their infiltration to a residence, *Mycol. Res.* **109**:1235–1242 (2005).

64. Raper, K. B. and D. I. Fennell, *The Genus Aspergillus*, R. E. Krieger Publishing, Huntington, NY, 1973.

65. Morton, F. J. and G. Smith, *The Genera Scopulariopsis Bainer, Microascus Zukal, and Doratomyces Corda*, Mycological Paper 86, CMI. Kew, Surrey, UK, 1963.

66. Morgan-Jones, G., and B. J. Jacobsen, Notes on hyphomycetes. LVIII. Some dematiaceous taxa, including two undescribed species of *Cladosporium*, associated with biodeterioration of carpet, plaster and wallpaper, *Mycotaxon* **32**:223–236 (1988).

67. Käpylä, M., Frame fungi on insulated windows, *Allergy* **40**:558–564 (1985).

68. Johanning, E., M. Gareis, C. S. Yang, E.-L. Hintikka, M.Nikulin, B. Jarvis, and R. Dietrich, Toxicity screening of materials from buildings with fungal indoor air quality problems (*Stachybotrys chartarum*), *Mycotoxin Res.* **14**:60–73 (1998).

69. Nielsen, K. F., Mycotoxin production by indoor molds, *Fung. Genet. Biol.* **39**:103–117 (2003).
70. Oppermann, H., C. Doering, A. Sobottka, U., Kramer, and B. Thriene, Exposure status of East and West German households with house dust mites and fungi, *Gesundheitswesen* **63**:85–89 (2001).
71. Hodgson, M. and R. Scott, Prevalence of fungi in carpet dust samples, in *Bioaerosols, Fungi and Mycotoxins: Health Effects, Assessment, Prevention and Control*, E. Johanning, ed., Eastern New York Occupational and Environmental Health Center, Albany, NY, 1999, pp. 268–274.
72. Scott, J. A., N. A. Straus, and B. Wong, Heteroduplex DNA fingerprinting of Penicillium brevicompactum from house dust, in *Bioaerosols, Fungi and Mycotoxins: Health Effects, Assessment, Prevention and Control*, E. Johanning, ed., Eastern New York Occupational & Environmental Health Center, Albany, NY and Mount Sinai School of Medicine, New York, 1999, pp. 335–342.
73. Horner, W. Elliott, A. G. Worthan, and P. R. Morey, Air- and dustborne mycoflora in houses free of water damage and fungal growth, *Appl. Environ. Microbiol.* **70**:6394–6400 (2004).
74. Pitt, J. J., *A Laboratory Guide to Common Penicillium Species*, Food Science Australia, 2000.
75. Wang, C. J. K. and R. A. Zabel, *Identification Manual for Fungi from Utility Poles in the Eastern United States*, American Type Culture Collection, Rockville, MD, 1990.
76. Klich, M. A., *Identification of Common Aspergillus Species*, Centraalbureau voor Schimmelcultures, Utrecht, Netherlands, 2002.
77. Kirk, P. M., P. F. Cannon, J. C. David, and J. A. Stalpers, *Ainsworth & Bisby's Dictionary of the Fungi*, 9th ed., CAB International, Egham, Surrey, UK, 2001.
78. Samson, R. A., Ecology, detection and identification problems of moulds in indoor environments, in *Bioaerosols, Fungi and Mycotoxins: Health Effects, Assessment, Prevention and Control*, E. Johanning, ed., Eastern New York Occupational & Environmental Health Center, Albany, NY and Mount Sinai School of Medicine, New York, 1999, pp. 33–37.
79. Ohtsuki, T., Studies on the glass mould. V. On two species of *Aspergillus* isolated from glass, *Bot. Mag.* **75**:436–442 (1962).
80. Gray, W. D. Food technology and industrial mycology, in *Biology of Conidial Fungi*, vol. 2, G. T. Cole and B. Kendrick, eds., Academic Press, New York, 1981, pp. 237–268.
81. Klich, M. A. and J. I. Pitt, *A Laboratory Guide to Common Aspergillus Species and Their Teleomorphs*, CSIRO, Division of Food Processing. North Tyde, N.S.W. Australia, 1988.
82. Vick, C. B., Adhesive bonding of wood materials, in *Wood Handbook, Wood as an Engineering Material*, Forest Products Laboratory, USDA Forest Service, Madison, WI, 1999.
83. Kerner-Gang, W. and H. I. Nirenberg, Identification of molds from particle boards and their matrix-dependent growth in vitro, *Mat. Ord.* **20**:265–276 (1987).
84. Moss, M. O., Morphology and physiology of *Penicillium* and *Acremonium*, in *Penicillium and Acremonium*, J. F. Peberdy, ed., Plenum Press, New York and London, 1987, pp. 37–71.
85. Butcher, J. A., Colonisation by fungi of *Pinus radiata* sapwood treated with a copper-chrome-arsenate preservative, *J. Inst. Wood Sci.* **28**:16–25 (1971).

86. Von Arx, J. A., J. Guarro, and M. J. Figueras, Ascomycete genus *Chaetomium, Beihefte zur Nova Hedwigia* **84**:1–162 (1986).

87. Smith, G., *An Introduction to Industrial Mycology*, 6th ed., Arnold, London, 1969.

88. Samuels, G. J., *Trichoderma*: Systematics, the sexual state, and ecology, *Phytopathology* **96**:195–206 (2006).

89. Bissett, J., Fungi associated with urea-formaldehyde foam insulation in Canada, *Mycopathologia* **99**:47–56 (1987).

90. Seaby, D., Differentiation of *Trichoderma* taxa associated with mushroom production, *Plant Pathol.* **45**:905–912 (1996).

91. Zuck, R. K., Isolates intermediate between *Stachybotryis* and *Memnoniella, Mycologia* **38**:69–76 (1946).

92. Li, D.-W., C. S. Yang, R. Haugland, and S. Vesper, A new species of Memnoniella, *Mycotaxon* **85**:253–257 (2003).

93. Andersen, B., K. F. Nielsen, U. Thrane, T. Szaro, J. Taylor, and B. B. Jarvis, Molecular and phenotypic descriptions of *Stachybotrys chlorohalonata* sp. nov. and two chemotypes of *Stachybotrys chartarum* found in water-damaged buildings, *Mycologia* **95**:1227–1238 (2003).

94. Li, D.-W. and C. S. Yang, Taxonomic history and current status of *Stachybotrys chartarum* and related species, *Indoor Air* **15**(Suppl. 9):5–10 (2005).

95. Lin, K. T., D.-W. Li, D. A. Dennis, R. Woodcock, and C. S. Yang, Qualitative identification of *Meruliporia incrassata* using real time polymerase chain reaction (real time PCR), in *Bioaerosols, Fungi, Bacteria, Mycotoxins and Human Health: Path-physiology, Clinical Effects, Exposure Assessment, Prevention and Control in Indoor Environments and Work*, E. Johanning, ed., Fungal Research Group Foundation, Albany, NY, 2005, pp. 335–345.

96. Pitt, J. I., Xerophilic fungi and spoilage of foods of plant origins, in *Water Relations of Foods*, R. B. Duckworth, eds., Academic Press, London, 1975.

97. Lacey, J., The aerobiology of conidial fungi, in *Biology of Conidial Fungi*, Vol. 1, G. T. Cole and B. Kendrick, eds., Academic Press, New York, 1981, pp. 373–416.

98. Van Asselt, L., Interactions between domestic mites and fungi, *Indoor Built Environ.* **8**:216–220 (1999).

99. Seeman, O. D. and H. F. Nahrung, Mites as fungal vectors? The ectoparasitic fungi of mites and their arthropod associates in Queensland. *Australasian Mycologist* **19**:3–9 (2000).

100. Ingold, C. T., *Spore Liberation*, Clarendon Press, Oxford, UK, 1965.

101. Deering R., F. Dong, D. Rambo, and N. P. Money, Airflow patterns around mushrooms and their relationship to spore dispersal, *Mycologia* **93**:732–736 (2001).

102. Tilak, S. T., Aerobiology and cereal crop diseases. *Rev. Trop. Plant. Pathol.* **1**:329–354 (1984).

CHAPTER 11

A RETROSPECTIVE AND FORENSIC APPROACH TO ASSESSMENT OF FUNGAL GROWTH IN THE INDOOR ENVIRONMENT

CHIN S. YANG

11.1. INTRODUCTION

Fungal spores are minute and invisible to the naked eye without the aid of a microscope. It may take days or weeks before a spore germinates and grows into a visible colony. Occupants of buildings are often surprised by the "sudden appearance" of fungal growth in the indoor environment without realizing that the process began days or weeks before. Spore germination and fungal growth require water and, therefore, are common in a damp or water-damaged environment. However, it is often difficult to determine when fungal germination and growth occur in an environment where long-term or repeated water damage may have occurred.

Because of interests due to litigations and insurance claims of water damage, fungal growth indoors, and the resulting health effects, issues are frequently raised on the timing of fungal growth and exposures. Although it is extremely difficult to determine the onset of fungal growth specifically and precisely, it is sometimes possible to retrospectively and forensically determine when observed fungal growth occurred in relation to the chronology of water damage events.[1]

This chapter discusses rationales, reasons, and processes useful in the retrospective and forensic determination of observed fungal growth in relation to the chronology of water damage events. A reasonable chronology of water damage and fungal growth can probably be estimated on the basis of an understanding of mycology literature, the ecology of indoor fungi (see Chapter 10 of this book and Ref. 2), the collection and critical analysis of environmental and building information and data, and the investigator's expertise, experience, and observations.

Sampling and Analysis of Indoor Microorganisms, Edited by Chin S. Yang and Patricia A. Heinsohn

11.2. MICROBIAL FORENSICS

Microbial forensics is a term that is applied to an analysis tracing a microbe to the source(s) of origin using molecular and genetic techniques.[3,4] The most common technique used is by comparing genetic variation and relatedness within and between species.[5] Although microbial forensics is relatively new, there are government surveillance systems and gene databanks that store and make DNA microbial fingerprints and gene sequences available for medical and economical investigations.

In response to the anthrax letter attacks in 2001 and other bioterrorism threats, many microbiologists and scientists in the United States have been looking for ways of using biologic or genetic markers to trace the origins of the microorganism. The scientific community in the United States as well as in many other countries has formulated a system of microbial forensics for investigation and attribution purposes in a bioterrorism event.[3,4] The basic principles of microbial forensics are useful not only in tracing bioterrorism agents but also in retrospectively and forensically determining fungal growth in relation to the chronology of water damage events indoors. The article by Budowle et al.[4] also highlighted the need for microbial forensic analysis to encompass sample identification, handling, collection, preservation and storage, method selection, casework analysis, interpretation of results, validation, and quality assurance.

The fungal genome is much larger and more complex than that of bacteria. In addition, much more research into bacterial genetics has been carried out as compared with fungal genetics, resulting in a greater understanding and knowledge of the bacteria genome. However, recent systematic studies of the fungal genome in important fungal groups, such as *Trichoderma*, are beginning to yield useful information, including authoritatively delineated species and their DNA sequences.[6] This makes future application of such information in fungal forensics possible. Although DNA-based forensic methods applicable to fungi are not fully ready at this time, this may become useful in the future. Nonetheless, retrospective tracing of fungal growth in the event of water intrusion in the indoor environment can be achieved using environmental and biological information and evidence. The same needs described by Budowle et al.[4] are applicable to the topic of this chapter.

11.3. RATIONALES AND MYCOLOGICAL BACKGROUNDS

Fungal spores are microscopic and invisible to the naked eye. Furthermore, developing fungal microcolonies may be small and inconspicuous in the early growth stages. Fungal growth may have occurred for a period of time and be hidden in the HVAC system, in wall cavities, behind baseboard, or underneath vinyl wallpaper. These make the detection of vegetative and reproductive growth in their early stage of development extremely difficulty, although not impossible. Some information can be obtained and used to determine whether the growth is relatively recent or has occurred for an extend period of time.

Indoor fungal growth is always the result and consequence of water damage or long-term moisture control problems. Information regarding any water damage history can provide an important clue to the approximate start of fungal growth. Observation of water-damaged building materials observed during inspection can also shed some light regarding a water damage history. Water is an excellent and common solvent and can dissolve many substances. A building material subjected to repeated water damage will show signs of varying degrees of deterioration. A ceiling tile may become stained, softened, and deformed because of repeated water damage. Nails and carpet strips may become rusty, stained, and weakened. Rusty steel beams and metal studs are common in buildings subjected to long-term water damage. The degree of rust development and the strength of such materials are good indicators of the history of water damage. If growth of cellulolytic fungi develops, cellulose-containing building materials can be weakened because of degradation of cellulose fibers and paper. In addition, wood-based building materials may be degraded and decayed by fungi to varying degrees. Severely decayed wood is not uncommon in wood-structured buildings suffering from long-term water issues from a few years to decades.

Fungal growth is usually concentric, and its colony is spherical.[7] This means that the greater the colony diameter, the longer the fungus has grown. Fungal growth is also affected by the fluctuation of moisture content or water activity (a_w) in the substrate, temperature, and other biotic and abiotic factors (see Chapter 10). Under optimal a_w and temperature conditions, fungal growth is fast. Under suboptimal conditions, the growth may be slow or may stop. Furthermore, different species may have different optimal growth conditions.[8] Therefore, the precise time chronology of a fungal colony growth can at best be estimated, and often depends on other environmental information.

Fungal succession in relation to resource utilization and penetrations by organisms such as oribatid mites and enchytraeid insects on Scots pine needles and fir needles have been studied and described.[9] Unfortunately, very little or no information on fungal and biological succession in the indoor environment is available, for a variety of reasons. The primary reason is that no owner of a building is willing to allow a moldy building to sit idle for many years. The author has, however, observed changes of fungal taxa in several water-damaged buildings over a time period. An example of changing fungal taxa over time occurred in an urban high-rise building that caught fire in winter in a northeastern U.S. city. Millions of gallons of water were used in firefighting. The entire building was soaked, including floors not directly damaged by fire, and no drying effort was implemented for many months. The building was also subjected to the prevailing weather conditions due to damage to the building envelope. Fungal growth was observed, and *Penicillium* and *Sporobolomyces* were detected in samples taken approximately one month after the fire. Additional sampling showed fungal genera *Aspergillus, Chaetomium, Penicillium*, and *Ulocladium* approximately 2 years later. A detailed assessment for fungal growth and contamination, approximately 5 years later, identified a wide variety of fungi, including species (except for *Sporobolomyces*) found in the first and second samplings as well as *Stachybotrys chartarum* and other uncommon fungi. This case study illustrates that changes and increased fungal diversity occurred in the building over a period of approximately 5 years.

In a chronically water-damaged environment, fungal populations and compositions are likely to change over time. Fungi are no different from any living organisms. They grow in an environment and compete with other fungi and organisms. The fungal population and composition as a whole evolve and change depending on various environmental factors, interaction with other fungi and organisms, and the biology of the fungi. The evolution of organisms in populations and biodiversity in an environment or ecological habitat is called *succession*. Knowledge of fungal succession in the indoor environment is poorly developed at present. However, some theories of indoor fungal succession have been discussed.[10,11] A knowledgeable mycologist can use a variety of information and means to evaluate fungal succession in a given indoor environment. The delineation of fungal succession in an indoor environment follows a forensic approach. Information and evidence gathering is important and should be as comprehensive as possible. A discussion follows on various observations, environmental information, and the biology of various fungi that are useful in conducting an assessment of fungal succession in a water-damaged environment.

Although fungal spores are often described as ubiquitous, there are significant differences in the ubiquitousness of each species or each genus. Yang et al.[12] compiled results of more than 2000 Andersen air samples, indoors and outdoors, collected on 2% malt extract agar in nonresidential buildings in the United States. By rank order, *Cladosporium, Penicillium, Alternaria*, basidiomycetes, and *Aspergillus* were the top five most frequently detected fungal groups outdoors. *Cladosporium, Penicillium*, basidiomycetes, *Aspergillus*, and *Alternaria* were the top five most abundant fungal groups outdoors. The indoor rank-order sequences in frequency and in abundance were slightly different from those outdoors. *Cladosporium, Penicillium, Aspergillus*, basidiomycetes, and *Alternaria* were the top five most frequently detected fungal groups indoors, while *Penicillium, Cladosporium, Aspergillus*, basidiomycetes, and *Alternaria* were the top five most abundant fungal groups indoors. Womble et al.[13] published findings of an USEPA-funded study including 86 randomly selected, public and private non-water-damaged office buildings in 10 climatic regions in the continental United States. Over 2000 time-integrated air samples for fungi were collected during winter and summer. Thirty genera of fungi were detected indoors and 28 outdoors. The five most commonly found fungal groups or taxa were the same for both indoor and outdoor samples. In descending order, based on the frequency of detection in a building, they were nonsporulating, *Cladosporium, Penicillium*, yeast, and *Aspergillus*. The findings are in general agreement with those of Yang et al.[12] A significant portion of nonsporulating fungi in the report[13] would likely have been reported as basidiomycetes if the laboratory had known how to identify them. Kendrick conducted a metaanalysis of 200 reports from around the world.[7] The results suggested that *Cladosporium* (mostly *C. herbarum*) accounted for an average of 33% of spore counts in air samples, basidiospores 22%, ascospores 14%, *Alternaria* 4.5%, and *Aspergillus*/*Penicillium* 3.5%.

Horner et al.[14] reported that *Cladosporium cladosporioides, C.* spp., and *C. sphaerospermum* were the top three fungal taxa in both indoor and outdoor air samples. Water indicator fungi (such as *Chaetomium, Stachybotrys*, and *Ulocladium*) were essentially absent from both air and dust samples from these houses.

In the same study,[14] a variety of *Penicillium* species were among the top 30 most abundant types of airborne and dustborne fungi. *Aspergillus niger* and *A. versicolor* were the only two aspergilli consistently found among the top 30 most abundant airborne and dustborne fungi. However, *Aspergillus niger* was consistently in the top 10 in air or dust samples, while *Aspergillus versicolor* was the 30th in abundance in indoor air samples, 25th in outdoor air, and 19th on MEA and 8th on DG-18 in dust samples. *Aspergillus versicolor* is moderately xerophilic and grows better on a moderately xerophilic medium, such as DG-18. Other fungi of consistent abundance in both air and dust samples includes *Epicoccum nigrum, Alternaria alternata*, yeasts, and nonsporulating fungi. This is in general agreement with the findings of Yang et al.[12] and Womble et al.[13] It may be concluded that after water damage fungal growth is more likely to be the common species rather than the rare species.

Some noteworthy and infrequently reported indoor fungi were described by Li and Yang,[15] Fernando et al.,[16] and Li and Yang.[17] Li and Yang[15] described seven new records or noteworthy fungi isolated from indoor environments in the United States: *Ascotricha chartarum, A. erinacea, Memnoniella echinata, Sporoschisma saccardoi, Stachybotrys microspora, S. nephrospora*, and *Zygosporium masonii*. All species were recovered from water-damaged building materials, such as drywall, wallpaper, wood, or in combinations of these substrates. Four species were reported for the first time from the United States. Fernando et al.[16] described nine infrequently reported fungi, including two ascomycetes and seven hyphomycetes, isolated from air samples collected in homes in the Syracuse, NY area: *Acremonium reseolum, Acrodontium intermissum, Acrodontium myxomyceticola, Aphanocladium album, Cladosporium macrocarpum, Gnomonia* sp., *Myxotrichum deflexum, Phialophora botulispora*, and *Tetracoccosporium paxianum. Acremonium reseolum, Acrodontium intermissum*, and *Acrodontium myxomyceticola* were very common in air samples collected from both indoors and outdoors in this study. The remaining six taxa were rare. Li and Yang[17] reported their study of several *Stachybotrys* species, including several rare or infrequently identified species, such as *S. elegans* and *S. yunnanensis*. A newly segregated species, *S. chlorohalonata*, was also included in the study. This shows not only that there are many species in the genus but also that the rare and infrequently identified species can be found in water-damaged indoor environments. These articles suggested that rare or uncommon fungi can be found growing indoors or in samples collected indoors. If rare or unusual fungi are found growing indoors, this strongly suggests that long-term water damage has occurred.

Fungi that produce abundant microscopic spores in a higher frequency have a better probability of encountering suitable environmental conditions for spore germination and growth. Fungi are likely to produce smaller spores in much more abundance than larger spores. Small-spored fungi, such as *Aspergillus versicolor* and *A. fumigatus*, produce spherical conidia (asexual spores) with a diameter of ~2.5 μm. On the other hand, *Stachybotrys chartarum* produces cylindrical conidia averaging ~5 × 9 μm.[17] On the basis of dimensions, *Aspergillus versicolor* is $\frac{1}{14}$th the size of *S. chartarum*. The same amount of energy and resources that are necessary to produce a single *S. chartarum* spore may be used to produce 14 *A. versicolor* conidia. A 2.5-cm-diameter colony of *Penicillium* was reportedly capable of producing 400 million spores[7]. In general, small-spored fungi are likely to have a wider distribution and better chance to encounter suitable environments for survival and growth.

Although fungi may produce abundant spores, many fungi may have only a portion of spores viable and thus capable of germination and growth. *Stachybotrys chartarum* reportedly has "only a small percentages of the conidia available in laboratory cultures."[18] In addition, growth of the fungus is inhibited by different species of *Penicillium*.[18] The second edition of the *Field Guide for the Determination of Biological Contaminants in Environmental Samples*[2] provides a table of longevity of spores of 18 fungi under air-dry conditions. Spores of *S. chartarum* are reported to have a half-life of approximately 0.8 year. This implies that the spores will die out in about 5 years if there is no moisture for germination, growth, and reproduction. Spores may lose viability over time if no suitable germination and growth conditions are available. Some fungal spores are known to remain viable for years.[2,7,19] A wide range of viability of fungi in laboratory storage, with several different storage methods, has been reported.[2,20,21] In general, basidiospores are considered short-lived, from a few days to a few weeks. Ascospores are long-lived, for a few years, and conidia of some species are long-lived but variable. For fungi, such as *S. chartarum*, a portion of their spores may be nonviable. This indicates that some fungi, in particular in spore form, are capable of remaining viable for an extended time period.

11.4. OTHER SIGNS AND INFORMATION

Environmental information and signs, such as a history of water damage in the environment, rust and the degree of rust in sheetmetals (such as galvanized sheetmetal) or metals (such as nails), deterioration of building materials (gypsum and paper layers), decay and degree of soundness in wood structures, and previous sampling data and reports, can also provide useful information for forensic analysis.

Galvanized sheetmetal and other metals are commonly used in a building. Metals, particularly iron, are oxidized and corroded when exposed to moisture. Oxidized iron is rust. Occasionally, blue-green oxidized copper (copper oxides) on copper or bronze structures in buildings is observed. Galvanized sheetmetal is iron sheetmetal coated with a thin layer of zinc in order to provide greater protection against corrosion. Zinc, when exposed to moisture and airborne pollutants (such as hydrogen sulfide and sulfur dioxide), becomes white powdery zinc carbonate, zinc oxides, and zinc sulfate (also known as "white rust"). Zinc carbonate and zinc oxides are not very water-soluble, and do not protect the metal from further corrosion because zinc carbonate and zinc oxides become crusty and split, exposing the iron metal for corrosion. Zinc carbonate and zinc oxides also react with moisture and acidic airborne pollutants to form water-soluble zinc sulfate. It is also important to point out that gypsum ($CaSO_4 \cdot 2H_2O$) in drywall wallboard may be a donor of SO_4 ion for chemical reaction with zinc.[22] On the other hand, zinc sulfate is water-soluble and will be dissolved and washed away on repeated moisture exposure. Zinc can chemically be corroded when exposed to plasters, cements (which contain chlorides and sulfates), acid rain, precipitation, and plant matter (such as wood shingles, moss, or lichens). Electrochemical corrosion between galvanized sheetmetal and dissimilar metals in the presence of an electrolyte, such as moisture, is also common.[22] Iron underneath the zinc coating will then be exposed to moisture to produce rust. Therefore, observation of rust on sheetmetal or iron nails is an indication of repeated exposures to moisture or acidic airborne pollutants.

Rust from metals, such as nails, can also produce chemical stains on wood. Many carpet wood strips or wood in buildings may show brown to black stains around nails due to exposure to moisture, whether from flooding, high humidity, or condensation. Iron stain, which chemically is iron tannate, is formed when iron reacts with water and tannin in wood.[23] The deeper the stain is, the longer it took to form. An intense black stain is likely due to repeated exposures or a long-term exposure to moisture. This chemical reaction can also degrade the wood. Investigators should be careful to distinguish between iron stain and fungal growth. If the investigator suspects growth of wood inhabiting fungi, it is important to collect a bulk sample and request the laboratory to conduct an evaluation of fungal growth.

Building materials, such as drywall, deteriorate as a result of repeated or long-term water exposures and mold growth. Gypsum drywall is essentially a thick layer of gypsum ($CaSO_4 \cdot 2H_2O$) sandwiched with starch between two layers of paper. Water slowly dissolves the gypsum mineral and creates pits or tiny holes. Repeated or long-term water exposures enlarge the holes. The larger the holes are, the longer the water exposure is likely. Furthermore, fungal growth on the paper layers by cellulolytic fungi, such as *Chaetomium globosum*, and *Stachybotrys chartarum*, causes paper to eventually disintegrate. The degree of paper disintegration is correlated with the duration of fungal growth.

Some fungi can cause wood decay, including brown rot, white rot, and soft rot. Wood decay by fungi is a slow and complex process because wood is composed predominantly of lignin and cellulose, which are complex polymers and require the involvement of complex enzymes to break them down.[11,23] Wood decay fungi are usually tertiary colonizers or latecomers in an ecological succession. They usually require $a_w \geq 0.90$ and can break down complex organics, such as cellulose or wood, for sugars. Both brown and white rots are caused by wood-decaying basidiomycetes. Soft rot is caused by microfungi and ascomycetes.[23,24] In the indoor environment, wood decay can be initiated only when moisture content in wood is greater than 20% and when spores or mycelia of suitable fungal species are present.[23,25] Wood structures close to soil can be colonized by wood decay fungi through mycelia. The degree of wood decay can be significantly affected by wood and fungal species. Some wood species, such as old growth baldcypress and redwood, are much more decay-resistant than are woods such as aspens and spruces.[26] Furthermore, heartwood is more durable than sapwood of the same species. Certain wood decay fungi are considered much more aggressive than others. For example, dry-rot fungi, *Meruliporia incrassata* and *Serupla lacrimans*, are considered aggressive wood decay fungi.[23] If wood members of a building show wood decay, this is an indication of a long-term issue due to chronic or repeated wet conditions. In general, the observation and detection of wood decay in a building is considered due to long-term moisture issues, generally years. If wood species and wood decay fungi can be positively identified, this will further narrow down the timing and assist in determining the duration and chronology of water damage leading to wood decay.

Other biological indicators, such as the presence of mycophagous mites, insects, or mycoparasitic fungi, are also excellent indicators of long-term (i.e., secondary or tertiary) colonization, in association with fungal growth. Mycophagous mites and insects (mites and insects that consume fungi) have been observed feeding on fungal matter in moldy insulation in the HVAC system[27] as well as on water-damaged moldy materials (see Chapter 10).[28–30] Observations of association between xerophilic fungi, *Aspergillus penicillioides* and *Eurotium halophilicum*, and common house-dust mites *Dermatophagoides pteronyssinus* have been reported.[31] Fungal growth and association of mites on a water-damaged drywall have also been described.[30] Mites and insects not only feed on fungal matter but also facilitate dispersal and spread of fungal spores when they crawl and move around.[28] Two common storage mites, *Acarus siro* and *Tyrophagus putrescentiae*, were reported to feed on common indoor mold, such as *Penicillium chrysogenum*.[18] When mycophagous mites, insects, and their exoskeletons (molts) and feces are observed and identified with fungal colonies in samples collected from the indoor environment, this is a very strong indication that their arrival must have been after fungal colonization and growth had occurred. Mycoparasitic fungi, such as *Trichoderma harzianum*,[6,32] are commonly isolated and identified from water-damaged, moldy, cellulose-containing building materials.[6,33] *Trichoderma harzianum* is also known as a "cellulolytic fungus."[32] *Trichoderma harzianum* is also a common soilborne saprophyte.[32] It is not clear whether *T. harzianum* grows saprophytically, as a mycoparasite, or both saprophytically and mycoparasitically indoors. Clearly, *T. harzianum* is likely a tertiary colonizer if it grows parasitically indoors. This also means that growth of *T. harzianum* follows the growth of preceding fungi and indicates long-term water problems and mold growth. One needs to be cautious in interpretating indoor *Trichoderma* findings because they can also be found in outdoor air.

Another rare form of parasitism was found and observed once in the indoor environment by the author. Some parasitic fungi are known to trap and utilize their prey as nutrient sources.[7] The author has in one instance, over 25 years of experience, observed that nematodes and *Harposporium anguillulae*, a nematode-trapping fungus, were found in water and sediment in a drain pan in an air-handling unit of an industrial building. This type of parasitic association has never been reported from a building environment in the literature, although there were two articles that reported nematodes in office buildings.[33,34] Nematodes are typically soilborne, although they may be transported via soil particles by foot traffic or in duststorms. *Harposporium anguillulae* is a well-known nematode-parasitizing fungus found in soils. The detection and identification of this rare parasitic association in a syntheitc (human-created) system suggested a long-term issue. Based on the roof location of the air-handling unit, which made it difficult for nematodes to reach, and the rarity of such parasitic association in the indoor environment, it was estimated that the nematodes and *Harposporium anguillulae* had been growing in the drain pan for more than a few years, possibly as long as 20 years when the system was installed on the building roof. In addition, it was confirmed that the drain pan and the HVAC had never been cleaned or serviced.

It is also important for investigators to work closely with highly qualified and experienced laboratories and scientists during a retrospective and forensic investigation. Careful observations and subtle signs under the microscope can be an important clue to the investigation. Detections of fragments and pieces of fungal structures, including spores, fruting spores, hyphae, and mycelia, are a strong indication that the fungal growth has died, most likely due to a period of prolonged drying. Investigators can also request that the laboratory look for signs of mycophagus mites and insects.

11.5. CASE STUDIES

Case 1

Background The new owners of a 20-year old townhouse took possession and found a decayed and rotten wood subfloor, between the first and the second floors, within the first month of ownership during a renovation. The decay was such that second floor could be seen from the first floor through two openings. In addition to the wood decay, *Aspergillus versicolor, Chaetomium globosum*, and *Stachybotrys chartarum* were identified from moldy drywall samples in two other locations in the townhouse. Questions were raised as to whether the moldy drywall, wood decay, and wood rot could have occurred within the month of new ownership.

Observations On a walkthrough inspection of the townhouse unit, three areas of water damage and fungal growth were evident. The wood structure between the first-floor kitchen area and the second-floor master bathroom was badly decayed. The subfloor plywood on the second floor was so badly decayed that two openings of approximately 16 × 6 in. and 12 × 4 in. were present. Some wood framing also showed varying degrees of wood decay. The extent of water damage and wood decay observed could be caused only by long-term water damage of probably 5 years or longer.

The other two areas of water damage and mold growth were on the second-floor bedroom and in the basement. Water stains and visible mold growth were observed on the wood framing and drywall of the chimney. Samples of mold growth on drywall collected on the second floor showed growth of *Aspergillus versicolor, Chaetomium globosum*, and *Stachybotrys chartarum*. The basement had a very strong musty, mold odor on entry. Visible mold growth was evident on the drywall as well as on cardboard boxes in the room. Some of the drywall was damp to the touch. A variety of fungi were identified from mold substrates.

Discussion The three areas of water damage appeared to be independent from each other. The water-damage on the second floor and the decayed subfloor on the second floor had no apparent direct link. The mold growth in the basement was limited to the bottom 3 ft of the drywall, suggesting wickup of water from the concrete floor.

Aspergillus versicolor is a late primary or secondary colonizer. It is not commonly found immediately after water damage. However, it appears rather quickly if the water-damaged condition is not addressed and dried. *Chaetomium globosum*, on the other hand, is a tertiary colonizer and a cellulolytic fungus and is known to cause soft rot in wood and degradation in paper products. Based on the author's extensive observations, it is often found to grow after approximately 6 months of prolonged wet conditions. *Stachybotrys chartarum* is a tertiary colonizer fungus and produces slimy spores, which indicates that their dispersal is not as effective as airborne dry spores. Because of its biology, the appearance and growth of *S. chartarum* usually requires at least 9 months of chronic or repeated wet conditions, provided that it starts with a dry, clean environment. Occasionally, new building materials, such as drywall and ceiling tiles, may be contaminated with spores, which will germinate and grow when wet. However, spores of *S. chartarum* are likely to die out rather quickly without water for germination because of their short half lives.[2] The 9–12-months timeframe may be modified if it occurs in a building with previous water damage and mold growth or in a new construction with contaminated building materials. This is based on the author's extensive field and laboratory experience as well as understanding and study of mycology. It was concluded that the water damage and mold growth on the second floor could not have occurred within a month of the real estate transaction.

Wood decay is a slow process. It requires the coexistence of water, spores of wood decay fungi, and wood products. The probability of these three factors occurring simultaneously is low. Furthermore, spores of wood decay basidiomycetes are usually short-lived, from a few days to a few weeks; therefore, the initiation of wood decay takes a long time. Wood decay also involves the process of digesting complex organic matter (cellulose and lignin) by wood decay fungi, which are late tertiary colonizers. For wood decay fungi to cause the degree of wood rot observed in this case, more than 5 years of long-term water damage was required, based on the best professional judgment. There is zero or little chance for wood decay of such magnitude to occur in a month.

In addition, it is highly unlikely that three independent events of long-term water damage and fungal growth occurred within one month after the real estate transaction.

Case 2

Background Samples of sediment and condensate water collected from the drain pan of several rooftop air-handling units (AHUs) were analyzed for fungal and bacterial growth. After a week of incubation at 25°C, many living and dead nematodes infected with a nematode-trapping fungus, *Harposporium anguillulae*, were observed on the agar medium under a dissecting microscope.

Observations It was reported that standing water and mudlike sediment were consistently observed in the sheetmetal drain pan, which had significant rust and corrosion. No record of HVAC maintenance was available during the inspection of the system.

Discussion *Harposporium anguillulae* is a well-known nematode-parasitizing fungus and is known to trap and utilize its prey as nutrient sources.[7] This is the only instance in which the author has ever observed nematodes and *H. anguillulae* in water and sediment in samples from the drain pan in AHUs in over 25 years of professional experience. The detection and identification of this rare parasitic association in a synthetic system suggested that this was a case of long-term neglect on cleaning and servicing the drip pans and the AHUs. Because of the roof location of the air-handling units (which is difficult for soilborne nematodes to reach except in an unusual airborne transport via soil particles in dusty conditions) rusty and corroded drip pans, and the rarity of such parasitic association in the indoor environment, it was estimated that the nematodes and *H. anguillulae* had been growing in the drain pan for years, possibly as long as 20 years, when the system was installed on the building roof. In addition, it was later confirmed, according to the building managers, that the drain pan and the HVAC had never been cleaned or serviced.

Case 3

Background A homeowner reported roof leaks and mold growth in a 20-year old house one month after a hurricane had passed through. A home inspector was hired to inspect and collect samples for laboratory testing. A laboratory reported *Stachybotrys chartarum* in the samples. A second inspector was hired to inspect, sample, and document reported water damage and mold growth.

Observations The second inspector found water stains on ceilings in a bathroom remote from the reported roof leaks. Two separate laboratories independently identified *Stachybotrys chartarum* from samples collected from the house. One of the laboratories detected signs of mites and their fecal particles in tape samples.

Discussion As discussed in Chapter 10, *Stachybotrys chartarum* is a hydrophilic and cellulolytic fungus, and a tertiary colonizer. Its appearance in a 20-year-old house within a month after hurricane-associated water damage is unlikely unless there is pre-existing contamination. Based on the author's own observations for over 20 years, it would take at least 9–12 months for *Stachybotrys chartarum* to appear and grow indoors as a result of chronic or long-term water damage, provided the building does not have a previously reported water damage and mold growth history. The possibility for slimy spores of *S. chartarum* to reach an area with long-term or chronic adverse water conditions and cellulose-containing materials is remote and requires an extended period of time. The observation of mites and their fecal particles extends the 9–12-month period further because additional time is needed for mites to find the fungal growth. It was concluded that *S. chartarum* was not the direct result of the hurricane water damage and was most probably due to preexisting water damage.

Case 4

Background A dispute arose between an insurance company and the owner of a five-story commercial office building in a major city on the Gulf of Mexico. The building owner claimed that mold growth, including *Stachybotrys chartarum*, in the building was due to various instances of water damage, which was covered by the insurance policy. The insurance policy had become effective 9 months before the discovery of mold growth, and the insurance company claimed that the water leaks and mold growth preexisted, before the effective date of the policy.

Observations On inspection, multiple roof leaks on the roofdeck and extensive mold growth was detected on the ground floor and the fifth floor, as well as the shaft of the outdoor air intake of the HVAC system. The sheetmetal roofdeck had significant loss of zinc coating and exposed rusty iron. The shaft was constructed with drywall and had obvious signs of *Stachybotrys chartarum* growth and water infiltration from the exterior. In addition, the interior walls of the building were covered with vinyl wallpaper. Bulk and sticky-tape samples collected from the building showed the presence of species of *Aspergillus, Chaetomium, Fusarium, Penicillium, Trichoderma*, and *Ulocladium*. The presence of insects and mites were observed. Dead oriental cockcoach was found on floors throughout the building. On close and detailed inspections, drywall underneath the vinyl wallpaper with mold growth was significantly weakened and deteriorated.

Discussion *Stachybotrys chartarum* and species of *Chaetomium, Trichoderma*, and *Ulocladium* are well-known hydrophilic and cellulolytic fungi and are tertiary colonizers. The use of vinyl wallpaper is known to serve as a moisture barrier and causes condensation in air-conditioned buildings in a hot, humid environment. On the basis of the inspections, the types of fungi identified, the detection of insects and mites among fungal growth, rusty sheetmetal roofdeck, significantly deteriorated drywall, and multiple water sources, it is highly unlikely and improbable that the water damage and mold growth occurred within the 9 months since the insurance policy had become effective.

Case 5

Background The roofdeck of a 25-year old school building was composed of gypsum wallboard. Visible black mold was found extensively covering the roofdeck during replacement of ceiling tiles. Tape and bulk samples were collected and analyzed with direct microscopic examination and culturing methods. *Stachybotrys chartarum* was identified as the predominant fungus by the direct microscopic examination method. The laboratory further reported that fungal structures appeared very brittle and broke into pieces and fragments during handling and preparation for the microscopic observation. However, repeated culturing failed to recover any fungi, in particular *Stachybotrys chartarum*, from the samples.

Observations On inspection, the roofdeck appeared dry; readings from a moisture meter confirmed this observation. Furthermore, there had been no report of active roof leaks or moisture problems associated with the roof or ceiling in previous years. There was a report that rainwater had penetrated the gypsum wallboards during roof construction.

Discussion Although *Stachybotrys chartarum* is a well-known hydrophilic and cellulolytic fungus and a tertiary colonizer, its growth is likely to slow and stop when a_w falls below the minimal threshold. It was possible that the growth occurred during roof construction when rainwater contaminated the gypsum wallboards. The growth was likely to die out after an extended drying period. The spores are also known to have a half-life of approximately 0.8 year and lose practically all viability in 5 years. That the fungus failed to grow during repeated culturing is another indication that the observed fungal colonies were dead. The microscopic observations that fungal structures appeared very brittle and broke into pieces and fragments on handling are another strong indication that fungal growth had ceased for an extended time period. Based on the inspections and moisture readings, no recent report of roof leaks or moisture problems, failure of the fungus to grow on culturing, and the microscopic observations of dead fungal fragments, the fungal growth appeared to occur during roof construction when rainwater contaminated the gypsum wallboards. An extended time period of dryness stopped the fungal growth, and the fungi eventually died out.

11.6. CONCLUSION

Based on the water damage history of a building, biological information of fungal growth indoors, and environmental signs and information, it is possible and feasible in many cases to estimate the chronology of fungal growth indoors both retrospectively and forensically. This is similar to the forensic thesis and approaches used in crime-scene investigations. Users of such information must have a clear understanding of mycology and fungal ecology in the building as well as keen observations of the environmental conditions conducive to fungal growth. An understanding of building materials, building structures, and building moisture dynamics is useful in such determinations. Experience in fungal assessment of the building environment is also important in educating professionals in defining the estimated time required for growth of certain species to occur. Individual professionals should be able to apply their practical field experience and calibrate with their geographic conditions to better define the chronology and timeline for establishment and growth of key moisture indicator fungi in the indoor environment.

REFERENCES

1. Li, D.-W. and C. S. Yang, Fungal contamination as a major contributor of sick building syndrome, in *Advances in Applied Microbiology*, Vol. 55, *Sick Building Syndrome*, D. Strauss, ed., Elsevier Academic Press, San Diego, 2004, pp. 31–112.
2. Hung, L-L., J. D. Miller, and H. K. Dillon, *Field Guide for the Determination of Biological Contaminants in Environmental Samples*, American Industrial Hygiene Association, Fairfax, VA, 2005.
3. Budowle, B., S. E. Schutzer, A. Einseln, L. C. Kelley, A. C. Walsh, J. A. L. Smith, B. L. Marrone, J. Robertson, and J. Campos, Building microbial forensics: As a response to bioterrorism, *Science* **301**:1852–1853 (2003).
4. Budowle, B., S. E. Schutzer, M. S. Ascher, R. M. Atlas, J. P. Burans, R. Chakraborty, J. J. Dunn, C. M. Fraser, D. R. Franz, T. J. Leighton, S. A. Morse, R. S. Murch, J. Ravel, D. L. Rock, T. R. Slezak, S. P. Velsko, A. C. Walsh, and R. A. Walters, Toward a system of microbial forensics: From sample collection to interpretation of evidence, *Appl. Environ. Microbiol.* **71**:2209–2213 (2005).
5. Cummings, C. A. and D. A. Relman, Microbial forensics—"cross-examining pathogens, *Science* **296**:1976–1979 (2002).
6. Samuels, G. J., *Trichoderma*: Systematics, the sexual state, and ecology, *Phytopathology* **96**:195–206 (2006).
7. Kendrick, B., *The Fifth Kingdom*, 3rd ed., Focus Publication, R. Pullins Co., Newburyport, MA, 2000.
8. Griffin, D. H., *Fungal Physiology*, 2nd. ed., Wiley-Liss, New York, 1994.
9. Ponge, J.-F., Fungal communities: Relation to resource succession, Chapter 8, in *The Fungal Community, Its Organization and Role in the Ecosystem*, J. Dighton, J. F. White, and P. Oudemans, eds., Taylor & Francis, Boca Raton, FL, 2005, pp. 169–180.
10. Grant, C., C. A. Hunter, B. Flannigan, and A. F. Bravery, The moisture requirements of moulds isolated from domestic dwellings, *Internal. Biodeter. Biodegrad.* **25**:259–284 (1989).
11. Singh, J., *Building Mycology*, Chapman & Hall, London, 1994.
12. Yang, C. S., L. L. Hung, F. A. Lewis, and F. A. Zampiello, Airborne fungal populations in non-residential buildings in the United States, *Proc. Indoor Air Conf.*, 1993, Vol. 4, pp. 219–224.
13. Womble, S. E., L. E. Burton, L. Kolb, J. R. Girman, G. E. Hadwen, M. Carpenter, and J. F. McCarthy, Prevalence and concentrations of culturable airborne fungal spores in 86 office buildings from the building assessment survey and evaluation (BASE) study, *Proc. Indoor Air Conf.*, 1999, Vol. 1, pp. 261–266.
14. Horner, W. Elliott, A. G. Worthan, and P. R. Morey, Air- and dustborne mycoflora in houses free of water damage and fungal growth, *Appl. Environ. Microbiol.* **70**:6394–6400 (2004).
15. Li, D.-W. and C. S. Yang, Notes on indoor fungi I: New records and noteworthy fungi from indoor environments, *Mycotaxon* **89**:473–488 (2004).

16. Fernando, A. A., S. E. Anagnost, S. R. Morey, S. Zhou, C. M. Catranis, and C. J. K. Wang, Noteworthy microfungi from air samples, *Mycotaxon* **92**:323–338 (2005).
17. Li, D.-W. and C. S. Yang, Taxonomic history and current status of *Stachybotrys chartarum* and related species, *Indoor Air* **15**(Suppl. 9):5–10 (2005).
18. Gravesen, S., J. C. Frisvad, and R. A. Samson, *Microfungi*, Munksgaard, Copenhagen, Denmark, 1994.
19. Yang, C. S. and R. P. Korf, *Ascorhizoctonia* gen. nov. and *Complexipes* amend., two genera for anamorphs of species assigned to *Tricharina* (Discomycetes), *Mycotaxon* **23**:457–481 (1985).
20. Smith, D. and A. H. S. Onions, *The Preservation and Maintenance of Living Fungi*, 2nd ed., CAB International, Oxon, UK, 1994.
21. Sussman, A. S., Longevity and survivability of fungi, in *The Fungi*, Vol. II, G. Ainsworth, and A. S. Sussman, eds., Academic Press, New York, 1968, pp. 12–20.
22. U.S. General Services Administration, *Galvanized Iron and Steel: Characteristics, Uses and Problems*, Historic Preservation Technical Procedures 05010-09 (accessed at http://w3.gsa.gov/web/p/hptp.nsf/0/ab5014b431e53490852565c50054b3e0?OpenDocument on Jan. 24, 2006).
23. Zabel, R. A. and J. J. Morrell, *Wood Microbiology*, Academic Press, San Diego, 1992.
24. Wang, C. J. K. and R. A. Zabel, *Identification Manual for Fungi from Utility Poles in the Eastern United States*, American Type Culture Collection, Rockville, MD, 1990.
25. Highley, T. L., Biodeterioration of wood, in *Wood Handbook: Wood as an Engineering Material*, USDA General Technical Report FPL-GTR-113, Madison, WI, 1999, Chapter 13, pp. 13-1–13-16.
26. Simpson, W. and A. TenWolde, Physical properties and moisture relations of wood, in *Wood Handbook: Wood as an Engineering Material*, USDA General Technical Report FPL-GTR-113, Madison, WI, pp. 3-1–13-24.
27. Yang, C. S., Fungal colonization of HVAC fiber-glass air-duct liner in the USA, *Proc. Indoor Air Conf.*, 1996, Vol. 3, pp. 173–177.
28. Samson, R. A. and J. Houbraken, Laboratory isolation and identification of fungi, in *Microorganisms in Home and Indoor Work Environments*, B. Flannigan, R. A. Samson, and J. D. Miller, eds., Taylor & Francis, London and New York, 2001, pp. 247–266.
29. Morey, P., Microbiological investigations of indoor environments: Interpreting sample data-selected case studies, in *Microorganisms in Home and Indoor Work Environments*, B. Flannigan, R. A. Samson, and J. D. Miller, eds., Taylor & Francis, London and New York, 2001, pp. 275–284.
30. Scott, J. A., *Studies on Indoor Fungi*, Ph.D. dissertation, Department of Botany, Univ. Toronto, 2001.
31. Samson, R.A. and B. v.d. Lustgraaf, *Aspergillus penicillioides* and *Eurotium halophilicum* in association with house-dust mites, *Mycopathologia* **64**:13–16 (1978).
32. Domsch, K. H., W. Gams, and T.-H. Anderson, *Compendium of Soil Fungi*, Vol. 1, reprinted by IHW-Verlag with supplement by W. Gams, 1993.
33. Arnow, P. M., J. N. Fink, D. P. Schlueter, J. J. Barboriak, G. Mallison, S. I. Said, S. Martin, G. F. Unger, G. T. Scanlon, and V. P. Kurup, Early detection of hypersensitivity pneumonitis in office workers, *Am. J. Med.* **64**:236–242 (1978).
34. Morey, P. R., M. J. Hodgson, W. G. Sorenson, G. L. Kullman, W. W. Rhodes, and G. S. Visvesvara, Environmental studies in moldy office buildings: Biological agents, sources and preventive measures, *Ann. Am. Conf. Gov. Indust. Hyg.* **10**:21–35 (1984).

CHAPTER 12

MICROBIAL REMEDIATION IN NONINDUSTRIAL INDOOR ENVIRONMENTS

PHILIP R. MOREY

12.1. INTRODUCTION

Considerable information is available in professional society and governmental publications on remediation of (1) mold growth on interior surfaces, (2) biofilms and legionellas in HVAC and water system components, and (3) sewage contamination in buildings. General principles for mold remediation first embodied in the 1993 New York City Department of Health *Guidelines on Assessment and Remediation of Stachybotrys atra in Indoor Environments*[1] have been updated in more recent consensus documents.[2–5] The second edition of Health Canada's *Fungal Contamination in Public Buildings: Health Effects and Investigation Methods*[6] has significantly expanded the definition of "remediation" to include not only the physical removal of mold growth from interior surfaces but also the elimination of the building defect (moisture problem) that allowed mold growth to occur. Methods to control *Legionella* amplification in heat rejection and potable hot-water systems, while well known,[7,8] are often underutilized as evidenced by the continuing occurrence of legionellosis.[9] Control of microbial growth in air handling unit drain pans and water sump humidifiers,[10,11] although straightforward, is sometimes neglected despite reaffirmation of general maintenance principles in documents such as ASHRAE Standard 62.[12] Standards from the Institute of Inspection, Cleaning and Restoration Certification[13] beginning in 1994 have provided guidance on cleanup of sewage contamination and restorative drying of water-damaged interior finishing and construction materials. General principles of mold remediation, biofilm and *Legionella* control, and sewage cleanup are reviewed in this chapter.

Sampling and Analysis of Indoor Microorganisms, Edited by Chin S. Yang and Patricia A. Heinsohn

12.2. MOLD REMEDIATION PRINCIPLES

It has been recognized for over 3000 years[14] that extensive mold growth on interior surfaces is unacceptable and that visually moldy materials should be discarded. The 1993 New York City *Guidelines*[1] recommended the physical removal of visually moldy materials, specifically contaminated with *Stachybotrys chartarum*, and also that cleanup be carried out by persons equipped with appropriate personal protective equipment including respirators and gloves. The use of containment barriers and negative pressurization was recommended when visually moldy materials with a surface area greater than that of a standard gypsum wallboard ($\sim$3 m^2) was removed.[1] These general principles have been reaffirmed by more recent guidelines,[2,3,4,15] although the threshold for use of depressurization and containment between guidelines varies from about 3 to 10 m^2 of visually moldy materials. The 1993 New York City *Guidelines*[1] also recommended that moisture incursion into building infrastructure be prevented and that small patches of visually moldy finishing and construction materials be removed by properly trained maintenance personnel.[14] These 1993 principles have also been reaffirmed in more recent guidelines.[2–6,15]

More recent mold remediation guidelines[2–6,15] all recommend increasingly stringent actions to contain dust and to remove visible mold growth as the extent of growth increases from just a little (a few square feet) to a large scale ($>$100 ft^2). The rationale behind these guideline recommendations is that more visible mold likely means greater exposure risk. Recent studies in Austrian residences provide evidence that exposure is related to surface area of mold growth observed during building inspection.[16] Inspection and air sampling were carried out in 66 households. The extent of mold growth on walls was divided into four classes as follows: (1) absence of mold growth, (2) small single spot, (3) medium ($<$1 m^2) growth, and (4) large ($>$1 m^2) growth.[16] Sampling results showed that the extent (surface area) of visible mold growth was significantly correlated with both the total concentration of mold spores as well as with the predominance of *Penicillium* and *Aspergillus* in room air. This Austrian study is especially significant in North America because mold remediation guidelines such as those published by OSHA, the USEPA, and the NYC Department of Health[2–4] all have assumed that the surface area of visible mold growth on interior surfaces is correlated with exposure risk to both remediators and building occupants. The Austrian studies[16] provide scientific documentation that the logic behind North American mold remediation guidelines is well founded.

Important objectives of mold remediation include (1) physical removal of mycelium (growth) from interior surfaces, (2) prevention of cross-contamination of occupied or clean zones by microbial contamination from areas undergoing remediation activities, and (3) removal of settled dusts containing spores and other microbial particulate that originated from zones where mold growth occurred.[14] Physical removal of mold growth from interior finishing and construction materials is the preferred remediation method.

The following section describes mold removal methodologies. A later section outlines quality assurance principles useful in monitoring the activities of the mold remediation contractor.

12.3. MOLD REMOVAL METHODS

Methods used to physically remove mold growth vary according to the porosity of the affected surface. Physical removal of moldy porous materials is recommended because attempts to superficially clean, treat, and seal visually moldy surfaces leave some residual mold propagules behind. Thus, the surface is primed for growth if moisture becomes nonlimiting. Consequently, visually moldy porous materials such as paper-faced gypsum wallboard (GWB), ceiling tiles, insulation, and carpet are discarded.[14] Because it is difficult to see hyaline (colorless) fungal microcolonies on porous surfaces such as GWB, it is prudent to remove some apparently sound material up to ~30 cm away from the visually colonized surface.[17] Physical removal of visually moldy porous materials is also the preferred remediation method because hidden patches of mycelium are often revealed, for example, on wall cavity surfaces during the removal process.

Mold growth that is present on nonporous surfaces such as sheetmetal, ceramic tiles, and glass is removed by cleaning.[14] Use of tap water with detergents should be effective for cleaning of nonporous materials.

The method recommended to remove visible mold growth from semiporous materials such as wood depends on the degree to which hyphae have penetrated into the substrate. Wood that is sound except for superficial colonization of the surface should be sanded, planed, refinished, and reused. Wood that is dry or wet-rotted is discarded.[18]

12.4. DUST CONTROL DURING MOLD REMEDIATION

Dust suppression methods, containment barriers, and depressurization techniques are used during mold remediation to prevent cross-contamination of clean or occupied areas and to prevent occupant or worker exposure outside of containment. The extent of mold growth on surfaces (small-scale, medium-scale, large-scale[15,19]) in a room or zone is important with regard to selection of dust suppression and containment methodology. For small-scale remediation, the positioning of the suction nozzle of a HEPA vacuum cleaner at the location where mold is being physically removed is generally adequate for dust suppression. For large-scale cleanups, a depressurized containment (−5 Pa) is recommended for isolation of remediation area dusts (including spores) from adjacent cleaned or occupied zones.

Spores from mold-colonized surfaces may have previously been dispersed on air currents into areas of the building unaffected by moisture problems. The IICRC S520 Standard[20] refers to this as "environmental condition 2." A combination of

HEPA vacuum cleaning and damp wiping (nonporous surfaces only) should be adequate to remove most of the settled dusts containing these dispersed spores. According to the IICRC,[20] the objective of HEPA vacuuming and damp wiping is the conversion of the condition 2 surface to environmental condition 1 (normal fungal ecology[20]) status. Documentation of condition 1 status or normal fungal ecology in an indoor environment is usually determined by the occurrence of a rank-order profile of mold propagules containing a significant amount of phylloplane (leaf-sourced[21]) molds[22,23] and always with the absence of evidence of actual mold growth (IICRC[20] condition 3). Pre- and postcleaning sampling and testing are required for rank-order comparisons and to document the change in condition. This sampling and testing should be conducted by competent third parties.

Restoration of interior surfaces to a clean and dry status (IICRC condition 1[20]) does not require sterilization or disinfection of interior surfaces. Thus, the use of disinfectants and biocides is unnecessary in mold remediation[24] except in healthcare environments, where disinfection of some surfaces (e.g., in surgical suites) is required to reduce the risks of nosocomial infections.

12.5. AIR SAMPLING FOR MOLDS

Air sampling is not required in most buildings in order to demonstrate that mold remediation has been effectively carried out.[3,4] However, air sampling can be useful as a part of the quality assurance sampling. For purposes of documenting effective remediation, it is generally adequate to verify that visually moldy materials (including hidden mold) have been removed according to industry-standard quality assurance measures[25] and that residual dusts have been removed by HEPA vacuum cleaning so that recleaning is unnecessary.[6] Air sampling is deemphasized as a quality assurance measure of the effectiveness of mold remediation, especially for small buildings (e.g., most residences) because of the difficulty associated with data interpretation when only a few outdoor and indoor samples are collected (see Chapter 3, this volume, and Refs. 26 and 27). If air sampling and analysis are required, a sufficient number of samples should be collected.

If air sampling is carried out in order to document a return to IICRC condition 1 status,[20] sample collection is recommended to occur after buildback and several days after the indoor building environment and its HVAC system have equilibrated with the mycoflora outdoors.[5,6] Table 12.1 (see also Chapter 3, this volume, and Ref. 28) provides an example where air sampling demonstrated that indoor mold spore types and concentrations had returned to a normal and typical condition. In the example in Table 12.1, chronic moisture problems and extensive visible mold growth occurred on the lower floor of a two-story building. *Aspergillus versicolor* (AV) and other nonphylloplane molds were dominant fungi on visually moldy surfaces in first-floor rooms as well as in the air on both floors prior to remediation (Table 12.1). Note that prior to remediation the collective concentration of *Penicillium, Aspergillus*, and *Eurotium* outnumbered the combined concentration of *Cladosporium, Alternaria*, and *Epicoccum* (phylloplane molds). Several weeks

TABLE 12.1 Air Sampling for Culturable Molds as an Indication of Effectiveness of Remediation

	Culturable Fungi (CFU/m^3)[a]			
Location of Samples	Total	AV only	P/Asp/E	C/Alt/Epi
Before remediation				
Floor 1 ($N = 10$)	650	140	560	45
Floor 2 ($N = 20$)	120	6	60	30
After remediation				
Floor 1 ($N = 20$)	450	4	75	300
Floor 2 ($N = 30$)	730	2	120	400

[a]Cellulose medium. AV = *Aspergillus versicolor*; P = *Penicillium*; Asp = *Aspergillus*; E = *Eurotium*; C = *Cladosporium*; Alt = *Alternaria*; Epi = *Epicoccum*; phylloplane molds (C/Alt/Epi) dominated the propagules collected in the outdoor air before and after remediation.

Source: Adapted from Ref. 28.

after the completion of mold remediation, which included a thorough HEPA vacuum cleaning throughout the building, air sampling showed that phylloplane molds dominated the mold spores on both floors of the building. The sampling data in Table 12.1 showed that restoration activities had changed the dominating profile of airborne fungi from nonphylloplane to phylloplane species. The presence of low concentrations of AV after restoration illustrates the difficulty associated with removal of all traces of the original mold spore contaminants. However, the overwhelming presence of airborne phylloplane mold spores after cleanup shows a return to normal and typical (IICRC status 1[20]) condition.

12.6. CLEARANCE BY MEASUREMENT OF RESIDUAL DUST

It is universally accepted in mold remediation guidelines[2–6] that both visible mold growth and associated dusts should be removed during the cleanup process. During final clearance, it is important to verify by inspection that visible mold growth has been removed and cellotape sampling can be used to assist in this documentation. The removal of fine dusts can be subjectively documented by a white or black glove test wherein the glove is moved across a surface and the amount of adherent dust on the glove may indicate a need for recleaning. It has been suggested that a quantitative measurement of residual dust left on nonporous surfaces after mold remediation is an appropriate final clearance for mold remediation.[6,22,25]

In 1992 the National Air Duct Cleaning Association (NADCA)[29] developed a standard in which the weight of the surface dust collected on a standard 37 mm polyvinyl chloride or mixed cellulose ester filter (three-piece cassette) was used to indicate whether nonporous surfaces in HVAC system ductwork needed additional cleaning. While this gravimetric measurement does not provide an indication of the microbial content of the collected dust, it does provide useful information on the

TABLE 12.2 Residual Dust on Various Surfaces after HEPA Vacuum Cleaning[a]

Material Surface	Residual Dust (mg/m^2)
Nonporous wood book shelves (cleaning satisfactory)	0.5[b]
Nonporous hardwood flooring (cleaning satisfactory)	7.2[c]
Semiporous concrete block with partially painted surface (cleaning satisfactory)	13[c]
Nonporous floor tile with semiporous grouting (cleaning satisfactory)	67[c]
Carpet before HEPA vacuuming	490[d]
Carpet after HEPA vacuuming (cleaning probably satisfactory)	200[b]

[a] Using a 37-mm minivacuum (three-piece) filter cassette with tared mixed cellulose ester (0.8 μm pore size) or polyvinyl chloride (5.0 μm pore size) filters; used closed face with 5 cm flexible tubing cut at a 45° angle on tubing inlet; approximately 0.2–1.0 m^2 surface area vacuumed for 4–6 mins.

[b] Dust samples collected 3–7 days after HEPA vacuum cleaning at a time when the interior space was occupied.

[c] Dust samples collected 3–7 days after HEPA vacuum cleaning prior to occupancy.

[d] Dust samples collected several days before mold remediation began.

efficiency of the cleaning process in reducing dust loading on interior surfaces. The measurement of residual dust in mold remediation is useful because it (1) is easily understood by cleaning personnel, (2) avoids difficulties associated with interpretation of microbial air and surface sampling data, and (3) provides a logical means of determining whether the dust mass per unit surface area is sufficiently low that regardless of microbial composition, the possibility of health risks is reduced to a lower, more acceptable level.

Examples are provided in Table 12.2 where the amount of residual dust collected from nonporous and semiporous surfaces such as bookshelves, floors, and painted concrete block walls during final clearance indicated adequate cleaning. In the example in Table 12.2, the residual dust level in carpet after HEPA vacuum cleaning was reduced by over 50% but not to the 100 mg/m^2 level.[29] Additional development of standard methodologies for collection and measurement of residual dust on various kinds of surfaces is needed before this method of clearance can be universally used during mold remediation.

12.7. HVAC SYSTEM MOLD REMEDIATION

Mold growth can occur in HVAC systems when airstream surfaces become dirty and moist. Dirt and mold growth on nonporous sheetmetal surfaces can be easily removed by physical cleaning.[14] However, mold growth on porous airstream surfaces cannot be removed by cleaning. Physical removal of the porous material is required once hyphae have penetrated into the porous substrate.[14]

When mold growth occurs in HVAC systems, it should be realized that air-handling units and supply air ductwork are designed to transport ventilation air directly to the occupant breathing zone.[14] Thus, during mold remediation in HVAC systems[2–4] it is necessary to prevent dust and mold propagules dislodged from airstream surfaces during cleaning from following ventilation pathways into occupied spaces. Because of the difficulty associated with mold remediation of airstream surfaces in HVAC systems, the manufacture of insulation with an airstream surface that is smooth, cleanable, and resistant to biodeterioration has been recommended.[14,30] Some success in preventing fungal growth by incorporating zinc-based encapsulants in HVAC system airstream surfaces has been reported.[31,32]

12.8. REMEDIATION OF ROOM CONTENTS IN MOLDY BUILDINGS

When visible mold growth occurs on walls, ceilings, or other room surfaces, spores from these growth areas will be dispersed and settle onto or into porous contents such as upholstered furniture, drapes, and clothing (IICRC condition 2)[20]. If the amount of mold growth on room surfaces is small-scale, then it is reasonable to assume that spores and other microbial particulate dispersed onto or into porous contents will be limited and HEPA vacuum cleaning should be successful in removing most of the contamination (returning to IICRC condition 1).[20]

However, when the amount of visible mold growth on room surfaces is large-scale, then cleaning of visually nonmoldy porous contents will be difficult. HEPA vacuum cleaning will have to be supplemented by other cleaning methodologies such as washing in aqueous or organic solvents.[33] Alternatively, porous contents that were present in rooms with large-scale mold growth may be discarded because of the high cost and difficulty associated with cleaning and documenting that the cleaning process was effective. Section 9 of IICRC Standard S520[20] contains detailed information on contents cleaning.

12.9. QUALITY ASSURANCE DURING MOLD REMEDIATION

An important endpoint in mold remediation is quality assurance that the moisture problem that caused mold growth has been eliminated. In addition, it is necessary to document that a thorough physical assessment or inspection (carried out before or during the remediation process) identified all significant mold growth sites, including those hidden in building construction. Quality assurance principles applicable to oversight of the methods used by the mold remediation contractor are reviewed elsewhere[5,15,25] and include the following:

- Document that appropriate methodologies or protocols were used during removal of visible mold and associated dusts identified during building assessment and inspections. Protocols and methodologies used during mold remediation become more complex as the extent of mold growth to be removed increases.[15,19] Protocols for mold remediation should be written by individuals with recognized professional training and experience in the investigation of mold problems following methods recommended in standard-of-care documents.[2–5,15,25] Generally the author of a mold remediation protocol is a certified industrial hygienist (CIH), professional engineer (PE), or a building scientist. Protocols for mold remediation may be prescriptive-based (tell the remedial contractor how to carry out the cleanup) or performance-based (the remedial contractor firm uses its own experience-based methods to carry out the cleanup). Regardless of whether the protocol is prescriptive- or performance-based, a clear set of endpoint guidelines should be available to independently judge whether the remediation was successfully carried out. Assurance by inspection that mold growth was removed and that affected space was HEPA vacuumed are important endpoints for all mold remediation protocols.

- Document that appropriate dust suppression, containment, and depressurization techniques were used by the cleanup contractor or cleaning personnel during mold remediation.
- Document that surfaces in remediated spaces have been HEPA-vacuumed and that the amount of residual dust remaining on interior surfaces does not indicate a need for recleaning.[6,25]
- Document for large-scale cleanups that a CIH, PE, or OS health and safety professional provided independent quality assurance that the remediation was appropriately carried out.[15] Independent or third-party oversight is required to avoid conflict-of-interest issues (the remedial contractor should not judge if its work was successful). Independent oversight is also beneficial for medium-scale and small-scale mold remediations.

12.10. MICROBIAL REMEDIATION OF LEGIONELLAS AND BIOFILMS

The presence of stagnant water in AHU water spray systems and drain pans and in humidifier water reservoirs provides conditions for the formation of biofilm (slime containing yeasts, bacteria, protozoa, and other microorganisms) on submerged or wet surfaces.[14] Procedures for removal of biofilm from AHUs and humidifiers include decommissioning of the HVAC component or device, physical removal of the biofilm aided by use of detergents and disinfectants, removal of cleaning chemicals, and recommissioning of the HVAC component or humidifier.[11,12,34] The use of biocides in place of routine physical removal of biofilm through cleaning is inappropriate. Prevention of biofilm development in HVAC components is promoted by sloping of drain pans for self-drainage, periodic cleaning of cooling coils and wet surfaces, and design of HVAC components for accessibility of potentially wet surfaces for ease of inspection and maintenance.

Legionella species, the causative agent of legionellosis (Legionnaires' disease and Pontiac fever) are found in outdoor water sources such as lakes and streams, as well as in heat-rejection systems in buildings (e.g., cooling towers) and in hot-water systems. Legionellas are found most frequently in or near biofilms where protozoa and the biofilm matrix itself can protect the bacterium from biocides or other means of inactivation. Legionellas grow most rapidly in lukewarm water with a temperature of about 85–100°F.

Excellent maintenance of heat-rejection and hot-water systems are appropriate actions useful in preventing the amplification of legionellas and in reducing the risk of legionellosis.[9,35,36] For hot-water systems, thermal or chemical disinfection of storage tanks and piping, avoidance of water stagnation in piping, and removal of sludge from storage tanks and reservoirs are actions effective against *Legionella* amplification.[7,8,35,36] *Legionella* amplification in cooling towers is controlled by attention to biocide usage and by scale or corrosion control. Where *Legionella* amplification has occurred, the addition of biodispersants and disinfectants and circulation of these chemicals in the water system/piping are important aspects of remediation.[14]

The finding of culturable legionellas in water systems does not in itself mean that legionellosis has or will occur. However, the occurrence of legionellas over and above levels found in piped water entering the building is indicative of a water system requiring more attention to sound maintenance procedures.[35]

12.11. REMEDIATION OF SEWAGE CONTAMINATION

Basic principles of sewage remediation include (1) prompt physical removal of sewage (black) water, (2) thorough and often repeated disinfection of interior surfaces that came in contact with sewage water, and (3) prompt drying of interior surfaces following disinfection so as to remove conditions that promote mold growth. Specific actions recommended during sewage cleanup are described in IICRC Standard S500[13] and by Berry et al.[37] Prompt removal of sewage water and disinfection of affected surfaces are recommended because of the presence of pathogenic microorganisms and invertebrates (e.g., hepatitis A, *Escherichia coli* 0157/H7, *Giardia lamblia, Ascaris lumbricoides*; see Ref. 38 for a description of sewage organisms).[14]

During sewage cleanup it is important to document which porous materials or contents were directly affected by sewage water. Materials and contents (e.g., carpet, gypsum wallboard, insulation, upholstered furniture, clothing) that came in contact with sewage water are discarded.[14,37] Semiporous materials such as wood framing and concrete can be cleaned and disinfected. Nonporous materials such as sheetmetal and glass can be easily cleaned and disinfected.[13,14,37] An important aspect of restoration is the identification of building components such as wall cavities, underflooring materials, HVAC ductwork, and crawlspaces that may have come in contact with sewage water. These areas should be disinfected, and this may require removal of undamaged finishes and construction material in order to gain access to contaminated materials.[14]

Microbial sampling to document the effectiveness of sewage remediation is seldom carried out because the primary emphasis is placed on physical removal of sewage water and disinfection and drying of affected interior surfaces. If microbial sampling is carried out, it is appropriate to compare the ecology of sewage-contacted surfaces with similar surfaces in clean, dry, and well-maintained buildings (see Table 3.9, in Chapter 3 of this book). The mere presence of a serotype *E. coli* in the soil around a building, in soil tracked into a building, and in niches such as cat litterboxes, does not in itself indicate the occurrence of sewage contamination. However, the occurrence of an *E. coli* serotype such as O157/H7 (a causal agent of hemorrhagic diarrhea) on building surfaces directly affected by sewage water is indicative of contamination needing remediation.

REFERENCES

1. *Guidelines on Assessment and Remediation of Stachybotrys atra in Indoor Environments*, New York City Department of Health, New York City Human Resources Administration, and Mount Sinai–Irving J. Selikoff Occupational Health Clinical Center, New York, 1993.
2. *Guidelines on the Assessment and Remediation of Mold in Indoor Environments*, New York City Department of Health, 2000.
3. U.S. Environmental Protection Agency, *Mold Remediation in Schools and Commercial Buildings*, EPA 402-K-01-001, Washington, DC, 2001.
4. OSHA, *A Brief Guide to Mold in the Workplace*, Occupational Safety and Health Administration, U.S. Department of Labor, Washington, DC, 2003.
5. AIHA, *Recognition, Evaluation and Control of Indoor Mold*, American Industrial Hygiene Association, Fairfax, VA (in press).
6. Health Canada, *Fungal Contamination in Public Buildings: Health Effects and Investigation Methods*. H46-2/04-358E, Ottawa, 2004.
7. HSE, *The Control of Legionellosis Including Legionnaires' Disease*, Health and Safety Executive, Health and Safety Booklet H.S. (G) 70, Sudbury, UK, 1994.
8. ASTM, *Standard D5952-96, Standard Guide for Inspecting Water Systems for Legionellae and Investigating Possible Outbreaks of Legionellosis (Legionnaires' Disease or Pontiac Fever)*, American Society of Testing and Materials, West Conshohocken, PA, 2002.
9. Cooper, A. J., H. R. Barnes, and E. R. Myers, Assessing risk of *Legionella, ASHRAE J.* 22–27 (April 2004).
10. Morey, P. R., M. J. Hodgson, W. G. Sorenson, G. J. Kullman, W. W. Rhodes, and G. S. Visvesvara, Environmental studies in moldy office buildings: biological agents, sources, and preventive measures, *Ann. Am. Conf. Gov. Indust. Hyg.* **10**:21–35 (1984).
11. Brundett, G. W., *Maintenance of Spray Humidifiers*, The Electricity Council, Research Center, Capenhurst, Chester, UK, 1979.

12. ASHRAE, *Ventilation for Acceptable Indoor Air Quality*, ASHRAE Standard 62.1-2004, American Society of Heating, Refrigerating and Air-conditioning Engineers, Atlanta, 2004.

13. IICRC, *Standard and Reference Guide for Professional Water Damage Restoration*, Standard S500, Institute of Inspection Cleaning and Restoration Certification, Vancouver, WA, 1999.

14. Morey, P., Remediation and control of microbial growth in problem buildings, in *Microorganisms in Home and Indoor Work Environmnents*, B. Flannigan, R. Samson, and J. Miller, eds., Taylor & Francis, London, 2001, pp. 83–99.

15. Canadian Construction Association, *Mould Guidelines for the Canadian Construction Industry*, Standard Construction Document CCA 82-2004, Ottawa, 2004.

16. Haas, D., J. Habib, I. Wendelin, M. Unteregger, J. Posch, G. Wüst, R. Schlacher, E. Marth, and F. F. Reinthaler, Concentrations of airborne fungi in households with and without visible mold problems, *Proc. Indoor Air* **2**(1):1505–1509 (2005).

17. Miller, J. D., Mycological investigations of indoor environments, in *Microorganisms in Home and Indoor Work Environments*, B. Flannigan, R. A. Samson, and J. D. Miller, eds., Taylor & Francis, London, 2001, pp. 231–246.

18. Lloyd, H. and J. Singh, Inspection, monitoring and environmental control of timber decay, in *Building Mycology*, J. Singh, ed., Spon, London, 1994, pp. 159–186.

19. ACGIH, *Bioaerosols Assessment and Control*, R. J. Shaughnessy and P. R. Morey, American Conf. Governmental Industrial Hygienists, Cincinnati, OH, 1999, Chapter 15, pp. 15-1–15-7.

20. IICRC, *Standard and Reference Guide for Professional Mold Remediation*, S520, Institute of Inspection Cleaning and Restoration Certification, Vancouver, WA, 2003.

21. Flannigan, B., Guidelines for evaluation of airborne microbial contamination of buildings, in *Fungi and Bacteria in Indoor Air Environments*, E. Johanning and C. S. Yang, eds., Eastern New York Occupational Health Program, Latham, NY, 1995, pp. 123–130.

22. Dillon, H. K., J. D. Miller, W. G. Sorenson, J. Douwes, and R. J. Jacobs, Review of methods applicable to assessment of mold exposure to children, *Environ. Health Perspect.* **107**(S.3): 473–480 (1999).

23. Horner, W. E., A. W. Worthan, and P. R. Morey, Air and dustborne mycoflora in houses free of water damage and fungal growth, *Appl. Environ. Microbiol.* **70**:6394–6400 (2004).

24. ACGIH, *Bioaerosols Assessment and Control*, E. C. Cole and K. K. Foarde, American Conf. Governmental Industrial Hygienists, Cincinnati, OH, Chapter 16, 1999, pp. 16-1–16-9.

25. AIHA, *Report of Microbial Growth Task Force*, American Industrial Hygiene Association, Fairfax, VA, 2001.

26. ACGIH, Developing a sampling plan, in *Bioaerosols Assessment and Control*, American Conf. Governmental Industrial Hygienists, Cincinnati, OH, 1999, Chapter 5, pp. 5-1–5-13.

27. AIHA, *Field Guide for the Determination of Biological Contaminants in Environmental Samples*, 2nd ed., L.- L. Hung, J. D. Miller, and H. K. Dillon, eds., American Industrial Hygiene Association, Fairfax, VA, 2005.

28. Morey, P. R., Microbiological investigations of indoor environments: Interpreting sampling data-selected case studies, in *Microorganisms in Home and Indoor Work*

Environments, B. Flannigan, R. A. Samson, and J. D. Miller, eds., Taylor & Francis, London, 2001, pp. 275–284.

29. NADCA, *Mechanical Cleaning of Non-Porous Air Conveyance System Components*, Standard 01-1992, Publication NAD-01. National Air Duct Cleaning Association, Washington, DC, 1992.
30. Morey, P. R., Control of indoor air pollution, in *Occupational and Environmental Respiratory Disease*, P. Harber, M. B. Schenker, and J. R. Balmes, eds., Mosby, St. Louis, 1996, pp. 981–1003.
31. Yang, C. S., Fungal colonization of HVAC fiberglass air-duct liner in the U.S.A., *Proc. 7th International Conf. Indoor Air Quality and Climate, Nagoya*, 1996, Vol. 3, pp. 173–177.
32. Yang, C. S. and P. J. Ellringer, Antifungal treatments and their effects on fibrous glass liner, *ASHRAE J.* **46**:35–40 (2004).
33. Moon, R., D.-W. Li, and C. Yang, Mold-contaminated fabrics, *Clean. Restor.*, **41**:30–38 (2004).
34. ISIAQ, *Control of Moisture Problems Affecting Biological Indoor Air Quality*, International Society of Indoor Air Quality, Helsinki, 1996.
35. ASHRAE, *Minimizing the Risk of Legionellosis Associated with Building Water Systems*, Guideline 12-2000. American Society of Heating, Refrigeration and Air-conditioning Engineers, Inc., Atlanta, 2000.
36. Dahlen, E., M. Shum, and M. Kerkhove, Control of *Legionella* growth in distribution systems, *Proc. Indoor Air.* **2**(2):2487–2491 (2005).
37. Berry, M. A., J. Bishop, C. Blackburn, E. C. Cole, W. G. Ewald, T. Smith, N. Suazo, and S. Swan, Suggested guidelines for remediation of damage from sewage backflow into buildings, *J. Environ. Health* **57**:9–15 (1994).
38. Fox, J. C., P. R. Fitzgerald, and C. Lue-Hing, *Sewage Organisms: A Color Atlas*, Lewis Publishers, Chelsea, MA, 1981.

APPENDIX

COMMON AIRBORNE AND INDOOR FUNGI AND THEIR SPORES

DE-WEI LI

Proper identification of fungi is challenging and the most important part in microscopic analysis. For spore count, only spore morphological characteristics are available. For microscopic analysis on other samples, additional fungal structures may be available. It is important for analysts to understand how to utilize available fungal structures for accurate identification and to recognize the significance of presence of certain fungi and fungal structures from perspectives of fungal biology and fungal ecology. The followings are common airborne, indoor fungi, and fungal structures for quick reference. Since colony characteristics are not available for direct microscopic analysis of most of the fungal samples, description of the colony will not be included in the following fungal descriptions. Readers should try to obtain and reference mycological books, journals, and monographs of important indoor fungi:

1. *Acremonium* and *Gliomastix* (Hyphomycetes). Conidia are usually single-celled, (they rarely have more than one cell) and may be colorless (*Acremonium*) or have dark pigmentation (*Gliomastix*). Their sizes and shapes vary from spherical, ovoidal, tear-shaped, subspherical, falciform, reniform, ellipsoidal, to cylindrical; their surfaces range from smooth to rough; they are usually thin-walled, developing mostly in a slimy mass, occasionally in dry conidial chains. Phialides range from being colorless to darkly pigmented, are erect, and occur mostly singly from hyphae with inconspicuous collorettes.[1,2] These two genera are morphologically very similar, and some mycologists, in fact, treat them under the genus *Acremonium*. The primary differences separating the two genera are pigmentation and roughness of tips of phialides. *Acremonium* is moniliaceous with smooth tips of phialides, while *Gliomastix* is dematiaceous with rough tips. *Acremonium strictum* (Fig. A.1) is a common indoor species, favoring high water activity, and

Sampling and Analysis of Indoor Microorganisms, Edited by Chin S. Yang and Patricia A. Heinsohn

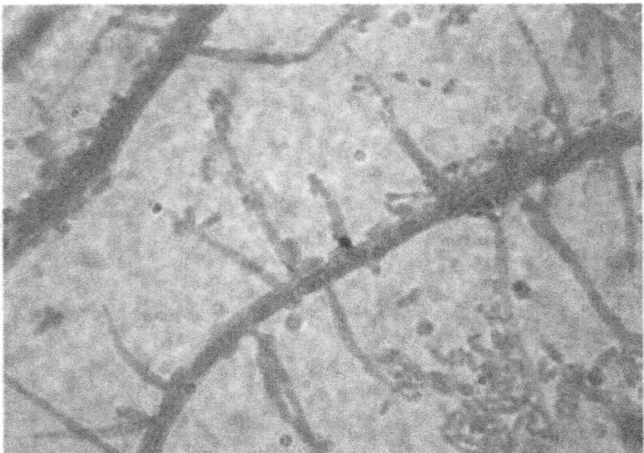

Fig. A.1. *Acremonium strictum.*

grows on drywall, cellulose-based ceiling tiles, wood materials, and other surfaces. Occasionally it may grow on building materials under high humidity without direct water damage due to repeated condensation. These genera can be identified with intact conidiophores and conidia, but it is very difficult to identify the conidia of this genus from air samples because of wide variations within the genera and similarity with other fungal spores. More often its airborne conidia were considered unidentified single-celled spores.

2. *Alternaria* (Hyphomycetes). *Alternaria* conidia are relatively characteristic because of their size, shape, and septation. Its conidia are large, light brown to brown, smooth or rough, multiple-celled with transverse and frequently longitudinal septation, (viz., brickwall pattern with flat bases and elongated tips), and range in size, up to 300 μm long.[3] Many species develop conidia in chains. Conidiophores are light brown to brown, simple or branched, and have multiple cells, developing conidia in sympodial mode. Mycelia are colorless to brown. Common indoor species are *A. alternata* (Fig. A.2) and

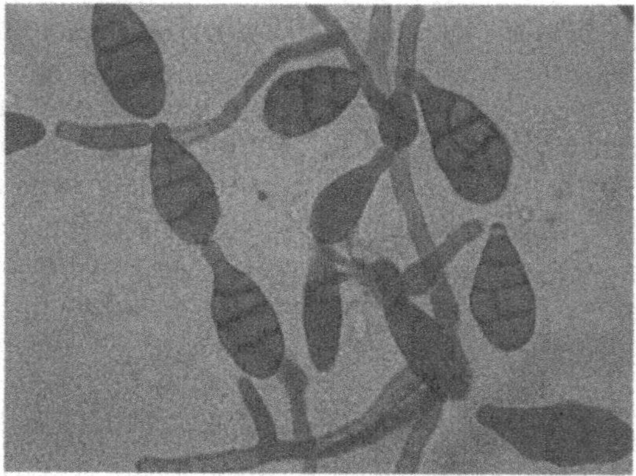

Fig. A.2. *Altenaria alternata.*

A. tenuissima. This genus comprises a large number of species, most of which are saprobic or plant pathogenic. Several other fungal species and genera (e.g., *Mystrosporiella, Phoma glomerata, P. pomorum, Ulocladium*) develop spores with some similar morphological characteristics.

3. *Aspergillus* (Hyphomycetes). Conidia colorless, green, bluish green to brown or black, spherical, subspherical to cylindrical, 2.5–7.5 μm, smooth to rough, in dry chains, forming columns or diverging. Some species develop very thick-walled cells (Hülle cells) or sclerotia. Phialides flask-shaped, without collorettes borne directly on vesicle (uniseriate) or on metulae (biseriate). Conidiophores erect, unbranched, with enlarged apical vesicles. Some species develop teleomorphs (sexual state) also. This is a very large genus and some species are very common contaminants and important indoor fungi. Some species are clinically significant to human health because of their pathogenic, mycotoxigenic, and allergenic nature. Species of this genus are mostly xerophilic to mesophilic, but rarely hydrophilic. This genus is easy to identify, due to the presence of vesicle and arrangement of phialides borne on it, but very difficult to identify to species. Several species may implicate human mycoses.[4] For instance, *A. flavus* and *A. fumigatus* (Fig. A.3) may cause pulmonary aspergillosis.[2] For direct microscopic analysis, it is relatively easy to identify the members of *Aspergillus* to genus. However, it is not practical to identify them to species without culturing them on required growth media and conditions. To identify common members of this genus to species, please refer to *Identification of Common Aspergillus Species* by Klich.[5]
4. *Aureobasidium* (Hyphomycetes). *Aureobasidium pullulans* is a very common black yeastlike fungus indoors and is often found in bathrooms, in kitchens, or on exterior building walls under shade. Conidia colorless, ellipsoidal, varying in size (mostly 9.0–11.0 × 4.0–5.5 μm) and shape, one-celled, in a slimy mass (Fig. A.4), and not very characteristic.[2] Hyphae

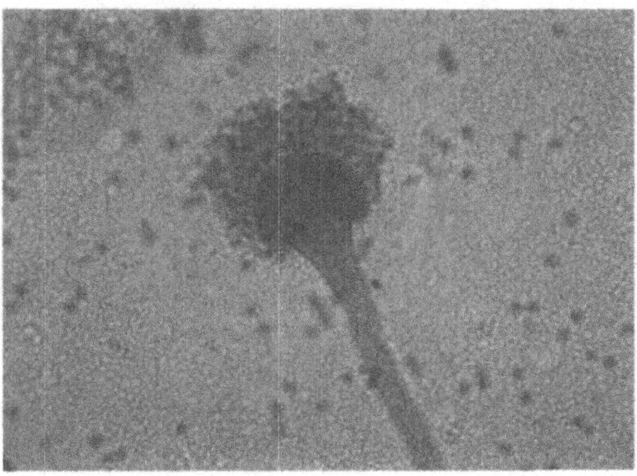

Fig. A.3. *Aspergillus fumigatus.*

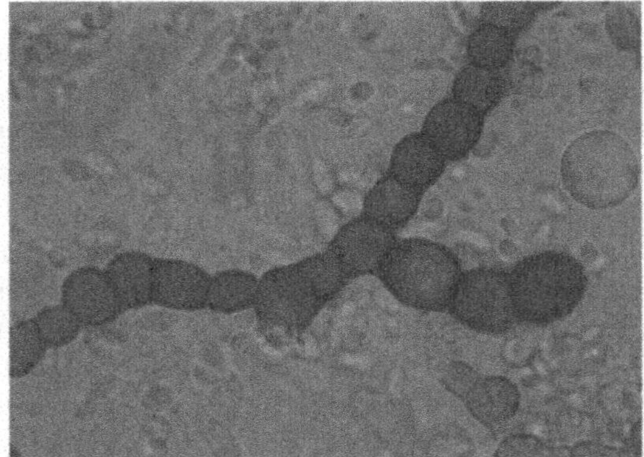

Fig. A.4. *Aureobasidium pullulans.*

colorless, 3–12 μm, locally developing into dark brown, thick-walled chlamydospores. Budding is often present and endoconidia are present in intercalary cells. Undifferentiated conidiogenous cells are in colorless hyphae intercalarily. It is not practical to identify spores of this fungus on spore trap samples without culture characteristics. For microscopic analysis on other samples, with more fungal structures, this fungus may be identified.

5. *Bipolaris* (Hyphomycetes). Conidia brown, mostly smooth, ellipsoidal, straight or slightly curved, multicellular, with a flat dark scar at base (Fig. A.5). Conidiophores brown, simple, erect, multicelled, developing conidia in sympodial mode. Germination is bipolar. Nearly all species are pathogens of grasses. Some species are saprobes. It is very difficult to differentiate the conidia of this genus from those of *Drechslera*, based solely on the morphology of conidia. Often the spores of these two genera are lumped together for spore count.

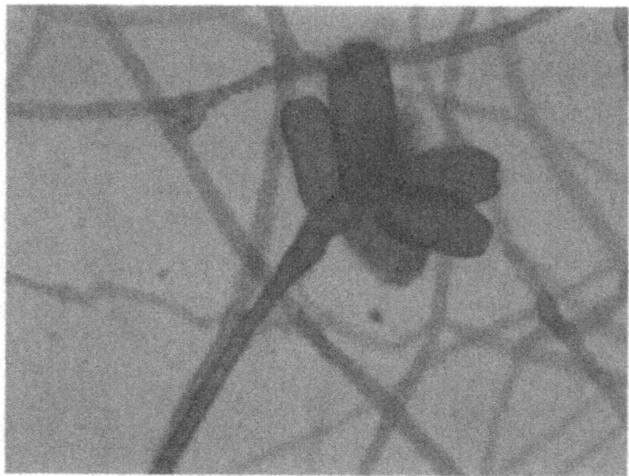

Fig. A.5. *Bipolaris specifera.*

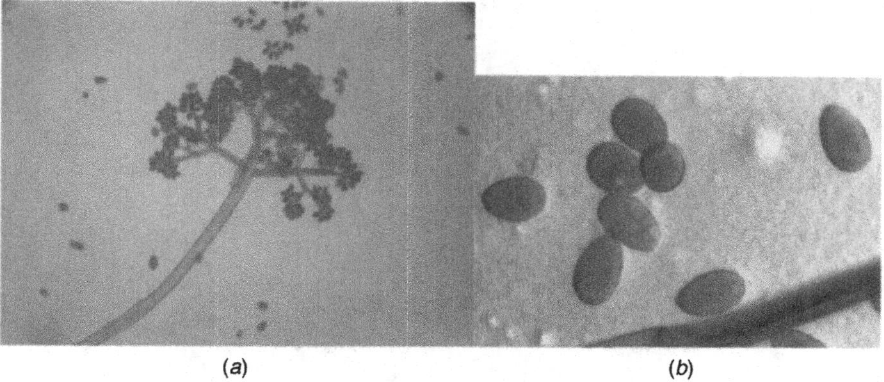

(a) (b)

Fig. A.6. *Botrytis cinerea*: (a) conidiophore; (b) conidia.

6. *Botrytis* (Hyphomycetes). Conidia colorless, gray to pale brown, smooth, ellipsoidal, obovoid, or subspherical, mostly nonseptate. Conidiophores gray to brown, straight or flexuous, smooth, branched, often dichotomously or trichotomously, with branch ends often enlarged. A conidiophore forms a long stipe and a dense head. *Botrytis* spp. are common fungi outdoors and not common indoors. The most common species is *B. cinerea* (Fig. A.6), which is a plant pathogen causing gray mold on a very broad range of hosts including some common ornamental plants, such as geranium, begonia, rose, lily, dogwood, rhododendron, dahlia, Magnolia, and camellia[6,7] and fruits and produce. This fungus is mainly of outdoor origin, although it may be from growth on fruits or flowers brought in from outdoors. Some houseplants can be infected by this fungus, such as cyclamen, poinsettia, chrysanthemum, and gerbera.[7] Other species of *Botrytis* may be present, such as *B. peoniae* on peonies, *B. squamosa* on onion, and *B. tulipae* on tulips. These species of *Botrytis* share some common characteristics in pathology and ecology. With the presence of a conidiophore, it is not difficult to identify this genus. However, it is a quite challenging to identify its conidia in the samples of spore count. Conidia of this genus are often described as unidentified fungal spores.
7. *Cercospora* (Hyphomycetes). Conidia colorless, filiform (scolecospores), multiple-celled, with a conspicuous thick dark scar at base (Fig. A.7). Conidiophores colorless or dark, simple, in clusters and bursting out of leaf tissue. Conidia develop successively on the new growing tips of conidiophores. Species of *Cercospora* are plant pathogens on a wide variety of higher plants, commonly causing leaf spots. Conidia of *Cercospora* are common in outdoor air during warm seasons. These spores do not have a significant effect on indoor air quality. The presence of these spores in indoor air are due to infiltration from outdoor sources. *Pseudocercospora*, and ascospores of *Balansia, Cochliobolus*, and *Gaeumannomyces*, and

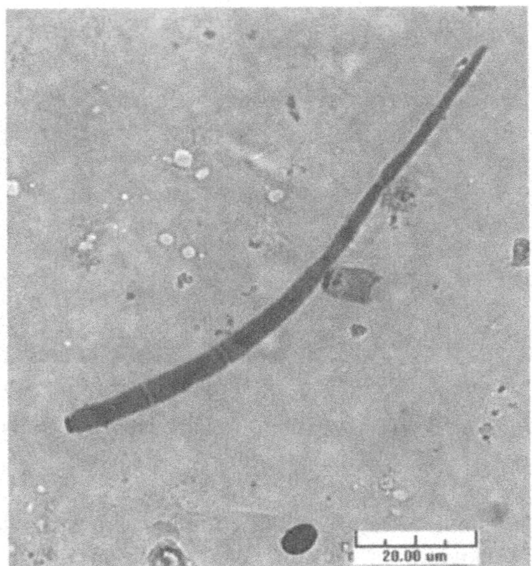

Fig. A.7. *Cercospora* conidium.

other genera develop spores sharing some similarities in morphology. Understanding the development of conidia and ascospores will be useful in differentiating them.

8. *Chaetomium* (Ascomycota). *Chaetomium* species produce ascospores, which are very characteristic and relatively easy to identify. Ascomata spherical to pyriform, covered with characteristic hairs, which are branched or unbranched and undulate or spiral coiled. Asci typically produce eight ascospores, cylindrical to clavate, slenderly stalked, and with deliquescing walls. Ascospores aseptate, smooth, dark, varying in size and shape, but limoniform in most indoor species, release in a dark mass. The most common species indoors is *C. globosum*, a hydrophilic species, often growing with in long-term water damage environments (Fig. A.8). A similar genus, *Chaetomidium*, produces similar spores. *Chaetomium* species grow on wood and paper products, and are hydrophilic (moisture-loving or have high water activity). They are common on water-damaged drywall, wood, or materials with significant cellulose content. Spores of *Chaetomium* detected indoors are excellent indicators of water damage.
9. *Cladosporium* (Hyphomycetes). Conidia dark, mostly one- or two-celled, varying in size and shape, smooth to mostly rough, ovoid to cylindrical, some subglobal, some limoniform with conspicuous scars at both ends, frequently in simple or branched acropetal chains. Conidiophores tall, dark, erect, unbranched or branched near the top, single or in clusters. *Cladosporium* spp. are weak phytopathogens or saprophytic. *C. cladosporioides* and *C. sphaerospermum* (Fig. A.9) are the common species indoors. Spores

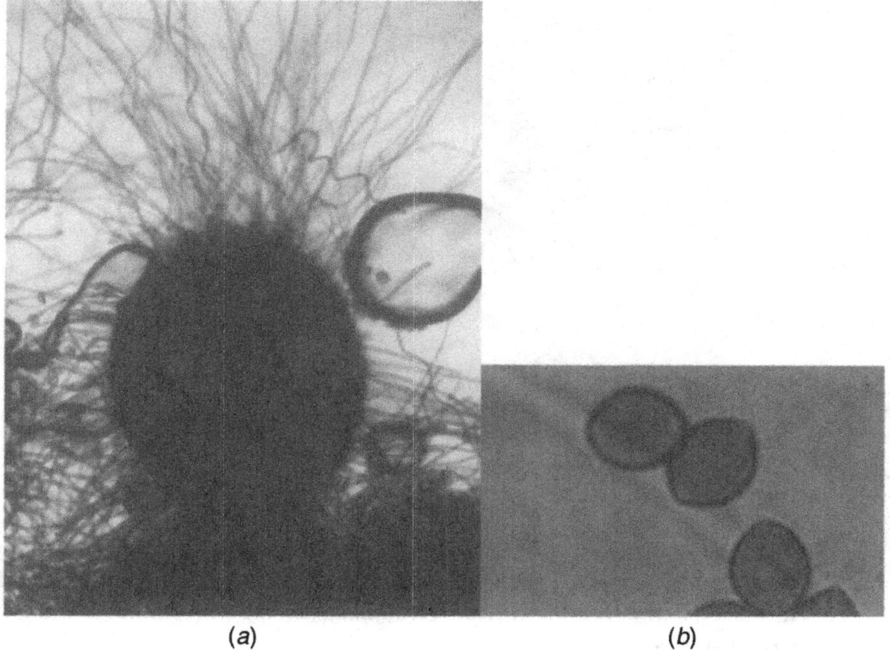

(a) (b)

Fig. A.8. *Chaetomium globosum* (*a*) and its ascospores (*b*).

of *Cladosporium* are the most common airborne spores on the earth. *Cladosporium* is the dominant airborne fungus outdoors in many areas around the world.[8] The spores of *Cladosporium* have well-defined seasonality with a peak in growing seasons. Their diurnal pattern has a midday peak.[9] However, species of *Cladosporium* can grow indoors, such as on fibrous glass

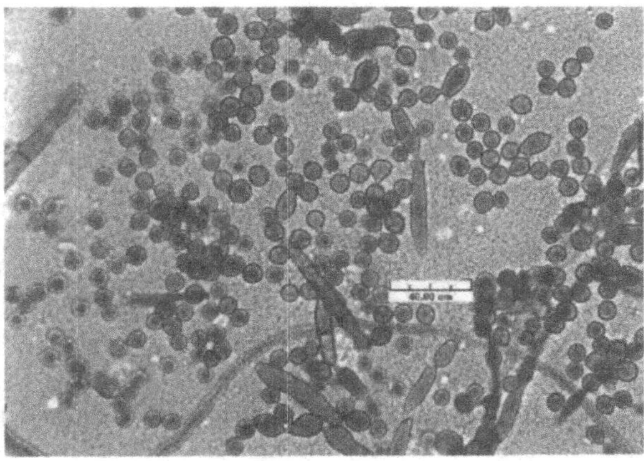

Fig. A.9. *Cladosporium sphaerospermum.*

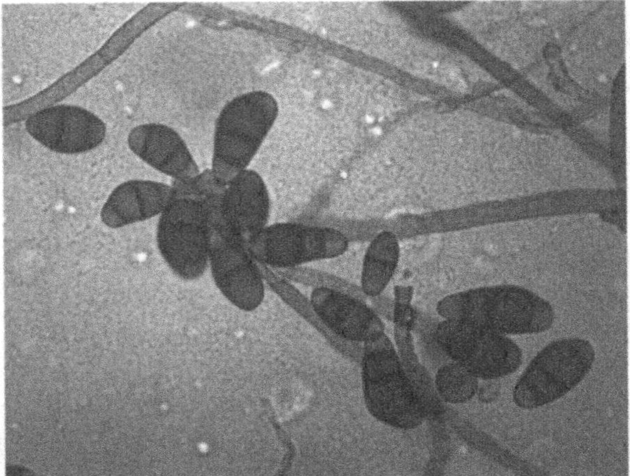

Fig. A.10. *Curvularia lunata.*

insulation materials in the HVAC and on cold, condensing surfaces. Many other fungal spores, such as spores of *Cladophialophora, Fulvia, Hormoconis* (*Amorphotheca*), and *Hyalodendron*, are very similar to *Cladosporium* spores.

10. *Curvularia* (Hyphomycetes). Conidia dark, three- to five-celled, fusiform, curved, with an enlarged central cell and paler cells at both ends; conidiophores brown, simple, developing conidia on new sympodial growing points or apically (Fig. A.10). *Curvularia*, *Drechslera*, and *Bipolaris* are closely related. Spores of *Curvularia* are very characteristic, as its name implies, and are easy to identify. Species of *Curvularia* are pathogens on a wide variety of plants and saprophytic. Their spores can be found outdoors and indoors.
11. *Drechslera* (Hyphomycetes). Conidia, light brown to midbrown, smooth to finely roughed, straight, broadly rounded at both ends, multicelled, with distoseptate (Fig. A.11). Conidiophores dark brown, simple, erect, thick, and smooth, often developed from dark brown stromata. This genus occurs mainly outdoors as plant pathogens. The airborne spore of this genus indoors originate outdoors. Its conidia geminate from any cells, while the conidia of *Bipolaris* are from polar cells.
12. *Epicoccum nigrum* (Hyphomycetes). Synonym is *E. purpurascens*. Conidia developed singly, golden brown to dark brown, rough, spherical to pyriform, 15–25 μm, with funnel-shaped base and broad attachment scar; conidial septation is obscuring in various directions and dividing conidia into a maximum of 15 cells (Fig. A.12). Conidiophores straight or somewhat flexuous, short, in clusters, colorless to pale brown. This is a common indoor fungus, often growing on water-damaged dry walls and paper products. Conidia of this species are very common in outdoor air, very characteristic, and easy to identify.

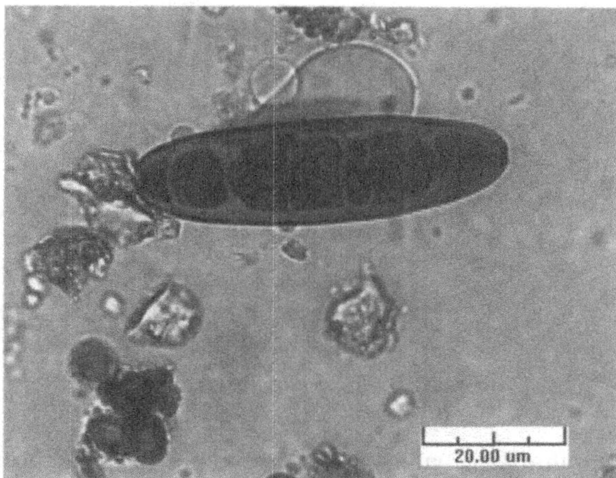

Fig. A.11. A conidium of *Drechslera* sp.

13. *Fusarium. Fusarium* may produce three types of spores: (a) macroconidia, fusiform, curved, multicelled, colorless, with a foot cell at the base end in most species (Fig. A.13); (b) microconidia much smaller, simpler, colorless, one- or two-celled, smooth, variable in size and shape, developed in chains or slimy heads in most species; (c) chlamydospores may develop or absent within hyphae intercalarily or terminally. Some species consistently

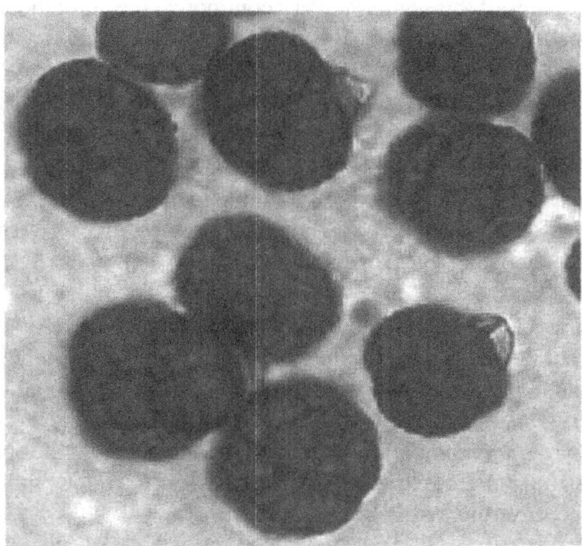

Fig. A.12. *Epicoccum nigrum.*

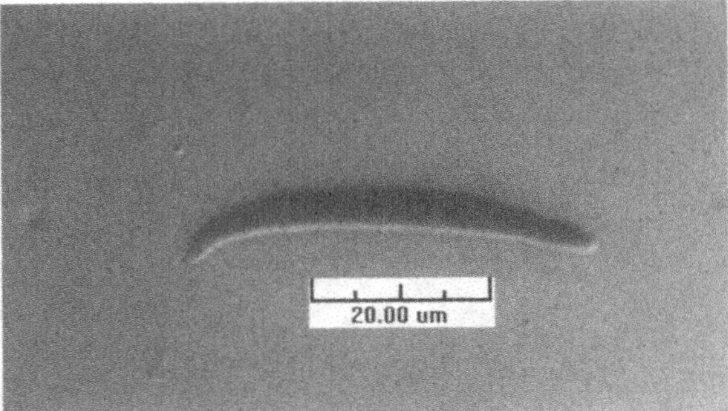

Fig. A.13. *Fusarium* macroconidia.

produce both macroconidia and microconidia. The macroconidia are easily identified, but the microconidia are difficult to identify. Microconidia may be similar to and confused with spores of some *Acremonium* species. There are two types of conidiophores: (a) densely branched conidiophores forming slimy pustules (sporodochia); (b) sparely branched or unbranched conidiophores. In both cases, phialides are slender and tapering, variable in length, colorless, and smooth. Several species of the genus may occur indoors under chronically wet or damp conditions (such as drain pans, around water faucets). However spores of *Fusarium* are often considered of outdoor origin.

14. *Ganoderma* (Basidiomycota). A wood decay fungus that produces bracket or conk on trees. It has never been reported growing in the indoor environment. Basidiospores of this genus are very common in the air outdoors and occasionally present in indoor air. Basidiospores of *Ganoderma* are thick-walled, rough, with a truncated end (Fig. A.14). The morphology of the basidiospores are very characteristic and are very reliable and useful for classification of this fungal genus.[10] The spores of *Ganoderma* are easy to identify from spore trap samples. This fungus is absolutely from outdoor sources.
15. *Memnoniella* (Hyphomycetes). Conidia, in chains, rough to warted, dark olivaceous to brownish black, globose to subglobose, 3.5–6 μm (Fig. A.15).[11] Phialides obovate, or ellipsoidal, colorless first, then turn to olivaceous, smooth, 7–10 × 3–4 μm, in clusters of 4–8 phialides. Conidiophores simple, erect, smooth to rough, colorless to olivaceous, slightly enlarged apically. The genus *Memnoniella* is differentiated from *Stachybotrys* by their different spore shapes and whether spores are in slimy conidial masses (*Stachybotrys*) or in dry chains (*Memnoniella*).[11] However, more recent studies indicate that *Memnoniella* may develop *Stachybotrys*-like spores.[12] *M. echinata* is the most common species and is an excellent indicator of water-damaged environments.
16. *Mucor* (Zygomycota). Sporangiophores are unbranched or irregularly branched, with basal rhizoids, ending with a sporangium, not originated

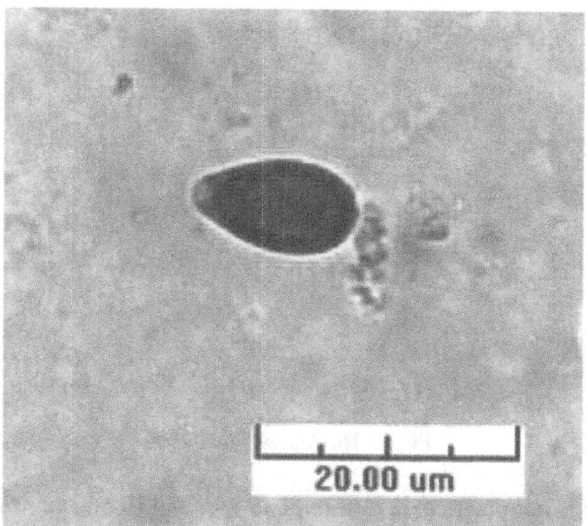

Fig. A.14. Basidiospore of *Ganoderma* sp.

from stolons. Sporangia vary in size, are spherical without an apophysis, have large columellae (versicle in a sporangium), rupture or persist on maturity, and develop many sporangiospores in it. Sporangiospores are single-celled, vary in shape, and are smooth or slightly ornamented. *Mucor plumbeus* is often present in indoor environments (Fig. A.16). For spore

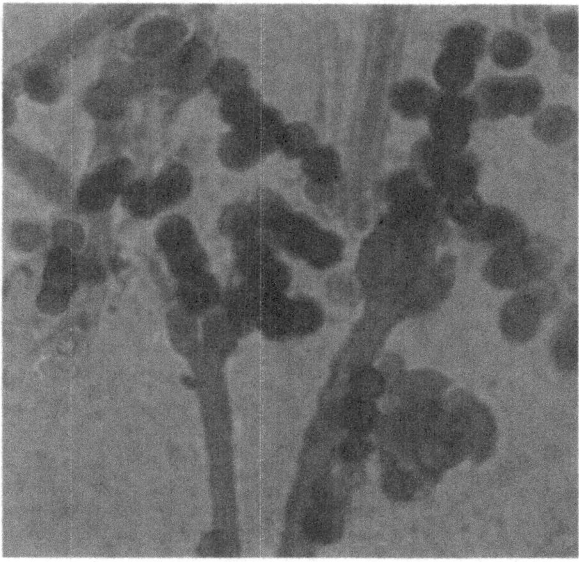

Fig. A.15. *Memnoniella echinata.*

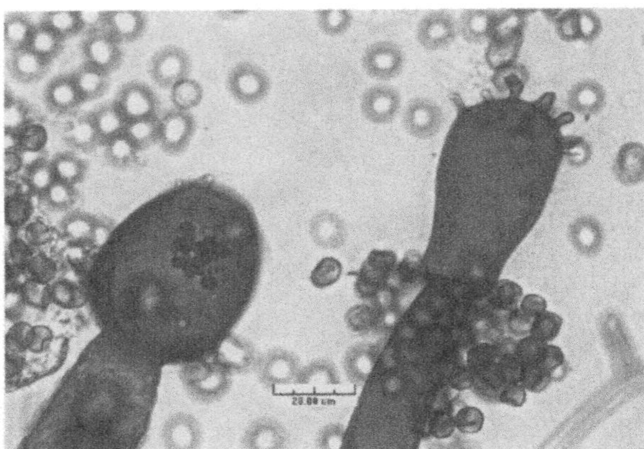

Fig. A.16. *Mucor plumbeus.*

count it is not practical to differentiate sporangiospores of *Mucor* from other spores with similar morphology.

17. *Nigrospora* (Hyphomycetes). Conidia spherical, large, black, smooth, 14–20 μm, and an equatorial furrow frequently develops (Fig. A.17). Conidiogenous cells develop on superficial hyphae terminally or laterally, swollen, ampulliform, colorless, develop conidium singly. Its conidia are very characteristic and easy to identify. With other fungal structures, it is not difficult to identify members of the genus. Species of *Nigrospora* grow on a variety of plants. Common species is *N. sphaerica*. They may occasionally grow on water-damaged materials.
18. *Penicillium* (Hyphomycetes). Conidia in dry chains, one-celled, colorless or very light colored in a shade of green, variable in size and shape, smooth to finely rough, 2.5–8 μm in length, globose, ellipsoidal, fusiform, or short cylindrical. Phialides acerose to flask-shaped, and have colorless to pale

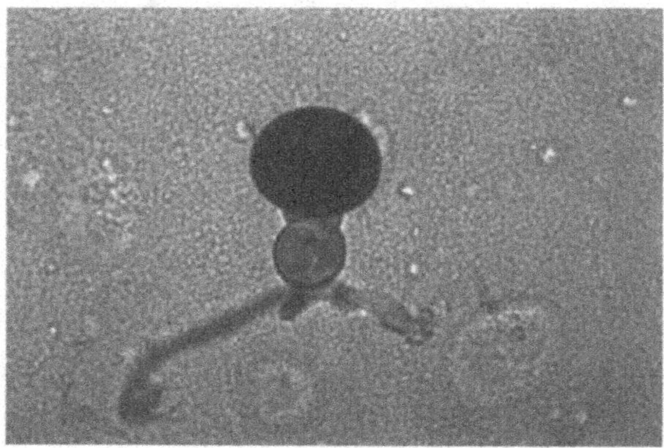

Fig. A.17. *Nigrospora sphaerica.*

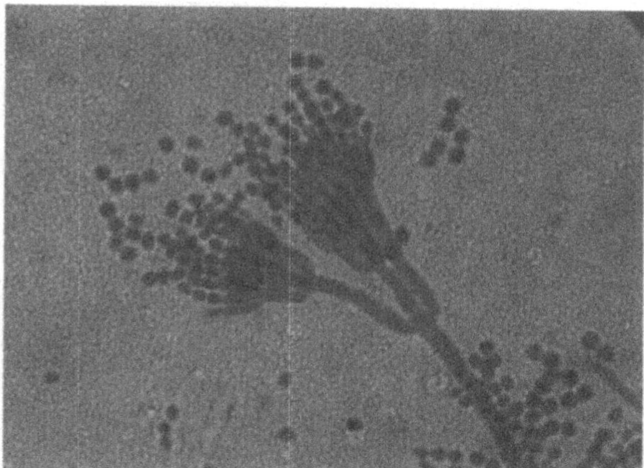

Fig. A.18. *Penicillium chrysogenum.*

pigmentation (Fig. A.18). Conidiophores, simple or aggregating into synnemata, erect, colorless or pale-colored, terminally bearing one or several layers of metulae that develop into a group of phialides. Stipes smooth or rough, colorless or pale-colored. A conidiophore with conidia resembles a broom shape. *Penicillium* is a large genus and is easy to identify to the genus, but very challenging to identify to species. Some species are common in indoor environments with water damage or under damp conditions. For common species identification, please refer to the laboratory guide to common *Penicillium* species by Pitt.[13] It is very difficult to differentiate the conidia of this genus from the conidia of *Aspergillus* or other similar genera. The conidia of these two genera are often grouped together as *Aspergillus/Penicillium* for spore count.

19. *Pithomyces* (Hyphomycetes). Conidia are dark brown, multicellular (phragmo- or dictyoconidia) develop on small peglike branches of the vegetative hyphae, broadly elliptical, pyriform, oblong, and commonly rough (Fig. A.19). Conidia of *Pithomyces* are cosmopolitan and are very characteristic, making them easy to identify. However, less experienced laboratory analysts may not be able to differentiate them from spores of *Ulocladium* with confidence. *Pithomyces chartarum* is the common airborne species. It is very common on dead leaves of a variety of plants and has also been found on paper.[3] *Ulocladium*, on the other hand, is an excellent indicator of a water-damaged environment.
20. *Rhizopus* (Zygomycota). Sporangiophores develop directly from opposite side of rhizoids, single or in groups, usually unbranched (Fig. A.20). Stolons are present and are an important characteristic of this genus. Sporangia are big, light-colored when young, later turning blackish-brown with aging, and have many spores. Columella are spherical or half-globose, are brown, and have an apophysis. Sporangiospores are ellipsoidal, often

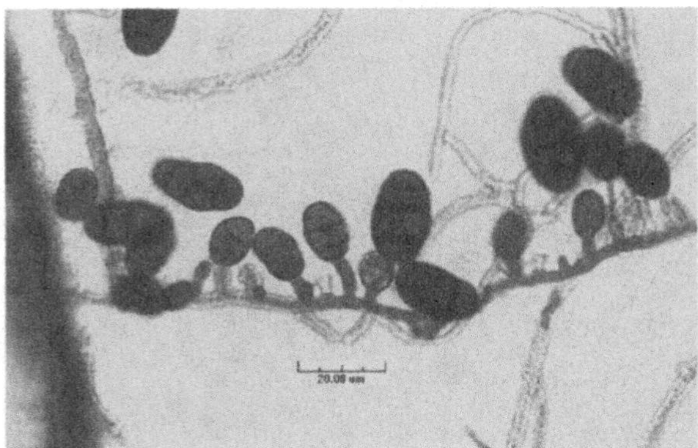

Fig. A.19. *Pithomyces chartarum.*

striate, and irregularly angled. Zygospores are occasionally produced in culture. Presence of rhizoids and apophysis will differentiate *Rhizopus* from *Mucor. Rhizopus* develops characteristic sporangiospores that are easily distinguishable by trained laboratory analysts. *Rhizopus stolonifer* is the most common species identified.

21. *Scopulariopsis* (Hyphomycetes). Conidia are in basipetal chains, one-celled, smooth or rough, spherical to ovate with a noticeable truncate end.

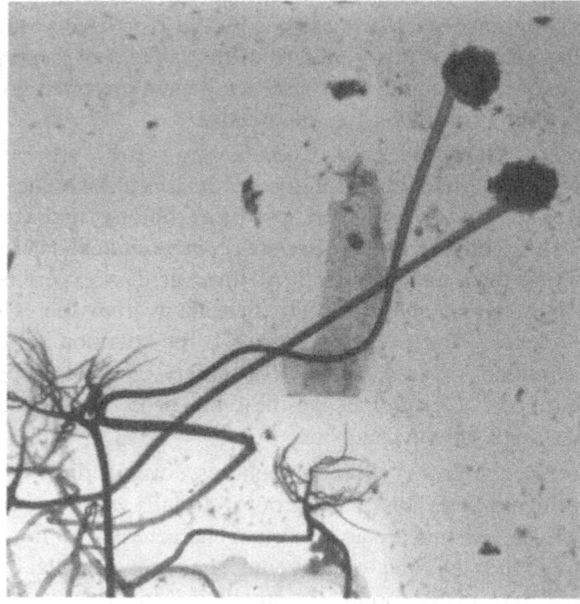

Fig. A.20. *Rhizopus stolonifer.*

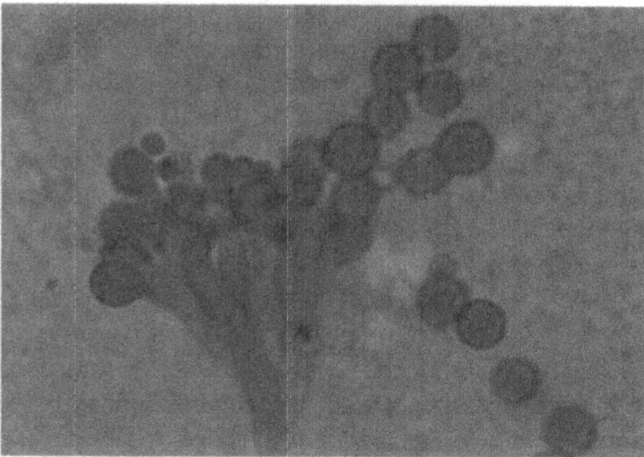

Fig. A.21. *Scopulariopsis brevicaulis.*

Conidiogenous cells are cylindrical or with a slightly swollen base (vase shape), annellate, single or in branched groups similar to the conidiophores of *Penicillium*. Several species may occur indoors. The most common species is probably *S. brevicaulis* (Fig. A.21). Conidia of *Scopulariopsis* are very characteristic with truncate ends.

22. *Spegazzinia* (Hyphomycetes). The species *Spegazzinia tessarthra* develops two types of spores. One is dark brown, spiny, four-celled, separated by a somewhat cross-septation. The other type is smooth, dark brown to black, four-celled, and with a well defined cross septation (Fig. A.22).

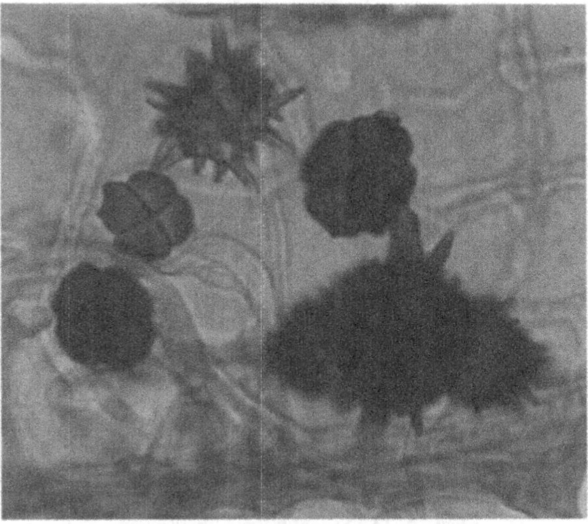

Fig. A.22. Two types of conidia of *Spegazzinia tessarthra*.

Conidiophores are slender, erect, solitary, pigmented, and develop apically. *Spegazzinia* conidia are very characteristic and easy to identify. There is no report of growth of this genus indoors. Its conidia are of outdoor origins, including plants, plant debris, and soils.

23. *Stachybotrys* (Hyphomycetes). Conidia are in slimy masses, smooth to coarsely rough, dark olivaceous to brownish black, obovoid, later becoming ellipsoid with age, 10–13 × 5–7 μm. Phialides are obovate or ellipsoidal, colorless early then turning to olivaceous with maturity, smooth, 12–14 × 5–7 μm, in clusters of 5–9 phialides (Fig. A.23). Conidiophores are simple, erect, smooth to rough, colorless to olivaceous, slightly enlarged apically, mostly unbranched but occasionally branched. Conidia of *Stachybotrys* are very characteristic and can be confidently identified in spore count samples. This genus is closely related to *Memnoniella*. Species of *Memnoniella* may occasionally develop *Stachybotrys*-like conidia, and vice versa.[12,14] *Stachybotrys* species grow on drywall and paper products, and are hydrophilic.[11] *S. chartarum* is probably one of the most common species of *Stachybotrys* found in buildings with long-term water damage.

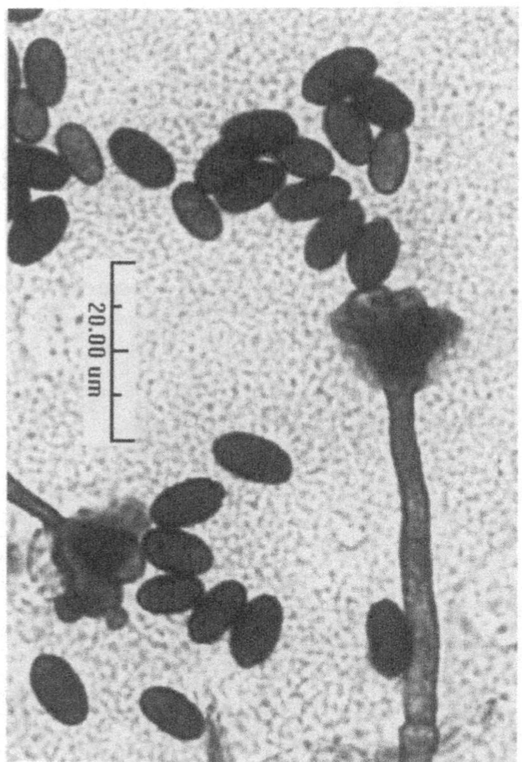

Fig. A.23. *Stachybotrys chartarum.*

Fig. A.24. *Trichoderma koningii.*

However, other species of *Stachybotrys* have been found in indoor environments.[14] *Stachybotrys* detected indoors is an excellent indicator of water damage. Precaution should be taken to interpret lab results with the presence of *Stachybotrys*.

24. *Trichoderma* (Hyphomycetes). Conidia appear colorless to green, smooth to rough, and are in moist conidial masses, variable in shape and size, small, 2.8–4.8 μm for common species. Phialides are flask-shaped, smooth, colorless. Conidiophores branch repeatedly, bearing clusters of phialides terminally in most cases. *T. harzianum*, *T. koningii* (Fig. A.24), and *T. viride* are reportedly found to grow in indoor environments. Recent research suggests that *T. harzianum* and *T. asperellum* are considered two true cosmopolitan species.[15] *Trichoderma* is a saprobe, and is very common in soil and on decaying wood. Some *Trichoderma* species are mycoparasites. These species are fast-growing in media and on building materials.
25. *Ulocladium* (Hyphomycetes). Conidia are black, rough, with pointed base when young, with both transverse and longitudinal septae, single or in a short chain (only in *U. chartarum*) (Fig. A.25). Conidiophores are pale brown, erect, multicelled, and develop conidia in a sympodial mode. The two common species of indoor *Ulocladium* are *U. botrytis* and *U. chartarum*. This genus is closely related to *Alternaria* and *Stemphyllium. Ulocladium* conidia are characteristic and can be identified by properly trained laboratory analysts, although spores of *Alternaria* and *Pithomyces* may be confused with *Ulocladium. Ulocladium* is an excellent indicator of water damage. However, *U. botrytis* was found to grow on ceiling tiles under high humidity (unpublished data).

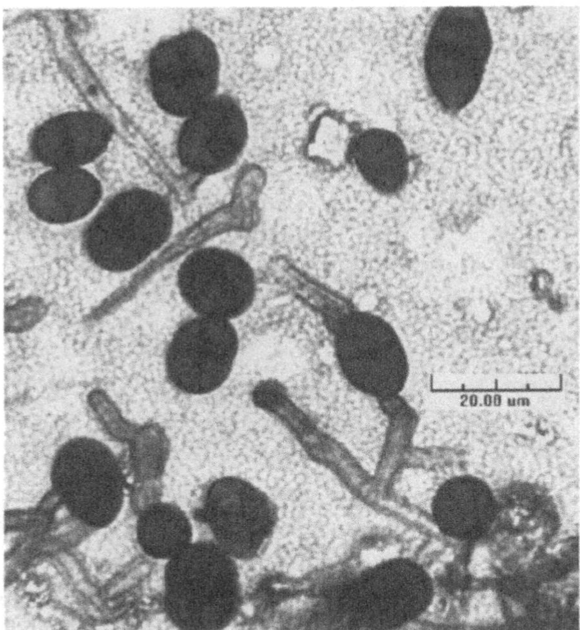

Fig. A.25. *Ulocladium botrytis.*

Important Fungal Structures

26. Amerospores (single-celled spores). Amerospores are defined as asexual spores that are nonseptate, single-celled spores with a length: width ratio not exceeding 15:1; if elongated, they have a single axis and are not curved through more than 180°; protuberances are less than one-fourth spore body length.[16,17] According to this definition, amerospores include a large variety of one-celled spores. Some laboratories may report amerospores as individual, small, rounded single-celled spores. *Aspergillus/Penicillium* spores are those in chains. In such a case, the category should be specified or defined to avoid confusion.
27. Ascospores and ascus (pl. asci). Ascomycota produce ascospores (sexual spores) in a saclike structure called an *ascus.*[17,18] In most cases eight ascospores are developed in an ascus. Only a few ascomycetes, are known to grow in indoor environments, due to water damage. Ascospores may be of indoor origins including those of *Chaetomium*, *Emericella*, *Eurotium*, and *Peziza*. Most ascospores are of outdoor origin. These ascospores include a large variety of different taxa of ascomycetes, which are saprophic or phytopathogens. Ascospores are highly variable and diverse in size and shape. They vary from single-celled to multicelled, from colorless to pigmented, from spherical to needle-shaped. Ascospores are abundant during warm seasons. There is a significant increase in airborne ascospore

concentration during and after a light shower or drizzling rain. Some members of the Ascomycetes class have well-defined diurnal patterns with early morning peaks.[9]

28. Conidiophores. Conidiophores are reproductive structures of anamorphic fungi and specialized hyphae on which asexual spores are developed. Morphological characters of conidiophores are very important to determine the mode of conidiogenesis so as to identify anamorphic fungi. Some anamorphic fungi are very common indoors. Conidiophores of some genera are quite characteristic. The presence of conidiophores can be a good indication of fungal growth indoors. Conidiophores can become airborne and appear in the samples of spore traps.
29. Hyphal fragments. A hypha (pl. hyphae) is a vegetative, filamentous fungal structure.[18] Its presence in air samples may indicate the presence of fungal infestation. It is unusual for hyphae or hyphal fragments to become airborne unless they are disturbed, such as in remediation. Elevated levels of hyphal fragments indoors may suggest fungal growth indoors and disturbance.
30. Mycelium (*pl.* mycelia) and mycelial fragments. Mycelium is a mass of hyphae.[18] Presence of mycelial fragment or mycelia most likely indicates the presence and growth of fungi. Unless they are disturbed (i.e., by removal of moldy materials or mold remediation), it is not easy for mycelial fragments to become airborne. Elevated levels of airborne mycelial fragments indoors suggest fungal infestation indoors.
31. *Aspergillus*/*Penicillium*-like. For spore count, it is impossible to differentiate conidia of *Aspergillus* from *Penicillium* only on the basis of morphological characteristics of conidia. These conidia are very simple, mostly small (2.5–5 μm; up to 7 μm), rounded or subspherical, and smooth to ornamented, colorless to lightly pigmented. Many fungal genera and species produce spores that are very similar or difficult to differentiate from *Aspergillus* and *Penicillium*. Spores of the following genera may be identified as *Aspergillus*/*Penicillium*-like spores: *Absidia, Acremonium, Aphanocladium, Beauveria, Chromelosporium, Phialophora, Gliocladium, Metarrhizium, Monocillium, Mortierella, Mucor, Paecilomyces, Thysanophora, Torulomyces, Trichoderma*, and *Verticillium*.
32. Basidiospores/basidiomycetes. *Basidiospores* are sexual spores developed on a basidium (pl. basidia) by basidiomycetes (including mushrooms, boletes, toadstools, bracket and conk fungi, polypores, jelly fungi, rusts, and smuts).[17,18] Some basidiospores are characteristic and can be differentiated by the hilum on the basidiospores. These spores are common and abundance from spring to fall in the temperate climate. They originate mostly from outdoor sources. Under very unusual situations, basidiospores may be from fruiting bodies or basidiomata of a few mushrooms and wood decay fungi indoors due to long-term water damage. Some members of the basidiomycetes have well-defined diurnal patterns with night or

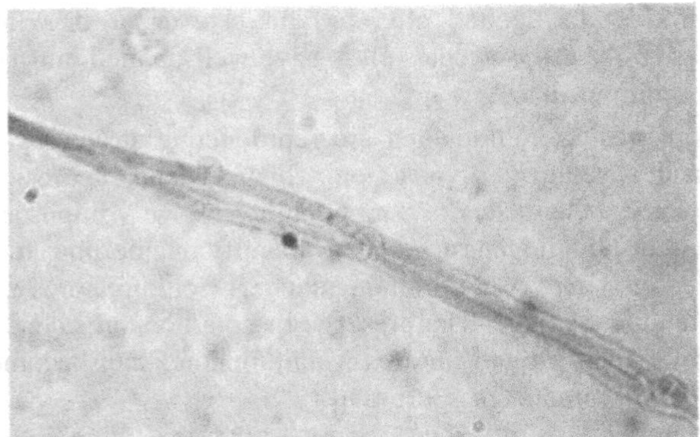

Fig. A.26. Clamp connections in basidiomycetous mycelium.

early-morning peaks.[9] For microscopic analysis of badidiomycetes, clamp connection is the unique characteristic of some basidiomycetous mycelia (Fig. A.26).

33. Myxomycetes. Myxomycetes are commonly called "slime molds". Mycologists study them but do not consider them to be true fungi. Spores of myxomycetes are variable, and their sources are mostly outdoors. Some fruiting bodies of myxomycetes have occasionally been identified from indoor samples. Without fruiting bodies, spores of myxomycetes are considered to be of outdoor origin. Their growth often occurs in hot, humid summer months.
34. Rusts. Rust spores are produced by a group of parasitic basidiomycetes. Rust fungi are obligate pathogens with a wide distribution.[19] They infect trees and a wide variety of plants. These plant pathogens depend on living hosts to survive. Some rusts require two different hosts to complete their lifecycle. Five kinds of spores may be developed in lifecycles of certain species.[19] They are called rusts because their spores (urediniospores and aeciospores) in mass appear rust in color. Their source is always outdoors.
35. Smuts. Smut spores are produced by a group of parasitic basidiomycetes.[18] Smut fungi mainly infect angiosperms, especially monocots, and produce most black (other colors are possible) spore masses in mid to late growing season (such as summer).[20] Members of the Gramineae family are the major hosts of smut. Cyperaceae and Compositae are important hosts also.[20] The well-known corn smut is a species associated with corn and is edible. Leaf smut is a disease on turfgrasses with a worldwide distribution[21] and may become airborne during mowing and other human activities. Their sources are always outdoors. Differentiation of smut spores from slime mold (myxomycetes) spores can be very challenging. This is the reason why some laboratory reports combine these two groups.

REFERENCES

1. Domsch, K. H., W. Gams, and T.-H. Anderson, *Compendium of Soil Fungi*, Vol. 1, IHW-Verlag, Eching, Germany, 1993.
2. de Hoog, G. S., J. Guarro, J. Gene, and M. J. Figueras, *Atlas of Clinical Fungi*, 2nd ed., Centraalbureau voor Schimmelcultures, Utrecht, Netherlands, 2000.
3. Ellis, M. B., *Dematiaceous Hyphomycetes*, Vol. 1, CAB Interantional, 1971.
4. Summerbell, R. C., *Aspergillus, Fusarium, Sporothrix, Piedraia* and their relatives, in *Pathogenic Fungi in Humans and Animals*, 2nd ed., D. H. Howard, ed., Marcel Dekker, New York, 2003, pp. 237–498.
5. Klich, M. A., *Identification of Common Aspergillus Species*, Centraalbureau voor Schimmelcultures, Utrecht, Netherlands, 2002.
6. Horst, K., *Westcott's Plant Disease Handbook*, 5th ed., Chapman & Hall, New York, 1990.
7. Sutton, J. C., D.-W. Li, G. Peng, H. Yu, P. Zhang, and R. M. Valdebenito-Sanhueza, *Gliocladium roseum*: A versatile adversary of *Botrytis cinerea* in crops, *Plant Dis.* **81**:316–328 (1997).
8. Li, D.-W., *Studies on Aeromycology in Kitchener-Waterloo, Ontario, Canada*, Ph.D. dissertation, Univ. Waterloo, Waterloo, Ontario, 1994.
9. Li, D.-W. and B. Kendrick, A year-round outdoor aeromycological study in Waterloo, Ontario, Canada, *Grana* **34**:199–207 (1995).
10. Zhao, J. and X. Zhang, *Flora Fungorum Sinicorum: Ganodermataceae*, Vol. 18, Science Press, Beijing, 2000.
11. Jong, S. C. and E. E. Davis, Contribution to the knowledge of *Stachybotrys* and *Memnoniella* in culture, *Mycotaxon* **3**:409–485 (1976).
12. Li, D.-W., C. S. Yang, R. Haugland, and S. Vesper, A new species of *Memnoniella, Mycotaxon* **85**:253–257 (2003).
13. Pitt, J. J., *A Laboratory Guide to Common Penicillium Species*, Food Science Australia, 2000.
14. Li, D.-W. and C. S. Yang, Notes on indoor fungi I: New records and noteworthy fungi from indoor environments, *Mycotaxon* **89**:473–488 (2004).
15. Samuels, G. J., *Trichoderma*: Systematics, the sexual state, and ecology. *Phytopathology* **96**:195–206 (2006).
16. Kendrick, B. and T. R. Nag Raj, Morphological terms in fungi imperfecti. in *The Whole Fungus*, Vol. 1, National Museum of Natural Sciences, Ottawa, Canada, 1979, pp. 43–61.
17. Kirk, P. M., P. F. Cannon, J. C. David, and J. A. Stapler, eds., *Dictionary of the Fungi*, 9th ed., CABI Publishing, Wallingford, CT, 2001.
18. Kendrick, B., *The Fifth Kingdom*, 3rd ed., Focus Publication, R. Pullins Co., Newburyport, MA, 2000.
19. Cummins, G. B. and Y. Hiratsuka, *Illustrated Genera of Rust Fungi*, 3rd ed., APS, St. Paul, MN, 2003.
20. Vánky, K., *Illustrated Genera of Smut Fungi*, 2nd ed., APS, St. Paul, MN, 2002.
21. Couch, H., *Diseases of Turfgrasses*, 3rd ed., Krieger Publishing, Malabar, FL, 1995.

INDEX

Sampling and Analysis of Indoor Microorganisms, Edited by Chin S. Yang and Patricia A. Heinsohn

Printed in the United States
By Bookmasters